The Resistance Dilemma

MW00774423

American and Comparative Environmental Policy
Sheldon Kamieniecki and Michael E. Kraft, series editors

For a complete list of books in the series, please see the back of the book.

The Resistance Dilemma

Place-Based Movements and the Climate Crisis

George Hoberg

The MIT Press
Cambridge, Massachusetts
London, England

© 2021 Massachusetts Institute of Technology

This work is subject to a Creative Commons CC-BY-NC-ND license.

Subject to such license, all rights are reserved.

The open access edition of this book was made possible by generous funding from the MIT Libraries.

The MIT Press would like to thank the anonymous peer reviewers who provided comments on drafts of this book. The generous work of academic experts is essential for establishing the authority and quality of our publications. We acknowledge with gratitude the contributions of these otherwise uncredited readers.

This book was set in Stone Serif and Stone Sans by Westchester Publishing Services. Printed and bound in the United States of America.

Library of Congress Cataloging-in-Publication Data

Names: Hoberg, George, author.
Title: The resistance dilemma : place-based movements and the climate crisis / George Hoberg.
Description: Cambridge, Massachusetts : The MIT Press, [2021] | Series: American and comparative environmental policy | Includes bibliographical references and index.
Identifiers: LCCN 2020048456 | ISBN 9780262543088 (paperback)
Subjects: LCSH: Environmentalism—North America. | Environmental sociology—North America. | Environmental policy—North America—Citizen participation. | Climate change mitigation—North America. | Climatic changes—Government policy—North America. | Renewable energy sources—Environmental aspects—North America. | North America—Environmental conditions.
Classification: LCC GE199.N73 H63 2021 | DDC 363.738/7460973--dc23
LC record available at https://lccn.loc.gov/2020048456

10 9 8 7 6 5 4 3 2 1

To Sophie and Sam, and their generation

Contents

List of Figures and Tables

Figures

Tables

Series Foreword

For decades, residents of communities around the world have fought against the imposition of energy production and related facilities that may impose local health or economic risks against their will. Sometimes the focus of the communities' wrath has been nuclear power plants or nuclear waste facilities and the public's understandable concern over radiation leaks and related hazards. This has been particularly so when government agencies have been less than forthright about such risks. At other times, the public has worried about the local impacts of producing oil and natural gas through both conventional drilling and hydraulic fracturing and its transportation across a region through pipelines for delivery to distant destinations.

These kinds of battles have been important in Canada and the United States, and increasingly they seem to divide the public along partisan and ideological lines. Even renewable energy projects that require the building of large solar arrays and expansive wind farms (on land and offshore) or high-voltage power lines have prompted community concern over aesthetic and environmental impacts, property values, and even public health. In such cases, the environmental community itself often has become divided between proponents of much-needed energy generation to replace fossil fuels and opponents concerned about specific local impacts.

These disputes raise fascinating questions about the public's role in governing, the options open to communities when a state or national decision adversely affects their residents or at least is perceived to do so, and the challenges that governments themselves face when they seek to develop new energy facilities as part of their efforts to ward off climate change disasters. Citizens might ask what opportunities they have to participate in decision-making and what forms of resistance to energy development projects have proven to be the most efficacious. That is, they may be looking for strategic guidance. Concerning the looming risks of climate change, governments need to know what they might do to overcome community resistance and build public support for the renewable energy projects that are now so essential if nations are to begin a serious movement toward sustainable energy systems.

George Hoberg's *The Resistance Dilemma* offers a fresh and intriguing examination of such crucial questions, drawing from a series of in-depth case studies of community resistance to energy development projects in Canada and the United States. He examines the origins, influences, and challenges facing this social movement strategy through a focus on resistance to new oil sands pipelines. In doing so, he addresses four core questions: (1) Has this kind of place-based resistance to fossil fuel development been effective, say in promoting climate action and reducing carbon emissions? (2) Does that strategy risk the unintended consequence of also feeding resistance to the clean energy transformation that is now so necessary? (3) Might more-innovative processes of governmental regulatory review and facility siting improve public acceptance of a transition to clean energy while avoiding the adverse consequences seen in fossil fuel resistance? (4) If such innovative approaches can reduce conflict, why are they not used more often?

The book employs a variety of policy theories to explore these questions, drawing from the most widely recognized work of the last several decades. These include the advocacy coalition framework, the multiple streams approach, punctuated equilibrium theory, and institutional theory. Hoberg distills them into an integrated policy regime framework that focuses on strategic actors both inside and outside government who interact within a context of ideas and institutional rules to pursue policy change. Doing so allows him to speak to the role of government institutions and rules, how key policy issues are framed over time, the impact of media coverage, the importance of legal issues, and the variety of actors who are critical to how these disputes are resolved. Both the book's theories and its dominant focus on oil sands development and resistance in Canada can be applied broadly to other energy challenges and other national settings.

Hoberg puts the challenge of climate change front and center in this analysis. In the first chapter, as well as later chapters, he reviews the evidence of climate change, and both national and international commitments to dealing with it, with a focus on Canada and the United States. He then proceeds in the following chapters to offer a rich and detailed history of specific cases of place-based resistance to fossil fuel and renewable energy development. That kind of coverage allows him to bring a variety of important data to the analysis, including energy production statistics, greenhouse gas emissions, public opinion, issue saliency, and keyword usage in media coverage, interest group strength, and trade and other key financial information.

Hoberg highlights the Keystone XL pipeline controversy and many others that have received less attention, and he analyzes the adverse consequences such resistance can have for clean energy projects, including wind and solar power; hence the book's reference in its subtitle to the dilemma of place-based resistance. At the end of the book, Hoberg gives his conclusions about how place-based resistance to clean energy projects might be overcome and thus aid in nations' responses to climate change.

At a time when environmental policies are increasingly seen as controversial and new approaches to address environmental issues are being implemented widely, we especially encourage studies that assess policy successes and failures, evaluate new institutional arrangements and policy tools, and clarify new directions for environmental politics and policy. The books in this series are written for a wide audience that includes academics, policymakers, environmental scientists and professionals, business and labor leaders, environmental activists, and students concerned with environmental issues. We hope they contribute to the public's understanding of environmental problems, issues, and policies of concern today and also suggest promising actions for the future.

Sheldon Kamieniecki, University of California, Santa Cruz

Michael Kraft, University of Wisconsin–Green Bay

Preface

The ideas for this book began germinating in 2006. I was just beginning an intellectual shift in direction from a focus on controversies in forest conservation to climate change. A field trip that year to Fort McMurray and the oil sands mines near there in northern Alberta sparked my alarm about the growing environmental footprint of that sector. Early research identified the centrality of expanded pipeline capacity to the future of that carbon-intensive oil resource. I became fascinated with an alliance of Indigenous and environmental activists who, while brought together by the forestry conflicts I'd been studying, were turning their attention to contesting oil sands pipeline projects as a way to advance their causes.

At the same time, a significant fraction of the British Columbia environmental community was mounting a campaign in opposition to new "independent power projects," small renewable energy projects being proposed by the private sector. That campaign spawned a rift in the environmental movement between those focused on combating climate change and those more focused on local environmental impacts. At first, I began analyzing these controversies separately, but by the early 2010s it became increasingly apparent that they were part of the same story. The climate crisis creates the imperative to move away from carbon-emitting sources of energy as quickly as possible. The process crisis—the challenges we face in getting social buy-in for significant new infrastructure projects—poses a direct challenge to the necessary solutions to the climate crisis. This book is the result of a scholarly exploration of this tension.

I should clarify my own positioning with respect to these conflicts. In 2011, as the alarming evidence of the threat of climate change grew, I abandoned my jealously guarded scholarly detachment and became engaged as an activist. I cofounded a small environmental group to mobilize my university community to engage in provincial and federal elections to promote climate action. The first several campaigns of the group targeted the two oil sands pipeline proposals crossing British Columbia. As a result, I have taken public positions against the Northern Gateway Pipeline and the Trans Mountain Expansion Project because of their climate impact and spoke at two anti-pipeline rallies on Burnaby Mountain in November 2014. I also applied to be an intervenor in the National Energy Board (NEB) hearings to discuss the climate impacts of the Trans Mountain proposal but was rejected by the NEB.

By 2017, I'd stepped away from that activist group to focus on completing this book. Nonetheless, I believe this combination of activism and scholarship helps inform the work. The activism allowed me to have deep appreciation for the strategy of mobilizing climate action by focusing on opposing new fossil fuel infrastructure. The academic viewpoint gave me a clear sense of the limitations of that strategy and the risks it carried, which I've depicted in the concept of the "resistance dilemma."

Along the way, I've gotten an enormous amount of help from graduate students who have worked with me. Xavier Deschênes-Philion played such a valuable role in chapter 7 that he is listed as coauthor of that chapter. I'm also indebted to Claire Allen, Sarah Froese, Alex Ash, Jessika Woroniak, Tracy Ly, Andrea Rivers, and Geoff Salomons. Guillaume Peterson St-Laurent has evolved from a student to a colleague during the process of writing this book and has been an invaluable source of support.

A number of colleagues, in and out of the academic sector, have supported, informed, or inspired me. Among them are Alasdair Bankes, Keith Brownsey, Angela Carter, Jennifer Ditchburn, Simon Donner, Monica Gattinger, Mark Jaccard, Andrew Leach, Shawn McCarthy, Keith Neuman, Martin Olszynski, Peter O'Neil, and Trevor Tombe. I was fortunate, for a portion of the project, to be supported by a Social Science Research Council of Canada Insight Development Grant led by Carol Hunsberger, whose support and collaboration contributed greatly to chapter 11. I gained invaluable insights from a number of conversations over the years with activists and advocates, many of whom are quoted in the book. I want to make special mention of Tzeporah Berman, Will Horter, Kai Nagata, Matt Price, and Keith Stewart. Beth Clevenger of MIT Press has been both supportive and patient throughout this process.

Earlier versions of some sections of this book were published elsewhere:

Hoberg, George. 2016. "Unsustainable Development: Energy and Environment in the Harper Decade." In *The Harper Factor: Assessing a Prime Minister's Policy Legacy*, edited by Jennifer Ditchburn and Graham Fox, 253–263. Montreal and Kingston: McGill-Queen's University Press.

Hoberg, George. 2018. "Pipelines and the Politics of Structure: Constitutional Conflicts in the Canadian Oil Sector." *Review of Constitutional Studies* 23 (1): 52–89.

Hunsberger, Carol, Sarah Froese, and George Hoberg. 2020. "Toward 'Good Process' in Regulatory Reviews: Is Canada's New System Any Better Than the Old?" *Environmental Impact Assessment Review* 82:106379.

My greatest debt is to my children. Their loving support has fueled me. Remarkably, over the duration of this project, they have evolved from youthful sources of inspiration to invaluable colleagues.

1 The Grand Challenge: Mobilizing to Address the Climate Crisis

Overview

It is increasingly understood that humanity faces a true climate crisis, where the pace and magnitude of climate change are on a civilization-threatening trajectory (IPCC 2018), yet the collective response by governments around the world to this emergency has thus far been too tepid to drive the pace and magnitude of energy-system change required to ensure a reasonably safe climate. Even if nations around the world carry out the commitments they made as part of the 2015 Paris Agreement, humanity will still be on a course toward a dangerous future climate.[1]

The explanations for this gap between the awareness of the crisis and the policy response are well understood by social scientists, and those explanations will be described shortly. This book focuses on a critical strategic choice by the North American wing of the global climate movement designed to address this gap. Frustrated with their inability to mobilize sufficient political pressure at the United Nations and at the national level, climate activists chose to ally themselves with place-based interests, including Indigenous groups, to block new coal plants, coal port expansion, fracking, and, more recently, oil sands pipelines (Klein 2014; Piggot 2018; Cheon and Urpelainen 2018). Organized resistance to new fossil fuel infrastructure has now become a formidable political force in North America as pipeline conflicts have become divisive national political issues in the United States and Canada, and in the relationship between the two countries.

This book examines the origins, influence, and challenges of this social movement strategy by focusing on the resistance to new oil sands pipelines. It addresses four core research questions: (1) Has the strategy of place-based resistance to fossil fuel development been effective at promoting climate action and the reduction of global warming emissions? (2) Does the strategy risk the unintended consequence of feeding place-based resistance to the clean energy transformation? (3) Is there hope in more innovative processes of energy infrastructure decision-making that can promote social acceptance of the rapid transition to the clean energy system but avoid the confrontational politics that have characterized fossil fuel resistance? (4) If innovative approaches have been demonstrated to reduce conflict, why are they so rarely used?

Before discussing how these four questions will be addressed, this chapter will explain why mobilizing to address the climate crisis has proven so challenging for humanity. The chapter will then examine how the climate movement's efforts to surmount these mobilization challenges produced the strategic shift to blocking fossil fuel infrastructure before looking at the analytical framework guiding the analysis. It is an actor-centered framework focused on strategic actors working through and on a particular context of institutions and ideas. The book uses this framework to develop hypotheses about the expected relative power and behavior of actors resisting new energy infrastructure in particular cases. The chapter concludes by describing the plan to address each of the four guiding questions.

The Atmospheric Tragedy of the Commons

The climate crisis creates the urgent imperative to transform the energy system to one that does not emit global warming emissions. There's also a political component to the climate challenge: the remarkable political difficulties in motivating concerted action on addressing climate change as a result of problem structure, psychological barriers, economic and cultural opposition, and, in many jurisdictions, dysfunctional governance structures. First, it has all the characteristics of a "wicked problem" (Lazarus 2009; Levin et al. 2012). The structure of the climate problem is characterized by three prominent features:

- immense uncertainty about the timing and magnitude of impacts;
- spatial inconsistency between local emissions, and the economic benefits that flow from them, and global impacts of climate impacts; and
- temporal inconsistency resulting from lags in the response of the climate system, in that the costs of climate action are in the present but the benefits of reducing emissions are uncertain and in the distant future. (Victor 2011)

All three of these problem characteristics aggravate the challenges of political mobilization and collective action (Olson 1965). Because the benefits of climate action are uncertain and far away in place and time, politicians have insufficient motivation to take the necessary short-term actions that inevitably come with some cost, frequently to politically powerful groups. As a result, there is a glaring mismatch between the incentives of policy-makers acting at the national or subnational level and the global community's shared goal of maintaining a safe climate. Paul Harris calls this dilemma "the atmospheric tragedy of the commons" (Harris 2013).

This problem structure also challenges our psychology as a species. Humans are "wired" to think about short-term, concrete issues, but climate change is long term and abstract. When faced with uncertainty, our psychology promotes optimism and wishful thinking rather than acknowledging the hard reality of the emerging crisis (Marshall 2014). Anthony Leiserowitz of Yale University has stated, "You almost couldn't design a problem that is a worse fit with our underlying psychology" (Gardiner 2012).

In addition to the organizational and psychological challenges of the problem structure, a further barrier to climate action is the fierce resistance of businesses and others who benefit from the status quo, especially the fossil fuel industry (Brulle 2014; Urquhart 2018). On top of the "privileged position of business" (Lindblom 1982) that gives business an enormous structural advantage over opponents, the fossil fuel industry has used its enormous wealth to fund campaigns and politicians opposed to climate action and to deliberately obfuscate climate science (Orsekes and Conway 2010). Economic resistance to climate action goes beyond big business, however, and includes many consumers (and voters) who are resistant to the price increases that would go along with a shift away from fossil fuels (Lachapelle and Kiss 2019). This resistance has spilled over into cultural politics. What began as strategic initiatives by fossil fuel companies to "manufacture doubt" about climate science has spawned a conservative social movement with a life of its own (McCright and Dunlap 2010). In January 2017, this social movement moved into the White House in the presidency of Donald Trump.

The Transformative Power of the Supply-Side, "Keep It in the Ground" Movement

In response to these formidable challenges, the climate movement has shifted strategies to focus on the supply side—blocking new fossil fuel infrastructure (Piggot 2018; Cheon and Urpelainen 2018; Green and Denniss

2018). In her book *This Changes Everything: Capitalism vs. the Climate,* Naomi Klein elevates local resistance movements to a hopeful progressive strategy to battle climate change. Klein characterizes this "keep it in the ground" movement (which she labels "Blockadia") as a "roving transnational conflict zone" that is provoked by "extreme extractivism," whose common characteristic is local resistance movements demanding local control. This movement, Klein notes, has quickly become remarkably effective: "It has taken the extractive industries, so accustomed to calling the shots, entirely by surprise: suddenly, no major new project, no matter how seemingly routine, is a done deal" (Klein 2014, 296).

Historically, these local conflicts were about local issues and disconnected from each other. Klein claims that, though reported in the mainstream press as isolated protests against specific projects, "these sites of resistance increasingly see themselves as part of a global movement" (303). An important catalyst in connecting these conflicts has been "widespread awareness of the climate crisis" (304). The emergence of the concept of a "carbon budget," discussed in chapter 2, has provided some scientific credibility for blocking new fossil fuel projects.

These themes have been echoed by Bill McKibben, the founder of 350 .org and perhaps the leader most influential in driving the shift to focusing on blocking pipeline infrastructure in North America (Klein and McKibben are close associates, and Klein is on the board of directors of 350.org). The *New Yorker* credited McKibben with transforming the politics of the Keystone XL pipeline (discussed in chapter 4) and with it the US environmental movement, saying that "McKibben has successfully made Keystone the most prominent environmental cause in America" (Lizza 2013). *Time* magazine referred to it as the "Selma or Stonewall" of the climate movement (Grunwald 2013).

In describing their strategy, McKibben emphasized the importance of allying the climate movement with place-based interests, saying, "After decades of scant organizing response to climate change, a powerful movement is quickly emerging around the country and around the world, building on the work of scattered front-line organizers who've been fighting the fossil fuel industry for decades" (McKibben 2013a). In the foreword to a book on the resistance to the oil sands, Klein and McKibben write, "The fight over the tar sands is among the epic environmental and social justice battles of our time, and one of the first that managed to marry quite explicitly concern for frontline communities and immediate local hazards with fear for the future of the entire planet" (Klein and McKibben 2014, xvii).

The great benefit of "keep it in the ground" as a political strategy is that it avoids many of the barriers to collective action that thwart mobilization on climate change. While climate change is complex, uncertain, abstract, and distant, fossil fuel infrastructure is comparatively straightforward and poses very specific and readily understandable risks to geographically specific locations (Piggot 2018; Cheon and Urpelainen 2018; Green and Denniss 2018). That doesn't mean it is *the* solution to the climate crisis, but as chapter 8 of this book demonstrates in addressing the book's first core question, it has empowered the climate movement in a new way that has helped force real climate policy actions.

The Resistance Dilemma

Addressing the climate crisis involves a rapid phaseout of carbon-emitting fossil fuels and the accelerated adoption of clean energy technologies. According to the seminal report of the Intergovernmental Panel on Climate Change (IPCC) about the implications of aiming to contain global warming as close to 1.5°C as feasible,

> Pathways limiting global warming to 1.5°C with no or limited overshoot would require rapid and far-reaching transitions in energy, land, urban and infrastructure (including transport and buildings), and industrial systems (high confidence). These systems transitions are unprecedented in terms of scale, but not necessarily in terms of speed, and imply deep emissions reductions in all sectors, a wide portfolio of mitigation options and a significant upscaling of investments in those options (medium confidence). (IPCC 2018)

The second core question of this book is: does the "keep it in the ground" strategy risk the unintended consequence of feeding place-based resistance to the clean energy transformation? Many renewable energy generation and transmission facilities have confronted determined opposition from local groups, leading to costly project delays or alterations and in some cases outright cancellations. Solar and wind power projects, vital to replacing fossil fuels for electricity generation, have generated pushback from local groups concerned about property values, changes to species habitats, landscapes, aesthetics, and human health. New high-voltage electric transmission lines have also attracted significant resistance. Renewable energy projects are frequently in quite different locations than fossil fuel infrastructure, so new transmission lines are usually required to supplement the buildout of new renewable energy sources. In addition, the integration of intermittent renewables into the electricity grid is projected to require significant new transmission capacity and deeper integration across larger geographic areas.

Concerns about renewable power infrastructure, as challenging as they have been, pale in comparison to the place-based resistance to nuclear power. Considered by many analysts to be critical to decarbonization (Morgan et al. 2018), there are virtually zero prospects for new nuclear energy plants in North America, in part as a result of vehement political opposition (Kinsella 2016). Place-based opposition has also thwarted the successful siting of nuclear waste repositories (Ramana 2018; McFarlane and Ewing 2006). While resistance to nuclear power is consistent with the arguments of this book, it has not been part of this research project and will not be addressed further.

Resistance to renewable energy is not a direct consequence of the movement to keep fossil fuels in the ground. In fact, as will be developed in chapter 10, the academic literature on the social acceptance of renewable energy emerged before the climate movement made the strategic pivot to blocking infrastructure. The resistance dilemma is that the "keep it in the ground" movement builds the institutional, social, and cultural muscles that strengthen the capacity of groups intent on resistance to renewable energy. Perhaps the most significant component of this dilemma is whether local governments should be granted veto power. If they are, it gives local authorities—Indigenous or not—the capacity to veto projects determined to be in the interests of the broader geographic political jurisdiction, but if that power is taken away, local groups may resent the disempowerment, and that can strengthen resistance.

Analytical Framework

To better understand the relationship between activist group strategies on the one hand and energy and climate policy outcomes on the other, this book applies an analytical framework from political analysis of public policies. Social scientists have developed a variety of theoretical perspectives designed to explain why governments adopt the policies they do. The advocacy coalition framework, developed by Paul Sabatier and his colleagues, focuses on the emergence of competing coalitions, one defending the status quo, the other change oriented. The groups are bound together because of their shared beliefs (Jenkins-Smith et al. 2014; see also Hochstetler 2011). The multiple-streams framework, arising out of John Kingdon's influential work, focuses on the interaction of three distinct streams—of problems, politics, and solutions—converging to produce windows for policy change (Kingdon 1995; Zahariadis 2014). Baumgartner and Jones's punctuated equilibrium model, inspired by biological theories of evolution, focuses on how different actors in the policy process work to alter policy images and institutional venues to generate change (Baumgartner and Jones 1993). The institutional analysis and development approach, built on the work of Elinor Ostrom, examines actors working in a particular action situation operating through a set of institutional rules to produce policy outcomes (Ostrom 2014).

Despite their differences in concepts and emphasis, there is a great deal of overlap in these approaches. All take the unit of analysis to be the policy subsystem defined by a particular policy domain, such as energy, climate, health, or foreign policy. All focus on the interaction of three core conceptual categories of strategic actors with interests and beliefs, institutional rules, and a set of ideas influential in that policy domain.[2]

This book employs an analytical framework, referred to as the policy regime framework,[3] distilled from these multiple theoretical perspectives (Hoberg 2001; May and Joachim 2013). The regime approach sees strategic actors, in and out of government, as the central agents of policy.[4] Each actor has their own interests, as well as political resources. They adopt strategies designed to best pursue their interests given their resources (Hoberg 2001). Strategic actors interact within a context of ideas and institutional rules but also work to change ideas through reframing or change institutional rules through venue shifting or other means (Pralle 2006a; Baumgartner and Jones 2010).

This book examines not only the strategic choices of environmentalists but also the battle over ideas by analyzing issues covered by the media and examining the conflict over the institutional rules of the game. Institutional design can be pivotal because when the location of authority changes, the balance of policy preferences could also change significantly. As a result, changing the venue can lead to different policy outcomes with quite different consequences for competing actors. That explains why institutional rules and venues are so frequently deeply contested. In this political struggle over institutions, government actors—whether individuals in positions of authority or organizational units—are important agents. As we explore the various infrastructure conflicts throughout this book, it will be important to keep in mind the distinction between government actors as actors within and on an institutional structure and the institutional structure of rules and venues themselves.

In earlier work, I developed a framework for analyzing the political risk to pipeline project proponents (Hoberg 2013). This framework is adapted here to analyze the strategic resources of project proponents and opponents, both for fossil fuel and renewable energy projects. This approach differs from the innovative work of sociologists McAdam and Boudet (2012), who built an explanatory model of opposition to energy facility siting inspired by the social movement literature. They use that framework to examine 20 cases in the United States, mostly liquefied natural gas (LNG) plants. Their model analyzes the relative importance of five factors: (1) the level of project risk, (2) whether the community has experience with similar projects, (3) whether the community experiences economic hardship, (4) the levels of civic capacity, and (5) political opportunity. Political opportunity is defined by whether the decision was made by elected officials, the proximity of the next election, and the level of government holding jurisdiction—with local government jurisdiction affording project opponents the greatest opportunities.[5]

The framework applied in this work argues that the relative power of project opponents is a function of four variables: (1) the salience of place-based, concentrated risks and benefits; (2) whether opposition groups have access to institutional veto points; (3) whether the project can take advantage of existing infrastructure; and (4) the geographic separation of risks and benefits. A detailed discussion of each of these variables follows.

1. *The salience of place-based, concentrated risks and benefits* The "logic of collective action" (Olson 1965) suggests that resistance to new projects is easier to organize if there are concrete, focused, place-based values at risk. By this logic, local concerns about risks to precious bodies of water are much more likely to galvanize opposition than more diffuse concerns such as global warming. The same can be said about local environmental impacts or the alteration of a sense of place resulting from deploying renewable energy infrastructure compared to the more diffuse concerns with decarbonization. The economic benefits of a project can be examined through the same lens. Jobs created in facility construction and operation are concrete and place based, whereas tax revenues and

corporate dividends are more diffuse. A variety of studies have found that close proximity to energy infrastructure tends to increase project support, because of both greater familiarity and greater expectation of economic benefit (Gravelle and Lachapelle 2015; Boudet et al. 2016; Boudet et al. 2018; Bishop 2014). As a result, a critical variable is the relationship between local risks and local benefits. Projects that provide local environmental benefits with minimal salient local risks are likely to face few opponents. When there are salient local risks and few local economic benefits, project opponents have a major advantage. The greater the local economic benefit, the more likely a project is to surmount resistance even if there are place-based risks. *The first outcome hypothesis is that the greater the placed-based risks in relation to local economic benefits, the more vulnerable the project is to resistance.*

2. *Whether opposition groups have access to institutional veto points* Veto points are locations of government authority that give a particular organization the ability to block approval of a project or policy (Immergut 1990; Tsebelis 2000). Examples are the organization granted formal decision-making authority (e.g., an independent regulatory body or the cabinet), whether the decision is subject to judicial review, and whether the approval of different levels of government is required.[6] In some cases, an organization can lack formal political authority but have sufficient power that they are equivalent to a veto point. These are referred to as "political veto points." *The second outcome hypothesis is that the more access opponents have to veto points, the more vulnerable the project is to resistance.*
3. *Whether the project can take advantage of existing infrastructure* Greenfield projects create more disruption to existing economic and residential patterns than projects that can take advantage of existing infrastructure.[7] *The third outcome hypothesis is that the more the project can take advantage of existing infrastructure, the less vulnerable it is to resistance.*
4. *The geographic separation of risks and benefits* All projects come with risks and benefits. If they occur in the same general area, it is more straightforward for affected interests and policymakers to consider both risks and benefits. The greater the geographic distance between those who benefit economically and those who face environmental risks, the more challenging it is to weigh risks and benefits. This situation is common in energy systems, where energy production is distant from its consumption—pipelines

and power lines being classic examples. This challenge is much greater when risks and benefits are separated by jurisdictional boundaries that represent veto points. *The fourth outcome hypothesis is that the greater the geographic separation of risks and benefits, the more vulnerable the project is to resistance.*

These outcome hypotheses are important to consider when predicting the level of opposition a new energy infrastructure proposal is likely to confront.

In addition to these hypotheses about outcomes, this analytical framework also yields several hypotheses about the anticipated strategic behavior of actors in this conflict.

1. To strengthen their leverage, climate activists will ally themselves with groups representing place-based interests when possible.
2. Strategic actors will focus their strategies on the institutional venue(s) most favorable to their interests.
3. Pipeline opponents will adopt framing that emphasizes place-based risks.
4. Decision rationales about pipelines will emphasize place-based risks far more than climate risks.

These hypotheses are important because they link the four-part analytical framework to predictions about the behavior of strategic actors within coalitions. They are explored through an analysis of actor strategies. In chapters 4–7, these hypotheses are applied to the four most contested oil sands pipeline cases of the 2010s: Keystone XL, Northern Gateway, the Trans Mountain Expansion Project, and Energy East. They are not a representative sample of pipeline cases. Rather, they were selected with the express purpose of examining the origins, implementation, and impact of the strategic choice by climate activists to shift from lobbying for policy reform to blocking fossil fuel infrastructure.

The core method is process tracing, a careful reconstruction of events within specific cases to identify causal process observations to draw inferences about explanatory hypotheses (George and Bennett 2005; Collier 2011; Mahoney 2010).[8] The four outcome hypotheses presented are the core causal hypotheses under investigation. They are further supported by the behavioral hypotheses, the first three of which point to mechanisms that help drive the case outcomes. The fourth behavioral hypothesis is what Mahoney (2010) calls an auxiliary outcome causal process observation, or an outcome that is consistent with the causal claims. Sources for this research include government documents, organizational websites, media accounts, peer-reviewed research, and personal interviews with key actors.

Organization of the Book

The chapters in part I examine the policy regime for the oil sands. Chapter 2 focuses on the characteristics of the oil sands resource, the most important background conditions influencing the oil sands policy regime, and the strategic actors that make up the competing oil sands and anti-pipeline coalitions. The chapter gives a brief history of oil sands development and its local, regional, and global environmental consequences. It also provides an overview of trends in markets influencing the Alberta oil sector as well as trends in electoral politics and public opinion in Alberta and across Canada. It then closes by looking at which actors constitute the oil sands coalition and the anti-pipeline coalition.

Chapter 3 examines ideas, institutions, and environmental policies. It examines the way the two competing coalitions frame the issues to best influence the public and policymakers. It delves into one particularly important idea, the concept of a carbon budget, which has become a scientific justification of sorts for the "keep it in the ground" movement discussed earlier. It also examines the macropolitical institutions in Canada, using the United States as a comparator both because it helps illuminate the structure of Canadian institutions and because US institutions are vital to understanding the Keystone XL case described in chapter 4 and several of the contested renewable energy cases described in chapter 10.

The chapters in part II examine the anti-pipeline campaigns and their impacts. Four pipeline cases are analyzed in depth, in the order they became prominent national or bilateral issues. The locations of these pipeline projects are shown on the map in figure 1.1. Chapter 4 examines the controversy over Keystone XL from Edmonton to Oklahoma, where the climate movement first adopted resistance to oil sands pipelines as a core part of its strategy to influence climate and energy policies. Chapter 5 looks at the Northern Gateway pipeline from Alberta to British Columbia's north coast. Chapter 6 examines the Trans Mountain pipeline from Alberta to the port of Vancouver. Chapter 7 analyzes the Energy East pipeline from Alberta to the east coast of Canada. With those cases described and explained, chapter 8 addresses the book's first core question directly, asking how effective the strategy of place-based resistance to fossil fuel infrastructure has been at promoting climate policy and reducing emissions. It summarizes the impacts of the four major anti-pipeline campaigns and shows how they directly influenced the adoption of more ambitious climate policies by both Alberta and Canada. It examines the changes that have occurred in the oil sands policy regime in the latter half of the 2010s.

The chapters in part III address the remaining three core questions. Chapters 9 and 10 address the "resistance dilemma" of whether place-based activism has the potential to threaten the much-needed transition to renewable energy. Chapter 9 examines resistance to one major clean energy megaproject, the Site C Dam in northeastern British Columbia. Chapter 10 examines

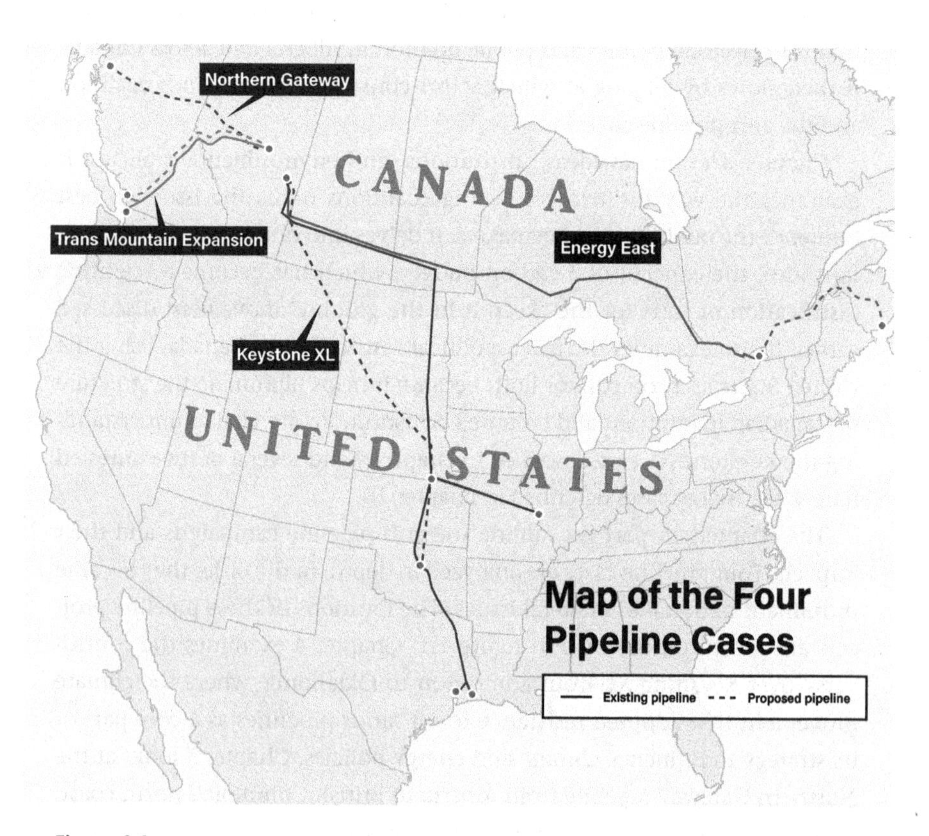

Figure 1.1
Map of the four pipeline cases.
Source: Joelle Lee.

a number of episodes of resistance to renewable energy infrastructure in eastern Canada and the United States.

It is helpful to think about these sorts of infrastructure conflicts as involving four stages that occur once the proponent proposes the project. First is the *project review stage*, where the project proposal is submitted for and undergoes regulatory review. Except in cases where an independent regulatory authority has final decision-making authority, the second stage is the *political stage*, where elected officials in government decide how to act on the results of the regulatory review. After that formal decision is made, in highly contested cases like the ones being examined here, there are two additional stages.

The third is the *legal stage*, where losers in the decision process challenge the legality of the decision in court and the courts hear and resolve those legal issues. Finally, the *on-the-ground stage* commences, where construction starts and physical resistance to the project emerges. Sometimes these stages overlap, especially the last two. Legal or physical conflicts can emerge during the project review and political stages, and physical conflict can emerge while legal proceedings are going on. Frequently, legal proceedings are battles over injunctions, either to prevent physical conflict from disrupting construction or to determine whether construction can proceed prior to the resolution of certain legal issues. Our case studies will trace the evolution of conflict through these stages.

Chapter 11 delves into the third question, whether we can build energy infrastructure decision-making processes that minimize the risk that unproductive place-based resistance will thwart projects that are in the broader public interest. Chapter 12 brings the book to a close by summarizing the results of the analysis and exploring the question of why more promising approaches to project review and approval have been used so rarely by decision-makers.

I The Oil Sands Policy Regime

2 The Oil Sands Policy Regime: Resource, Markets, and Politics

This book focuses on campaigns to resist the expansion of the oil sands, a major unconventional oil resource in northern Alberta, Canada, by blocking the development of new pipelines designed to increase access to markets in Canada, the United States, and abroad. This chapter and chapter 3 use the policy regime framework to provide an overview of Canada's oil sands and how they are governed by provincial and federal rules. This chapter examines the oil sands resource, its economic significance, and its environmental impacts. It also examines background conditions that provide the context for the oil sands policy regime, particularly changes in markets and politics. The final sections examine the strategic actors that form the core of the policy regime: the oil sands coalition and the anti-pipeline coalition. Chapter 3 examines the ideas and institutions of the oil sands policy regime and how government policies have responded to environmental concerns.

The Oil Sands Resource and the Political Economy of Canada

The oil sands are a massive deposit of unconventional oil in northern Alberta. Their size gives Canada its rank as the world's third-largest holder of oil reserves, behind only Saudi Arabia and Venezuela. The oil sands represent 165 billion barrels of proven oil reserves (2016 estimate), a remarkable 96% of Canada's total oil reserves. Their rapid growth over the past 15 years has been a major driver in the economies of Alberta and of Canada as a whole. Production was 2.9 million barrels per day in 2018 and is projected to increase by over 50% to 4.3 million barrels per day by 2035(Canadian Association of Petroleum Producers 2019).[1]

There are two production methods for the oil sands. For bitumen that is relatively close to the surface, the product is mined with shovels and trucks. For deeper deposits for which mining is not feasible, the bitumen is accessed through wells injected with steam and/or solvents. In this so-called in situ process, the heat from the steam reduces the viscosity of the bitumen so it can be brought up to the surface. All the early production facilities were mines, but by 2012 production from in situ facilities exceeded that of mines. Of the total oil sands reserve, only one-fifth is recoverable through mining; the remainder requires in situ processes. The product of these extraction operations is bitumen, which can either be shipped as "dilbit" (bitumen diluted with condensate) or upgraded to a quality equivalent to that of crude oil (synthetic crude oil) (Royal Society of Canada 2010, chapter 4).

The first commercial operation in the oil sands opened in 1967, but it wasn't until shortly after the new millennium that the region experienced explosive growth. Figure 2.1 tracks the growth in production of the oil sands from 1971 to 2019. It took until 1979 for production to reach 100,000 barrels per day. Production did not exceed 500,000 barrels per day until 1997 and did not top 1 million barrels per day until 2004. From 2004 to 2019, the average annual rate of growth in production was 8.7%. In 2009, oil sands production exceeded conventional oil production in Canada for the first time. In

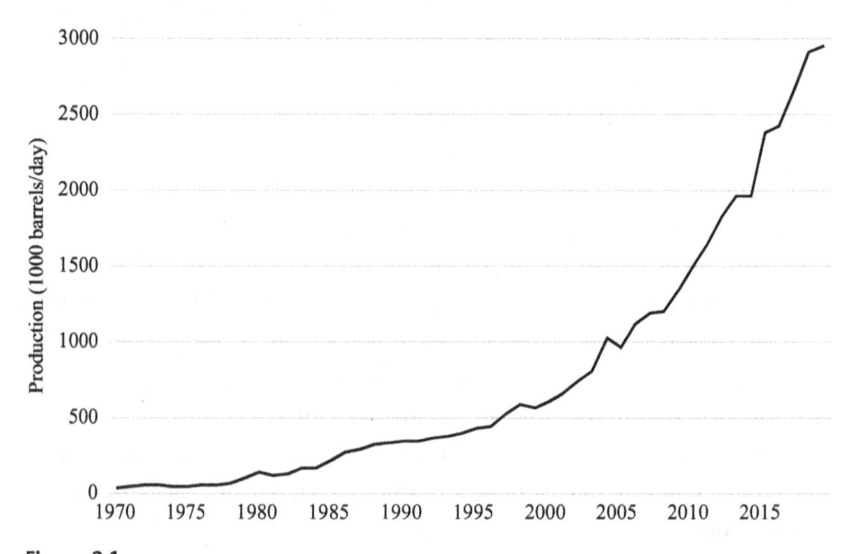

Figure 2.1
Canadian oil sands production, 1971–2019.
Source: *Statistics*, Canadian Association of Petroleum Producers, https://www.capp.ca/resources/statistics/.

2016, the oil sands constituted 78% of Alberta's oil production[2] and 62% of Canada's. Over the period 2010–2016, an average of 46% of production was bitumen upgraded to synthetic crude oil and 54% was produced as raw bitumen (Canadian Association of Petroleum Producers 2017).

The overwhelming majority of oil sands are exported to the United States, making the sector highly dependent on US markets and politics. In 2016, 79% of the 3.9 million barrels per day of Canadian oil production was exported, and between 97% and 99% of those exports go to the United States (the remaining 1%–3% going out through the Trans Mountain pipeline to Burnaby, British Columbia). While trade data are not collected specifically for the oil sands component of the Canadian oil sector, it is possible to estimate, based on a combination of production and trade data produced by the National Energy Board, that approximately 85% of oil sands production is exported to the United States. The remainder finds its way to refineries in Alberta and other Canadian provinces.[3]

The geography of North American energy production and transportation has created a distinctive problem for the oil sands. Petroleum products receive different prices depending on their quality and location. Alberta's bitumen is priced as part of the Western Canadian Select (WCS) index. Most North American oil gets the West Texas Intermediate (WTI) price, and international oil is typically priced according to the Brent crude index. WSC has always faced a price discount because it is of lower quality; it needs to be upgraded to have the characteristics of the "light, sweet" crude oil. But that differential can vary depending on market conditions, one of the most important being transportation capacity (Heyes, Leach, and Mason 2018; Walls and Zheng 2020). There is also a variable differential between West Texas Intermediate and Brent prices. While the two benchmarks historically have tracked each other closely, in the early 2010s a substantial discount for WTI emerged relative to Brent. Figure 2.2 shows these trends. This discount emerged because of oil transportation constraints between the midwestern United States and coastal markets, where the products would be exposed to the Brent price (Borenstein and Kellogg 2014). For 2007–2010, the WTI and Brent benchmarks were, on average, less than a dollar (US) per barrel apart.[4] Between 2011 and 2014, the price discount averaged $13 (National Energy Board 2016, 89). The inability to get oil sands products the global price resulted in billions of dollars per year in lost revenue to the sector, creating the urgent sense within the oil sands coalition that getting new pipelines and access to tidewater was imperative.

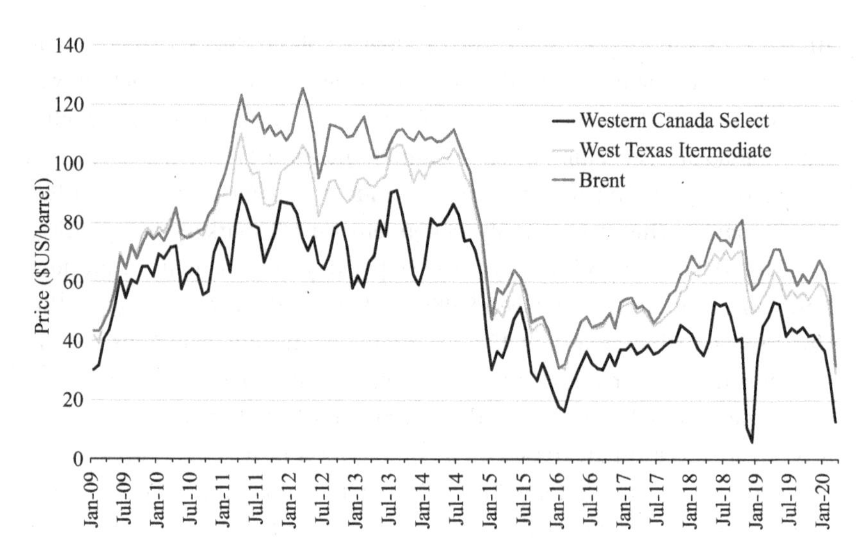

Figure 2.2
Brent, WCS, and WTI prices, 2009–2020.
Sources: WCS and WTI: Government of Alberta Economic Dashboard (Government of Alberta, n.d.a); Brent: Europe Brent Spot Price (Energy Information Administration, n.d.).

Alberta's economy is heavily dependent on the oil sector, a dependence manifested in statistics on employment, GDP, and provincial revenues. In 2017, 140,300 people were employed in the upstream energy sector in Alberta, the majority of whom worked in the oil sands (Government of Alberta, n.d.b). The oil and gas industry overall accounted for 16% of GDP in 2018, with the lion's share of that coming from the oil sands (Statistics Canada 2018). Crude oil makes up 43% of the province's exports, with the oil sands making up about four-fifths of that. With respect to government revenues, figure 2.3 shows total nonrenewable resource revenues and their bitumen royalty component from 2011 to 2018. The contribution of oil sands royalties to provincial budget revenues from 2011 to 2017 averaged 8.6%.[5]

This dependence on resource revenues is a double-edged sword for the province. On the one hand, it helps foster the "Alberta advantage" for attracting capital and labor. The province has chosen to fund government programs with these resource revenues, creating the fiscal space to avoid having a sales tax and keeping income and corporate taxes low, making Alberta the least taxed Canadian province (Government of Alberta 2017).

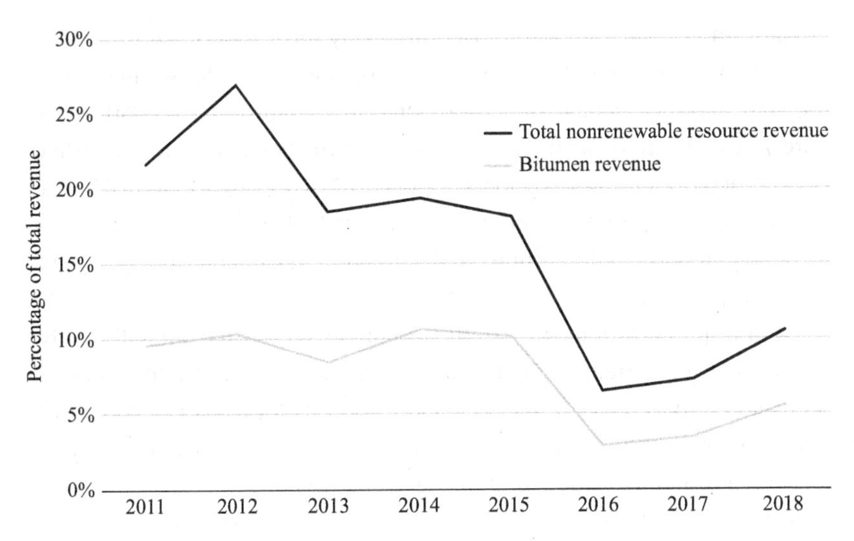

Figure 2.3
Alberta's nonrenewable resource revenues.
Source: Government of Alberta, n.d.d, "Resource Revenue Collected," Schedule 1 (Revenues).

On the other hand, the boom and bust nature of commodity markets means that provincial revenues fluctuate significantly. In the 2010s, oil sands revenues varied widely, from a low of $1.2 billion in fiscal year (FY) 2015–16 (2.8% of government revenue) to a high of $5.2 billion in FY 2013–14 (11.2% of government revenue).

The Environmental Challenges of the Oil Sands

As an unconventional source of crude oil, the oil sands have massive environmental impacts on land, water, and air. According to an expert panel assembled by the Canadian Council of Academies, "The environmental footprint of oil sands operations on air, water, and land is wide-ranging, significant, and cumulative, and will grow as production using current methods increases" (Canadian Council of Academies 2015, xiv). The disturbance to the land varies by production method, with mining being by far the most intense. The mining process involves stripping away the boreal forest and soil and then gathering the bitumen with mammoth shovels and trucks. The resulting removal and fragmentation of the boreal forest has had significant impacts on wildlife habitats. Impacts are likely most significant for wildlife dependent on old forest habitat, for migrating species, and for wildlife requiring large areas for habitat (Alberta Biodiversity Monitoring Network 2014). Boreal caribou, an iconic threatened species of the boreal region, are particularly vulnerable. Of the seven local populations of boreal caribou in Alberta's oil sands region, population trend data are available for four, and all four of those populations are in decline (Environment and Climate Change Canada 2017c, appendix A).[6]

Land impacts from in situ production are less intensive but still involve significant fragmentation of the forest from seismic lines and roads. Oil sands companies are required to reclaim land disturbed for development but at present only a miniscule fraction of the disturbed land has been reclaimed (Canadian Council of Academies 2015).

Significant land and water impacts are also created by the massive tailings ponds that are created in the process of managing the water and chemical waste when removing the bitumen from the material dug up with it. Some of the most enduring images of environmental damage caused by the oil sands were the jarring photos, promoted by the media and environmental critics, of those tailings ponds amid the northern boreal forest. In 2008, the deaths of 1,600 ducks that landed on one of the ponds created a dramatic international media moment. The structures are massive: the Syncrude Tailings Dam is the largest dam in the world by volume (US Bureau of Reclamation, n.d.). Oil sands tailings are different from other mining tailings because of the prevalence of fluid fine tailings, which take a very long period to settle, so their treatment and disposal are more challenging. By 2011, the tailings ponds and their containment structures covered 182 square kilometers of land, and by 2016 they held a volume of 1.3 billion cubic meters of tailings (equivalent to 400,000 Olympic swimming pools) (McNeill 2017a).

These tailings ponds are of concern because of both the size of their land disturbance and the fate of the contaminants in the ponds and their impact on reclamation and its costs. Tailings have now been shown to have seeped into groundwater, and there is also concern about the safety and stability of the very large dams that have been constructed to contain them (Canadian Council of Academies 2015, 42–43). There is also concern about oil sands companies closing down and walking away from their obligations to reclaim the areas, saddling the government of Alberta with the obligation and cost of doing so (McNeill 2017b). According to the Royal Society of

Canada, "Current government policy on financial security for reclamation liability leaves Albertans vulnerable to major financial risks" (Royal Society of Canada 2010, 279).

Freshwater withdrawals from rivers in the region are substantial and have prompted significant concern about maintaining downstream flow, especially during low flow periods in the Athabasca River area, where the big mines operate. Over time, however, operators began recycling more water and using saline groundwater. Until now, the rate of withdrawals has been limited to quite small percentages of annual flows. There are concerns about whether climate change will significantly reduce regional flows and further constrain water use in the future (Canadian Council of Academies 2015, 34–35).

Oil sand operations and upgraders produce significant quantities of air pollution, including sulfur dioxide, nitrogen dioxide, ozone, particulate matter, polycyclic aromatic hydrocarbons (PAHs), mercury, and volatile organic compounds. While these air pollutants have a significant environmental impact, thus far they have rarely exceeded air quality standards set by Alberta and the Council of Ministers of the Environment (Canadian Council of Academies 2015, 14).

While impacts on water, land, and regional air quality are of great concern in northern Alberta, the large greenhouse gas footprint of the oil sands is what has prompted the most international concern and mobilization of the "keep it in the ground" movement against new pipelines. Oil sands create carbon dioxide emissions from the massive trucks and shovels in the mines and from the use of natural gas for upgrading and generating steam for in situ production. The high consumption of natural gas for steam makes the in situ process more emission intensive than mines (Canadian Council of Academies 2015, chapter 2).

The energy intensity of production makes oil sands more emission intensive than most sources of oil. Estimates of the emission intensity vary, but there is general agreement that over the "well-to-wheels" life cycle of oil sands production—from their extraction and production in northern Alberta to their final consumption as combusted fuel—emissions from oil sands are higher than for most other types of crude oil (Masnadi et al. 2018). One widely cited finding is that of the US State Department in the Keystone XL pipeline environmental assessment. It concluded that the emission intensity of the oil sands is "an estimated 17% more GHGs on a life-cycle basis than the average barrel of crude oil refined in the United States" (US

Department of State 2015, 10; see also Masnadi et al. 2018). The Pembina Institute uses an estimate of "31 percent more emissions than the average North-American crude" (Israel 2017).

Canada's greenhouse gas (GHG) inventory report shows that 2018 oil sands emissions were 84 million tonnes, or 11.5% of Canada's 729 million tonnes. The sector's emissions have grown by a factor of 2.2 since 2005, when they constituted 5.1% of national emissions. Alberta is Canada's highest-emitting province. Its 273 million tonnes of emissions are 37.4% of Canada's total, and oil sands constitute 29.7% of Alberta's emissions, a significant increase compared to the 15.1% of Alberta's emissions they accounted for in 2005 (Environment and Climate Change Canada 2019). Figures 2.4 and 2.5 show oil sands emissions in the context of other economic sectors in Canada and Alberta.

Current oil sands emissions are not as troubling as the projected increases in the sector's emissions. If anything like the 50% increase in production by 2030 estimated by the Canadian Association of Petroleum Producers (CAPP) comes true, the oil sands will be the fastest-growing sector in Canada and

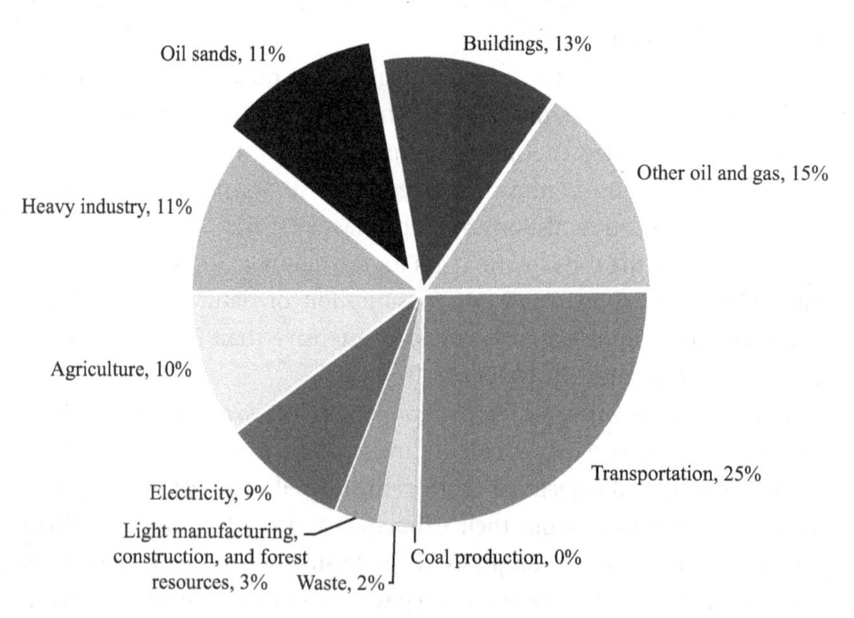

Figure 2.4
GHG emissions for Canada by Canadian economic sector, 2018.
Source: National Inventory Report 1990–2018: Greenhouse Gas Sources and Sinks in Canada. http://data.ec.gc.ca/data/substances/monitor canada-s-official-greenhouse -gas-inventory/.

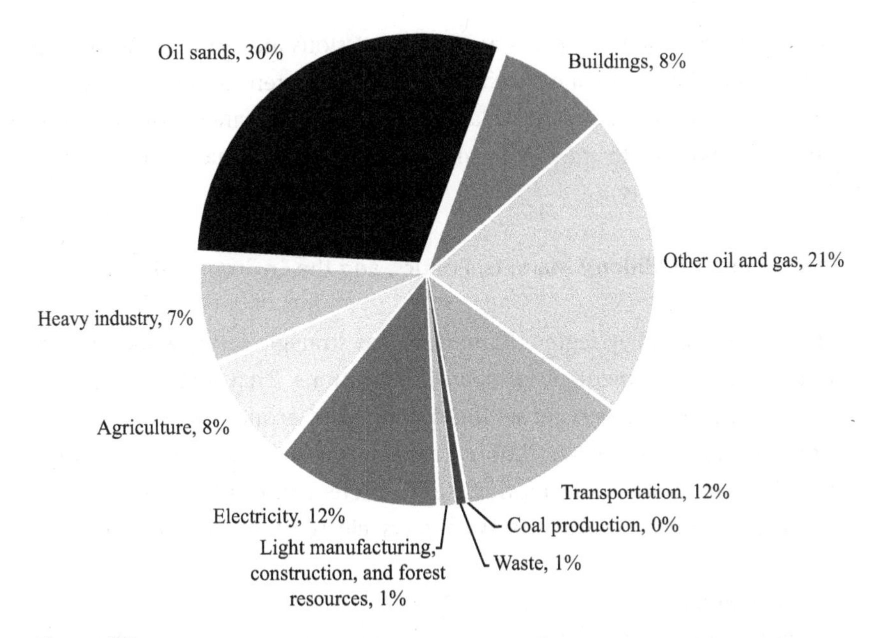

Figure 2.5
GHG emissions for Alberta by Canadian economic sector, 2018.
Source: National Inventory Report 1990–2018: Greenhouse Gas Sources and Sinks in Canada.

take up an increasingly significant fraction of Canada's emissions. Total emissions nationwide are projected to be 742 million tonnes in 2030 compared to 732 million tonnes in 2014, a 1% increase. The projections for most sectors are either flat (transportation, agriculture) or declining (electricity). The building sector's emissions are projected to increase by 8% and heavy industry's by 28%. The oil and gas sector is projected to experience an increase of 21% by 2030, but projected oil sands growth is much higher. (In fact, emissions in other segments of the oil and gas sector are projected to decline.) Government projections show oil sands emissions between 2014 and 2030 increasing by 40 million tonnes, or 59%. In 2030, oil sands are projected to constitute 15% of Canada's emissions (Environmental and Climate Change Canada 2017b). The expected rapid growth in oil sands emissions is a major challenge to Canada's capacity to reach its Paris Agreement target of reducing its emissions to 30% below 2005 levels by 2030.

Given their scale and the pace and intensity of their development, the oil sands have had a wide variety of significant environmental impacts. The two most prominent concerns are GHG emissions and tailings. According to the Canadian Council of Academies, "Under current trends, GHG emissions and tailings disposal and related land disturbance are the most significant contributions to the environmental footprint" (Canadian Council of Academies 2015, xiv).

Background Conditions: Markets, Politics, and the Environment

At its core, the policy regime framework has strategic actors working with and in an environment of ideas and institutions. That regime of actors, ideas, and institutions is set within a context of economic, political, and biophysical background conditions that can be powerful sources of change regarding the strategies and influence of different strategic actors. This section focuses on changes in energy markets, elections, and public opinion.

Energy Markets

The critical market driver for the oil sands is the price of oil. Oil price trends are shown in figure 2.2. The dramatic expansion in oil sands production that occurred in the middle of the first decade of the new millennium was fueled by a rapid runup in oil prices from early 2002, when the WTI price hovered around $20 per barrel, to July 2008, when the price peaked at $135. Then the impacts of the Great Recession took hold and by February 2009 WTI had dropped to $39 before it began another recovery that lasted for more than five years.

In late June 2014, the price of oil began another steep decline despite an economic recovery. On June 20, 2014, the WTI price was $108. By late January 2015, it had fallen to $48. There was a brief recovery over the next six months, but then another steep decline occurred, reaching its nadir in February 2016 at $29. WTI recovered to $50 by June and remained between $40 and $70 from mid-2016 through November 2017. This unexpected oil price collapse was driven by a variety of factors, including the dramatic rise in US production resulting from the shale revolution, and the failure of Saudi Arabia–led OPEC to restrain production once the price softened.

The price shock had a tremendous impact on the Canadian oil sector. The economic implications were felt across Canada, but Alberta was hit hardest. Spending by the oil patch dropped by one-third from 2014 to 2015 and in 2016 was two-fifths of the spending level two years earlier. It continued to drop during 2017 and 2018 (IHS Markit 2019; Canadian Association of Petroleum Producers 2019).

While much of the North American oil industry benefited from the rebound in the WTI index, the Alberta oil sector was plagued by a skyrocketing discount for Western Canada Select. The price differential between WCS and WTI, after several years of being in the teens, jumped to $27 in February 2018. After a brief decline, it resurged in the second half of 2018, reaching a stunning $45 in October 2018. At the time, the WCS price was only $11 per barrel. For 2018, the differential was more than double what it was the previous year. This price crisis led to a remarkable response from the government of Alberta—a decision to curtail oil sands production. That decision reduced the discount for oil sands crude back to historical levels (IHS Markit 2019), at least before the COVID-19 crisis roiled markets further.

Alberta's unemployment rate was 4.5% in early 2015, but by early 2016 it exceeded the Canadian average and then peaked at 9% in late 2016 before recovering somewhat to below 8% by May 2017. In June 2019, it stood at 7%, above the national average of 5.7%. In 2015, capital investment in the province was 37% less than in 2013. The province's 2015 GDP was 7% lower than two years earlier and at the end of 2018 still sat below the 2013 peak.[7]

The value, profitability, and growth of the oil sands are heavily influenced by the price of oil, more so than conventional oil, because of the high costs associated with the complexity of the operations required in extracting and converting bitumen into synthetic crude oil. A 2015 study found that the breakeven price justifying new oil sands facilities was between $85 and $95 per barrel for new oil sands mines (without an upgrader). For new in situ sites, the breakeven price was lower, between $55 and $65 per barrel, but still marginal given recent price trends (IHS Markit 2015). According to one source, these costs make Canada the producer with the third-highest cost in the world, far above major sources of conventional oil such as the Middle Eastern countries (Knoema, n.d.).[8]

Studies exploring oil supply cost curves, which examine how much supply is available at different oil price ranges, routinely rank the oil sands high on the curve, meaning that there are many global sources of oil that can be competitive at far lower oil prices (Aguilera 2014). In addition to making the Canadian oil sector more sensitive to global oil market fluctuations, high costs also make the oil sands highly vulnerable to future trends in international climate policy. As the international community finally comes to grips with the need to stay within a global carbon budget, future increases in sands production may not be economical (McGlade and Ekins 2015; Jaccard, Hoffele, and Jaccard 2018; Heyes, Leach, and Mason 2018).[9]

Politics

The Canadian political environment was relatively stable from 2005 to 2015 but underwent quite a dramatic change in 2015 as a result of the election in Alberta and in Canada's federal election. Until that tumultuous year, the Alberta oil sector was fortunate to have the stability of Alberta's electoral politics for almost the entire history of oil sands development. The Progressive Conservative Party of Alberta had held power in the province since 1971, an exceptional period of one-party dominance. That political stability ensured that any pressure to slow oil sands expansion or increase regulatory costs was met with strong resistance from the government in power.

Cracks in that dominance began to show in 2009, when a party to the Conservatives' right, the populist Wildrose Party, emerged to seize advantage of growing discontent with the ruling Conservatives. Despite leading the Conservatives in opinion polls prior to the election, Wildrose was unable to take the government in the 2012 election but did break through to become the official Opposition. The emergence of the new party as a threat to Conservative dominance helped trigger a leadership crisis in the ruling party, whose leadership changed hands three times in four years after Premier Ed Stelmach resigned in 2011.

While the Conservatives' dominance had slipped, no one expected its crushing defeat by the left-leaning New Democratic Party (NDP) in the May 2015 election. Going into the election, the Conservatives held 61 seats, Wildrose 17, and the NDP 4. When the election results were counted, the stunned Conservatives had been reduced to 9 seats. Wildrose would continue as the official Opposition with 21 seats. The NDP won the majority, taking 54 seats, more than it had won in every Alberta election since 1955 combined. The stunning victory has been credited to a combination of an overly entitled establishment party that had lost touch with how the province was changing and NDP leader Rachel Notley's effective and inspiring campaign (Bratt et al. 2019).

The substance of Notley's platform was not particularly threatening to the oil sector, but the thought of a social democrat running the oil province did create deep fears within the business community, which had grown to expect pro-business governments in power. Conservative leader Jim Prentice played those fears for all they were worth in the election, but Notley deflected his attacks as fear-mongering rumors. Her platform promised a review of oil and gas royalties and a modest increase in the corporate tax. A promise to phase out the use of coal for electricity was a hint of things to come, but there was no indication of the comprehensive climate plan that emerged later that year. The platform contained no positions on pipelines, but during the campaign Notley did express support for Energy East and Trans Mountain and skepticism regarding the political feasibility of Northern Gateway and Keystone XL given the quagmire both had entered by 2015 (Bratt et al. 2019).

Notley's premiership only survived one term, as she was replaced by Jason Kenney of the United Conservative Party in April 2019. The implications of this change will be addressed in chapter 8.

Canada's national electoral landscape was not quite as stable as Alberta's during the surge in oil sands development, but it still posed little threat to oil sands' emergence, and in the first decade and a half of the twenty-first century it has been quite conducive to it. Jean Chretien's Liberals took advantage of a fractured right wing to govern from 1993 to 2003. One of his legacies was entering and then ratifying the Kyoto Protocol, which did anger Alberta and the oil sector, but the fact that Chretien and his successor, Paul Martin, never developed a plan to implement it softened the blow. Alberta and the oil sector were relieved when its stalwart champion, Calgary-based Stephen Harper, united the Right and then won three consecutive federal elections to hold government for over nine years, from 2006 to 2015, the final four with a solid majority.

The 2008 election campaign featured Liberal leader Stephane Dion's ill-fated "Green Shift," a proposal for a revenue-neutral carbon tax.[10] Dion's wobbly communication skills were no match for Harper's relentless criticism. The 2008 world economic crisis further undermined the political viability of Dion's proposal, and Harper made the most of it: "It is like the [1980] national energy program in the sense that the national energy program was designed to screw the West and really damage the energy sector—and this will do those things. This is different in that this will actually screw everybody across the country" (CBC 2008). Harper went on to win his second minority government. Harper's success against Dion changed the politics of climate, as he came to believe that opposition to carbon taxes and aggressive climate policy was a winning political strategy. As global climate policy momentum began to increase, Harper adopted the strategy of limiting policy change and castigating proposals that were more ambitious.

Six months after Notley's shocking victory, the oil sands coalition was dealt another blow when Harper's decade-long reign as prime minister was brought to an end by the election of Justin Trudeau's Liberals. The 2015 federal election campaign was a tightly fought three-way race between Harper's ruling Conservatives, Trudeau's Liberals, and Tom Mulcair's NDP, with the NDP holding the lead when the election was called in August. But the youthful Trudeau's strong debate performances, stellar campaigning, and shrewd positioning on economic policy contributed to the Liberals' resurgence. Trudeau's party topped Harper's by 7 percentage points in the voting, and Mulcair's lackluster campaign relegated his NDP to a distant third. Trudeau won a majority with 184 seats to the Conservatives' 99 and the NDP's 44 (Pammett and Dornan 2015).

The oil patch was deeply concerned about Trudeau's rise, given the legacy of his father's 1980 National Energy Program, which poisoned the Liberal brand in Alberta, not to mention a Liberal Party run by a Trudeau. While his prospects for gaining seats in Alberta were slim, he understood that he needed Alberta's cooperation to be able to forge any sort of effective Canadian climate policy. To his credit, Justin Trudeau confronted this legacy head on and delivered one of his first major energy and climate policy speeches in the heart of the Canadian oil patch, to the Calgary Petroleum Club, proclaiming: "I'm the Leader of the Liberal Party of Canada, my last name is Trudeau, and I'm standing here at the Petroleum Club in Calgary. I understand how energy issues can divide the country. But I also know that strong leadership can see us through the challenges we face" (Trudeau 2015).

In that speech, he explicitly criticized his father's National Energy Program for being pushed through over the vehement opposition of Alberta. He built the case that the failure to get new pipelines built during Harper's tenure was a consequence of having weak environmental policies and poor relations with Aboriginal peoples, which undermined trust domestically and internationally. He proposed federal leadership on climate policy in target setting and carbon pricing but would leave the implementing mechanisms up to the provinces (Trudeau 2015).

In addition to the commitment to enact carbon pricing, the Liberal platform in 2015 promised to review the environmental assessment process and ensure that assessments include "upstream impacts and greenhouse gas emissions." The Liberals pledged to "modernize the National Energy Board, ensuring that its composition reflects regional views and has sufficient expertise in fields like environmental science, community development, and Indigenous traditional knowledge." While the platform didn't mention pipelines specifically, during the campaign Trudeau expressed strong support for Keystone XL and opposition to Northern Gateway. His positions on Trans Mountain and Energy East were similar: he expressed openness to them but could only decide after a review under a reformed, more rigorous process. The platform did say that the Liberals would implement the UN Declaration of the Rights of Indigenous Peoples, and it repeated a campaign refrain about social license: "While governments grant permits for resource development, only communities can grant permission" (Liberal Party of Canada 2015b, 42).

While these two 2015 elections transformed the political context for Canadian energy and environmental policy, there's no evidence that they were precipitated by a surge in public concern about the environment generally or climate change in particular. One of the best indicators of public concern is what polls reveal when they ask respondents what the "most important problem facing Canadians" is. When looking at the available data on this question, shown in figure 2.6, from 2005 to 2020, there was

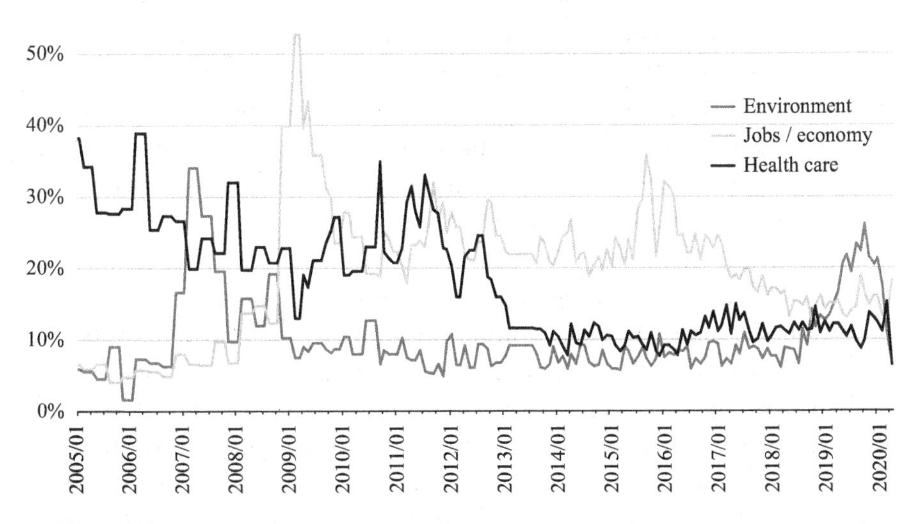

Figure 2.6
National issue of most concern according to public survey data, 2004–2020.
Source: Nanos Research.

in fact a strong surge in the salience of the environment, but it occurred in 2007, when the environment briefly exceeded both health care and the economy as the most important issue. That surge in salience did strongly influence policy development in the provinces of British Columbia and Ontario, but it made little dent in the Harper government's actions (Harrison 2012). That surge was subsequently crushed by the Great Recession and did not return in advance of the 2015 Canadian elections.

With respect to climate change in particular, despite accumulating evidence and a variety of extreme weather events domestically and internationally, there had been no measurable increase in concern about or salience regarding climate change among the Canadian public over the period 2007–2015. Leading up to the 2015 elections, Canadians were concerned about the environment and climate change, but there is no evidence that the breakthrough election results of 2015 were fostered by a surge of environmental concern. That changed in 2019, however. The 2019 election and its implications for the policy regime are discussed in chapter 8.

Actors—Oil Sands Coalition

Strategic actors work within the context of these background conditions. It is very common for actors within policy subsystems to be organized as coalitions, one in defense of the status quo and the other challenging it (Jenkins-Smith et al. 2014; Hochstetler 2011). At the core of the oil sands coalition are the oil industry, pipeline companies that ship their products, and the government of Alberta. The Alberta oil industry consists of both upstream producers in the oil sands and the vast network of affiliated suppliers and service companies that the large firms rely on. Naturally, their core interest is in profits.

The fundamental challenge for the sector is that it is landlocked in northern Alberta, and as the rapid growth of the industry occurred around 2005, pipeline capacity did not keep pace and was constraining the sector's access to markets. Transportation capacity constraints had two impacts on industry profits. First, without access to new and expanding markets, growth and the resulting revenues potentially available with expanded market access were constrained. Being shackled by existing routes to the United States tied the sector to a country whose domestic oil production was increasing substantially and whose oil consumption had plateaued at that time (Energy Information

Administration 2017, 9). Second, as described earlier in this chapter, the particular configuration of North American oil markets created a price discount for oil sands producers whereby their products could not fetch the same price that they would if the sector had greater access to international markets. As a result, getting "access to tidewater," where their products could fetch the global price, became the rallying cry of the oil sands coalition in motivating pipeline expansion.

The largest firms operating in the oil sands are Suncor, Canada Natural Resources, Imperial Oil, Syncrude, and Cenovus. These companies are all active in government and public relations but are also members of the industry's trade association, the Canadian Association of Petroleum Producers (CAPP), which represents the sector in the media and in policy discussions (Hussey et al. 2018).

The pipeline companies are naturally eager to benefit from this tidewater imperative, and a flurry of pipeline proposals emerged late in the first decade of this century and early the following decade. The three largest oil pipeline companies are represented in this book's case studies. Enbridge, proponent of the Northern Gateway Pipeline, was the largest company, with US$25 billion in revenue in 2016. Kinder Morgan, the Houston-based proponent of the Trans Mountain Expansion Project, had revenues of US$12.9 billion in 2016. TransCanada, the Calgary-based proponent of both Keystone XL and Energy East, was the third-largest oil pipeline company, with US$8 billion in revenues in 2016. (In May 2019, the company's name was changed to TC Energy.) For these companies, new pipelines promise greater revenues and profits. While they share an interest with the oil sands production companies in expanding the pipeline network, there are differences between the two subsectors in what the price for shipping oil sands products, or tolls, should be. For oil sands production companies, lower tolls promise lower costs of getting the product to market and therefore higher profits. For the pipeline companies, being able to charge higher tolls brings more profit. Like the oil sands producers, these companies conduct their own lobbying and media relations but are also represented by a trade association, the Canadian Energy Pipeline Association (CEPA).

The oil sands coalition has been strengthened by social mobilization of resource-dependent workers and communities. This more recent development will be addressed in chapter 12's examination of recent changes in the oil sands regime.

The government of Alberta has also been a reliable member of the oil sands coalition because of the dominant role the oil sands play in the provincial economy. This heavy dependence on the oil sector and the oil sands in particular has led a number of scholars and commentators to label the province a "petro-state" (Carter 2016a; Shrivastava and Stefanik 2015; Adkin 2016). The term is usually reserved for developing countries whose economies are dominated by oil, resulting in a series of dysfunctional economic and political consequences (Karl 1997; Ross 2012). Canadian writers who refer to Alberta as a "first world petro-state" emphasize the dominance of the oil industry and its interests in Alberta politics, usually emphasizing the relatively low royalty rates charged by the province and lax environmental regulations designed to minimize the costs of doing business (Carter 2016a). The dominance of the Progressive Conservative Party in Alberta elections from 1971 to 2015—virtually the entire period of the establishment and explosive growth of the oil sands—certainly contributes to this impression. Research has demonstrated how the oil industry and the Conservative government worked together to protect the industry from escalating demands for environmental regulation and greater recognition of Aboriginal rights (Hoberg and Phillips 2011). Despite a different ideology and more-ambitious environmental policies, the NDP government of Rachel Notley could also still be considered a core member of the oil sands coalition.

The government of Canada can also be considered a member of the oil sands coalition, although the situation became somewhat more complicated with the election of Liberal Justin Trudeau as prime minister in 2015. The oil sands boom coincided with the emergence and subsequent dominance of Stephen Harper as prime minister of Canada. Harper is from Calgary, and his political roots are with the Reform Party, a western, conservative populist party that emerged in the late 1980s. He has been a stalwart champion of the oil sector and a relentless opponent of climate policies he thought would threaten the industry (Hoberg 2016). Some commentators have even gone so far as labeling Harper's Canada a petro-state (Nikiforuk 2013). Harper's energy and environmental policies are discussed in more detail in chapter 3. Whether it's fair to say that Trudeau's federal government is still a core part of the oil sands coalition is something that will be assessed later in this book.

Actors—Anti-pipeline Coalition

The two core elements of the anti-pipeline coalition are environmentalists and Indigenous groups, joined by some municipalities and provincial governments. Resistance to oil sands pipelines has been a major focus for the Canadian environmental movement since 2011. In the United States, the Keystone XL pipeline was arguably the country's biggest environmental controversy of the 2010s, and the same can be said about the Canadian pipelines. These resistance campaigns frequently involve a variety of groups. Some are local groups concerned about local impacts to environmental values, whereas others are large regional, national, or even international groups concerned about broader issues, including climate change.

Environmentalists

Environmental groups' attention to the environmental risks of the oil sands really began in 2005, when the Pembina Institute began publishing reports emphasizing the risks to regional air quality, regional water quality and quantity, habitat loss, Indigenous rights, and global climate change posed by oil sands expansion (Pembina Institute 2005; Hoberg and Phillips 2011). Greenpeace Canada opened an Edmonton office in 2007, and Environmental Defence and Équiterre joined the fray shortly thereafter.

Table 2.1 lists the major North American environmental groups involved in the oil sands and associated pipeline disputes, where they are headquartered, what their annual budget (Canadian or US dollars) was in fiscal year 2016, and their social media status in 2017.

The challenge for the environmental community was how to contain the growing environmental footprint of the oil sands. Different groups had different priorities, depending on their focus, but after 2009, climate change became an increasing priority for the Canadian environmental movement. Activists had grown increasingly impatient with the reluctance of Jean Chretien's Liberal government to develop a meaningful plan to implement its commitment under the Kyoto Protocol to reduce emissions to 6% below 1990 levels by 2012 (Simpson, Jaccard, and Rivers 2008). Given the decentralized federalism in Canada and the alliance of the government of Alberta with the oil industry, environmentalists knew that any significant policy change on the oil sands or climate policy more generally would need to be driven from outside Alberta. Yet even before Harper took over the reins in

Table 2.1
Major environmental groups involved in oil sands and pipeline resistance

Group	Headquarters	Annual budget ($ millions) FY16	Facebook likes (March 2017)	Twitter followers (March 2017)
350.org	Washington	US$10.6	569,020	340,513
Canadian Parks and Wilderness Society	Ottawa	C$5.1	11,117	12,695
Ecojustice	Vancouver	C$5.4	58,270	26,281
Environmental Defence	Toronto	C$3.1	84,780	24,010
Equiterre	Montreal	C$3.7	94,684	20,708
Greenpeace Canada	Toronto	C$12.2	209,151	43,051
Natural Resources Defense Council	New York	US$140.4	912,983	277,813
Pembina Institute	Calgary	C$4.5	6,181	21,825
Sierra Club Canada	Ottawa	C$0.4	6,897	14,154
Sierra Club (US)	Oakland	US$63.4	910,943	314,301
Stand (formerly ForestEthics)	San Francisco	US$2.7 C$0.8	274,089	13,557
West Coast Environmental Law	Vancouver	C$2.1	7,050	14,262

the January 2006 federal election, the federal government was clearly resistant to more-aggressive regulation (MacDonald 2020). The environmental movement needed a new and different source of leverage.

The pipeline resistance strategy began to take form when a group of environmentalists began meeting in 2005, under the name Upstream Strategy Working Group, to discuss how to address oil sands expansion.[11] Its February 2006 strategy document provides an early glimpse into the vision at the root of the "keep it in the ground" movement, which energized resistance to the expansion of fossil fuel infrastructure. The Upstream Strategy Working Group declared that its overriding goal was "to transform Canada's fossil fuel sector into a sustainable energy sector through a combination of '*full cost*' and '*full stop*' strategies, and by working with others on the development of a positive energy *vision* for Canada." Their core target was the oil sands, and their core motivation was climate. As they put it, "Our focus will be on Canada's West and North, where the vast majority of energy development is taking place, with special attention to the tar sands, the centre of the fossil fuel spider web and possibly the largest climate disaster in North America" (Canadian Upstream Strategy Working Group 2006).

The most important lever for power was their "full stop" campaign, designed to "stop or radically delay three pipeline projects that bottleneck the industry (MVP,[12] Enbridge, and Keystone) in the period between now and 2008" (Canadian Upstream Strategy Working Group 2006). Jessica Clogg, a participant in the working group discussions and a lawyer with West Coast Environmental Law, describes the thinking that led to the strategy:

> A number of people said that if we want to get meaningful climate policy in Canada we have to show that we can stop a pipeline. At that moment, we did not have the power. No one could picture a pathway to addressing climate change unless we could tackle the tar sands. And the tar sands were too big. The players were too big. If you think about the tar sands as this spider at the centre of a web, what could we do to show that we could be powerful and to change the policy conversation around climate in Canada? And that was the genesis of the idea that we have to stop the pipeline. (Clogg 2017)

While their credible threat took several more years to achieve than they hoped, their strategy constitutes the birthplace of the oil sands pipeline resistance campaign. Will Horter, then executive director of the Dogwood Initiative and one of the key architects of the strategy, elaborates: "The idea was to kill a project. The industry has gotten everything it wants from the federal government and the Alberta government. We need to psychologically stop one thing, so they are willing to talk to us about something meaningful. And Enbridge rose to the top of that. And also what became Kinder Morgan" (Horter 2016).

The group was appropriately attuned to the need to do more than block infrastructure, saying: "Ensuring the formation of a politically powerful coalition promoting positive policy change to achieve a sustainable energy vision for Canada is the necessary backdrop to our 'no' work—we can't just be against everything. The positive vision will fill the space created by our meta-communications project that creates a public demand for change by highlighting the negative impact of the current energy paradigm (dirty energy serving US and not Canadian interests)" (Canadian Upstream Strategy Working Group 2006).

The group's 10-year strategy seems remarkably prescient:

1. Build a credible threat (by 2008)—go from current powerlessness to being on industry's radar screen as directly relevant to their business interests

2. Become a player (2009–2014)—being able to set a good part of the agenda, rather than just responding
3. "Sun Nation" (2015)—the majority of actors are on the same side as us in implementing a positive energy vision for Canada (Canadian Upstream Strategy Working Group 2006)

The document also listed 10 "sources of power," several of which have become foundations of the movement's influence:

- Working with Canadian First Nations to exercise their legal rights
- Smarter targeting of elected decision makers
- Using EA [environmental assessment] and permitting hearings as a vehicle for communications and to highlight the need for regulatory reform
- Capitalizing on events as a focal point for communications (e.g., climate change forums; tragic accidents like spills)
- Grassroots organizing, particularly with students, to build an engaged and focused group of future energy activists (Canadian Upstream Strategy Working Group 2006)

Strategy in hand, the group went to seek funding from philanthropic foundations. They did not have success with Canada's relatively small and conservative foundations. As one interviewee who wished to remain anonymous told me in 2017:

> One component is there wouldn't have been enough money, just from the Canadian funders, relative to the task at hand. Second, even among the Canadian environmental funders, they tended to be more conservative. They tended to be older money. Some of it was tied to historic or ongoing resource extraction. Their number one funding priority was conservation. They weren't yet funding climate change. Even after some of the US funding started to kick in, it very quickly became politically polarized. Then a lot of the Canadian foundations were afraid to touch it because they felt it would be too polarizing. They didn't want to be associated with it.

With Canadian foundations not interested, the activists turned to the network of large American foundations that had supported many of the groups in their forest conservation initiatives. These American foundations, including Pew, Rockefeller Brothers, Tides, Moore, and Bullitt Foundation, had already invested substantially in conservation initiatives in Canada's north and west, including the Canadian Boreal Initiative, the Great Bear Rainforest, and protecting wild salmon in the North Pacific. As a result, they were quite receptive to the arguments that oil sands expansions and their associated pipelines and tanker projects posed a significant threat to the very regions and natural processes that were already the foundations' priorities. American foundation funding for the resistance strategy became a controversial and divisive issue in Canadian energy and environmental politics, which will be described in chapter 3's discussion of framing.

By 2008, the Tar Sands Solutions Network had quietly emerged as a mechanism for financial support and strategic coordination of the pipeline resistance campaigns. Michael Marx of Corporate Ethics International was hired as coordinator, along with Canadian (initially Dan Woynillowicz of the Pembina Institute) and US (Kenny Bruno of Corporate Ethics International) deputy coordinators (Bruno 2017). The organization has remained very low profile. While it made a brief appearance online in 2013 (Uechi 2013), its website has been taken down, as its organizers have chosen to remain behind the scenes. It still has a Twitter account, @TarSandsSolns, which began tweeting in July 2013, but it hasn't posted since November 2015.

While each pipeline campaign has had its distinctive issues and local actors involved, the foundation-backed Tar Sands Solutions Network surreptitiously provided a surprising degree of both capacity and coordination to the different campaigns. According to Clogg:

> It was all one campaign. A better way of saying it was that we all made concerted efforts to stay in communication. There has never been on our side of the border a [formal] coalition. A coalition has common positions. A network is probably a better word. There was a commitment to work collaboratively. It was more that there were good communication channels between the teams than that there was one big coalition. It was a network of groups with shared goals, diverse strategies, and a shared commitment to work through strategic differences to understand where other groups were coming from. (Clogg 2017)

Most of the funding through the network went to existing groups, but the missing piece was an organization dedicated to promoting the positive vision of a clean energy economy. The first effort to fill this void was PowerUp Canada, created in 2009 by Tzeporah Berman and Chris Hatch, the powerful husband-wife team who emerged as prominent activists in Canada during the Clayoquot Sound campaign in the 1990s (Berman 2011). But that organization didn't survive, and the work of constructing and communicating the positive vision for change was taken over by Clean Energy Canada, initially a project of the Tides Canada foundation and now affiliated with Simon Fraser University in Burnaby, British Columbia. In 2012, Berman replaced

Michael Marx as coordinator of the Tar Sands Solutions Network, a position she held until 2018.

While Canadian environmental groups took a lead role in initiating the resistance networks, leading American environmental groups were also involved from the start. The most active group has been the Natural Resources Defense Council (NRDC), whose mammoth US$140 million budget dwarfs those of all the other groups involved. Its BioGems initiative was designed to identify conservation hotspots on which the group could focus its campaign, and the Peace-Athabasca Delta in northern Alberta had been identified as one of the world's most significant breeding grounds for migratory birds (Price 2017). When oil sands expansion was identified as a risk to the area, the NRDC became involved, stating that "the birds and their wetland rest stops are downstream from the world's largest industrial project—Alberta's tar sands mines. As the dirty oil industry grows, it is already having devastating impacts on the delta" (Casey-Lefkowitz 2009). The Sierra Club in the United States also became deeply involved, especially in the Keystone XL campaign, but on the Canadian pipelines they tended to defer to Sierra Club Canada and its regional chapters.

Indigenous Groups

Indigenous groups have also been critical members of the anti-pipeline coalition. Not all Indigenous groups have been opposed to pipeline projects; some see economic benefits and few environmental risks in pipelines and have willingly cooperated with pipeline companies. But many Indigenous groups have been adamantly opposed to pipelines through their traditional territories, either because of concerns about environmental risks or as assertions of their decision-making rights. Individual Indigenous groups have played important roles in many of the pipeline conflicts, but regional coalitions, to be discussed in the context of specific pipeline cases, have also played a critical role.

Cross-border alliances among groups in Canada and the United States have emerged as a powerful political force in these disputes. In January 2013, at the initiative of Pawnee and Yankton Sioux whose territories would be affected by Keystone XL, there was a gathering of nations to sign the International Treaty to Protect the Sacred from Tar Sands Projects (Indigenous Environmental Network 2013).

In September 2016, an even larger group of Indigenous nations across North America signed the Treaty Alliance Against Tar Sands Expansion. The organization was initially formed by 50 Indigenous groups but, as of August 2017, had expanded to more than 120 Indigenous nations in Canada and the United States. Signatories include Indigenous nations from all Canadian provinces and American states in the paths of the proposed new oil sands pipelines. The Treaty Alliance aims "to prevent a pipeline/train/tanker spill from poisoning their water and to stop the Tar Sands from increasing its output and becoming an even bigger obstacle to solving the climate crisis" (Treaty Alliance 2016a).

The text of the treaty states that the signatories, under their inherent legal authority, ban pipelines supporting tar sands expansion:

> Therefore, our Nations hereby join together under the present treaty to officially prohibit and to agree to collectively challenge and resist the use of our respective territories and coasts in connection with the expansion of the production of the Alberta Tar Sands, including for the transport of such expanded production, whether by pipeline, rail, or tanker.
>
> As sovereign Indigenous Nations, we enter this treaty pursuant to our inherent legal authority and responsibility to protect our respective territories from threats to our lands, waters, air and climate, but we do so knowing full well that it is in the best interest of all peoples, both Indigenous and non-Indigenous, to put a stop to the threat of tar sands expansion. (Treaty Alliance 2016a)

More details about the role of Indigenous nations in natural resource decision-making in the two countries will be addressed in the section on institutions in chapter 3.

Conclusion

The oil sands are a vast oil reserve, making Canada the third-largest holder of oil reserves worldwide. Their economic importance to Alberta and Canada more broadly has led to the creation of a powerful industry-government alliance making up the oil sands coalition. Until about 15 years ago, the coalition seemed to have a monopoly over relevant policy, but the scale and intensity of oil sands development became an issue. Whether looking at habitat disturbed, area covered in tailings ponds, or greenhouse gas emissions, its environmental footprint is massive. Environmentalists and Indigenous groups became alarmed and within several years were able to arm themselves with enough information and campaign funding that they turned the environmental footprint of the oil sands into a significant regional, national, and even international controversy. These are the roots of the anti-pipeline coalition that animates the intensive conflicts depicted in chapters 4–7.

Chapter 3 examines the arenas within which the struggle between the anti-pipeline and oil sands coalitions has taken place. The arena of ideas has witnessed a battle of words and images designed to influence the public and policymakers. The arena of institutions sets the rules by which decisions are made, although as the cases to follow will demonstrate, in policy conflicts strategic actors frequently work to change the rules to benefit their interests.

3 The Oil Sands Policy Regime: Ideas, Institutions, and Environmental Policies

The strategic actors making up the oil sands and anti-pipeline coalition work within an environment of ideas and institutions. Ideas shape both the normative and causal beliefs of actors, and institutions influence the strategies and power of actors. But strategic actors certainly don't take ideas and institutions as given. They employ framing and other forms of discourse to mold ideas to better suit their interests. If the rules of a particular institutional venue aren't working in their favor, strategic actors will try to change those rules or shift decision-making to a venue more to their liking.

The first two sections of this chapter examine ideas, first by looking at framing strategies and then by looking at the emergence of a powerful scientific idea that provides a rationale for the "keep it in the ground" movement. The third section examines institutions—first the basic rules for policy-making in Canada and the United States and then the rules governing the role Indigenous groups play in natural resource policy decisions. The rules specific to pipeline and other energy project decision-making will be examined in the case study chapters. The final section of this chapter examines how these actors, ideas, and institutions have combined to produce environmental policies relevant to the oil sands by the governments of Alberta and Canada. While that section addresses policy beyond climate change, the book's focus is on climate policy. This chapter deals with climate policy up through the elections of Rachel Notley and Justin Trudeau. Chapter 8 will address how those policies changed from 2015 to 2020.

Ideas—Strategic Actor Framing

The conflict over oil sands expansion is not just about a conflict of interests among strategic actors. It is also a struggle over ideas, a framing contest over who can construct images that most influence the minds of policymakers and the public. There are few issues in which framing disputes are more apparent: in this case, the very label one uses to describe the energy resource, oil sands versus tar sands, reveals which side you are on. The framing conflict goes well beyond labels to clashing worldviews about how best to advance human progress. The oil sands coalition constructs the resource as essential to Canadian prosperity; the anti-pipeline coalition constructs the resource as destructive to the environment and a threat to the cultures and governance of Indigenous people.[1]

These competing images are reflected both in the values to which the two coalitions appeal and in the factual arguments they make. Oil sands advocates advocate what Shane Gunster calls "petro-nationalism," emphasizing the resource's contributions to material values through employment, economic growth, energy security, and prosperity, not just in Alberta but in all of Canada (Gunster 2019; Gunster et al. forthcoming). Values of patriotism and nationalism have also been reflected in the oil sands coalition's promotion of pipeline projects as nation building and Stephen Harper's aspiration to make Canada an energy superpower (Hoberg 2016). Critics tend to appeal to environmental values and fairness and castigate the sector as dirty oil destroying wildlife habitat, poisoning downstream communities, threatening the global climate, and violating Indigenous rights (Environmental Defence 2008; Nikiforuk 2010). Some have gone as far as likening the oil sand mines to Hiroshima and (J. R. R. Tolkien's fictitious) Mordor (Berman 2011, chapter 14).

While much of the time the two coalitions seem to be talking past each other, they've also developed arguments and images to counter their opponents' core frames. The oil sands coalition counters the rights-violating dirty oil frame by emphasizing responsible resource development of ethical oil (Lavant 2010; McGowan and Antadze 2019). The anti-pipeline coalition challenges the oil sands coalition's rhetoric of prosperity and ethical oil by portraying Canada as a petro-state, in which big oil corrupts democracy (Nikiforuk 2013; Taft 2017), afflicted with "Dutch disease," where one region's oil wealth undermines the economies of the country's manufacturing regions (Lemphers and Woynillowicz 2012). The oil sands coalition counters environmentalists' depiction of the oil sands as a "carbon bomb" undermining Canada's inability to meet its UN emission reduction obligations by finding other forms of oil that are even more carbon intensive by pointing to the modest contribution of the oil sands to global emissions and by shifting the focus to end users' continued demand for and use of fossil fuels (Lavant 2010).

Both coalitions also appeal to the logic of necessity and inevitability but use it to draw dramatically different conclusions. For the oil sands coalition, oil is an essential fuel for economic development and will remain so for some time, and if global demand is not met by the oil sands, other sources of oil will simply fill the void, usually from less well-governed countries with lower environmental standards. The oil sands will find their way to markets, the oil sands coalition claims, one way or another. If pipelines are blocked, oil by rail will increase, with its attendant elevated risk of accidents.

For the anti-pipeline coalition, the global shift away from fossil fuels is necessary and inevitable so humanity can have a safe climate. More recently, the accelerated market penetration of electric vehicles has brought the idea of peak oil demand into the discourse. For the anti-pipeline coalition, since the shift away from fossil fuels is inevitable, it's best to stop fighting and plan for an orderly and just transition. The logic of inevitability also appears in the anti-pipeline coalition's discourse on pipeline and tanker spills. In response to industry and government's emphasis that there is a very low risk of major spills, pipeline opponents frequently simply state: "It's not a matter of if, it's a matter of when."

Another framing strategy is to use dismissive or disparaging labels about the people in the opposing coalition, in an effort to undermine their credibility or legitimacy. Many environmentalists cast aspersions on the motives of leaders of the fossil fuel industry, suggesting that their behavior, even when it's in compliance with laws, is criminal. One strategy of the climate movement has been to cast the industry as an enemy in order to inspire and mobilize a resistance movement. According to Bill McKibben, the fossil fuel industry "has become a rogue industry, reckless like no other force on Earth. It is Public Enemy Number One to the survival of our planetary civilization." He justified this construction by drawing an analogy to Bull Connor, the notorious city official in Alabama who ordered that civil rights protesters be attacked with fire hoses and police dogs: "A rapid, transformative change would require building a movement, and movements require enemies. As John F. Kennedy put it, 'The civil rights movement should thank God for Bull Connor. He's helped it as much as Abraham Lincoln.' And enemies are what climate change has lacked" (McKibben 2012).

The oil sands coalition has also at times chosen to demonize elements of the anti-pipeline coalition. One of the most infamous instances was Natural Resources Minister Joe Oliver's open letter (discussed in more detail in chapter 5). Oliver claimed that pipeline opponents "threaten to hijack our regulatory system to achieve their radical ideological agenda" with the help of "funding from foreign special interest groups" (Oliver 2012). This framing of Canadian environmentalists originated with Vancouver-area blogger Vivian Krause and was then amplified by the oil sands advocacy group Ethical Oil. Krause has been a thorn in the side of the Canadian environmental community since 2010 by persistently reporting the extent of environmental group funding in Canada that comes from US foundations, a number of whose money originally came from oil wealth. She has argued that the US foundations' influence is either to protect American oil interests by preventing Canadian access to foreign markets or simply inappropriate foreign influence on Canadian domestic policy issues (Hislop 2019).

The actual level of support from US foundations to Canadian environmental groups active in pipeline resistance campaigns is not nearly as substantial as the framing suggested by defenders of the oil sands coalition. An analysis of the budgets of 10 environmental groups active in anti-pipeline campaigns, combined with data from tax filings of US foundations, reveals that the average fraction of US foundation funding of Canadian environmental groups was 18.4%.[2] That is enough support to strengthen the capacity of these groups, but it does not seem like enough to justify the reaction from the oil sands coalition, especially given the very substantial involvement of foreign oil companies in the oil sands. Nonetheless, the frame of "foreign funded environmentalists" has remained prevalent in Canadian discourse. This rhetorical strategy was taken to new heights by Alberta politician Jason Kenney, as will be discussed in chapter 8.

Ideas—Scientific Rationale for "Keeping It in the Ground"

Much of the contest over ideas has involved the dynamic interaction of competing frames of oil sands development and its consequences. One fundamentally important related development has been the emergence of the concept of a "carbon budget" (Carbon Tracker Initiative 2011)—the idea that we need to keep a substantial majority of fossil fuels in the ground in order to stay within safe limits of global warming—to provide some scientific credibility for blocking new fossil fuel projects. The concept was first conceived by environmental activists but was then given scientific and establishment credibility. The scientific concept seems to have first appeared in environmental discourse as far back as 1997. Bill Hare, Greenpeace International's climate policy director, wrote a report that year, *Fossil Fuels and Climate Protection: The Carbon Logic,* that explicitly used the carbon budget concept. The focus of this report was to calculate the cumulative carbon dioxide (CO_2) emissions to the year 2100 that would be consistent with limiting the magnitude of global warming to within defined ecological limits. This calculation can be seen as a global "carbon budget," which if exceeded would most likely mean that ecological limits would be breached (Hare 1997, 1). Hare's analysis concluded that the carbon in fossil fuel reserves is "far greater than the total allowable carbon budget" (Hare 1997, i).

It took another decade for this idea to gain momentum and get tied to the mobilizing rhetoric of blocking new fossil fuel infrastructure. *Guardian* columnist George Monbiot introduced a new discourse with a column published shortly after the close of the 2007 United Nations Framework Convention on Climate Change (UNFCCC) meeting in Bali. Monbiot wrote: "Ladies and gentlemen, I have the answer! Incredible as it might seem, I have stumbled across the single technology which will save us from runaway climate change! From the goodness of my heart I offer it to you for free. No patents, no small print, no hidden clauses. Already this technology, a radical new kind of carbon capture and storage, is causing a stir among scientists. It is cheap, it is efficient and it can be deployed straight away. It is called . . . leaving fossil fuels in the ground" (Monbiot 2007). In making his case, Monbiot explicitly focused on the need to move beyond strategies to reduce demand to those that restrict supply, saying, "Most of the governments of the rich world now exhort their citizens to use less carbon. They encourage us to change our lightbulbs, insulate our lofts, turn our televisions off at the wall. In other words, they have a demand-side policy for tackling climate change. But as far as I can determine, not one of them has a supply-side policy" (Monbiot 2007).

The carbon budget concept came into use in peer-reviewed science in 2009. The first appearance of the concept in science journals appears to be a special issue of *Nature* on "The Coming Carbon Crunch" on April 30, 2009 (see Allen et al. 2009; and especially Meinshausen 2009). An article by Meinhausen (2009) appears to be the first peer-reviewed journal article to explicitly adopt the term "carbon budget." The analysis estimated how much more carbon humanity could afford to burn and still remain within the 2°C limit (the target set by the EU and soon to be adopted by the UN later that year) and concluded that more than three-quarters of the remaining proven fossil fuel reserves would need to stay in the ground in order to stay within the temperature target. Zickfeld et al. (2009) added momentum to the concept with a similar analysis of carbon budgets later that year in the *Proceedings of the National Academy of Sciences*.

By 2009, the carbon budget concept, initiated by activists, had been certified with scientific credibility. Within several years, the concept was picked up by think tanks such as Carbon Tracker Initiative, which began warning of a "carbon bubble" in 2011 by issuing reports about the financial implications of "unburnable carbon," fossil fuel reserves that become stranded assets as a result of the need to adhere to a carbon budget (Carbon Tracker Initiative 2011). The concept garnered additional establishment credibility when it was adopted by the International Energy Agency in its 2011 World Energy Outlook (International Energy Agency 2011, 236), by the World Bank in 2013 (World Bank 2013), and by the IPCC in its fifth assessment report, in 2014 (IPCC 2014).

The significance of the carbon budget concept is that it provided a scientific justification of sorts for blocking new fossil fuel infrastructure. Environmental writers, and soon environmentalists, began picking up the mantra, initiated by Monbiot, of "leave it in the ground" or "keep it in the ground." The LINGO (Leave it in the Ground) coalition was formed to advance the concept of supply-side mitigation at COP 17 (Durban) in 2011 (LINGO, n.d.). Greenpeace began using the hashtag #keepitintheground in their anti-coal campaign in early 2012. *The Guardian* made "keep it in the ground" a major focus of its climate reporting beginning in 2015 (Rusbridger 2015), and in January 2015, 350.org launched its own "keep it in the ground" campaign, building on its 2012 "do the math" campaign.

Born at Greenpeace as far back as 1997, justified scientifically in the world's leading science journals by 2009, and adopted by establishment institutions in the early 2010s, the carbon budget frame has bolstered the legitimacy of the supply-side strategy of blocking new fossil fuel infrastructure. "Keep it in the ground" became one of the signature slogans of the anti-pipeline resistance.

Institutions

This struggle among interests, between the oil sands and anti-pipeline coalitions, occurs within a particular set of institutional structures that establish the rules of the game that shape the resources and strategies of different actors. These strategic actors also try to alter the rules to enhance their position and power. Despite their similarities in culture, Canada and the United States have remarkably different political structures. The core institutional features of greatest significance are those that affect the fragmentation and diffusion of authority, both vertically and horizontally (Harrison and Sundstrom 2010, 16–18).

Canada has adopted the Westminster style of parliamentary democracy, whereas the United States developed its distinctive system of separation of powers and checks and balances. As a result, the American system is highly fragmented horizontally: Congress, the president, and the courts all play critical roles in public policy, including energy, environmental, and climate policy. In Canada, power is far more concentrated at the horizontal level, meaning at each level of government. Over time, the evolution of institutions in the two countries accentuated these differences. The exceptionally polarized nature of contemporary American national politics and the post-1968 tendency for Congress and the presidency to be controlled by different parties have combined to take the fragmentation of authority to extremes (Fiorina 1996). The emergence of a leader-centered form of the Westminster system has concentrated enormous powers in the prime minister at the federal level and in the premiers at the provincial level, leaving little role for legislatures except in the case of a minority government (Savoie 1999).

These broader institutional differences have fostered quite different styles of legislation, which in turn have had significant implications for the role of courts in policy-making. In the Canadian system, legislation tends to be broad and enabling, with few limits on the discretion of implementing agencies. In the United States, the separation of powers fosters a different style of lawmaking, where legislation tends to contain far more specific directions and timelines and thus constrains the exercise of executive authority (Moe and Caldwell 1994; Hoberg 2000). These differences in statutory design have direct implications for the role of the courts. The Canadian system provides fewer legal hooks for interest groups to challenge decisions, especially in the domain of administrative (nonconstitutional) law. In contrast, the American system creates seemingly endless opportunities to challenge administrative decisions, fostering a culture of what's become known as "adversarial legalism" (Kagan 2001).

The one major exception to this characterization of Canadian courts as playing a limited role in policy disputes is Aboriginal law. Because Aboriginal rights are enshrined in the Canadian constitution (section 35), First Nations have formidable legal tools to challenge government decisions in the courts. The evolution and significance of Aboriginal law for energy decision-making will be discussed later.

The two countries' differences in fragmentation are the reverse when looking at the vertical dimension, how they divide authority between the federal government and the states or provinces. Canada has evolved into a far more decentralized federation than the United States. That difference is particularly apparent in energy and environmental policy, where the US government has traditionally played a much stronger role. In Canada, provinces play the lead role on energy and the environment through their jurisdiction over lands and resources. Notwithstanding clear provincial jurisdiction, the Canadian government does have broad powers to act on energy and the environment, including in the areas of interprovincial and international trade, criminal power, fisheries, navigable waters, and issues involving First Nations, among others, but with few exceptions it has, for political reasons, deferred to provincial authority even in areas where it has jurisdictional authority (Harrison 1996; Harrison 2013).

The year 1980 was a watershed year for the Canadian government in the energy area. In the face of the second oil price spike, the Liberal government of Prime Minister Pierre Trudeau introduced the National Energy Program (NEP), designed to strengthen Canadian ownership and control of the oil sector and also channel to other provinces some of the windfall profits the western oil and gas sector was making as a result of the sudden rise in oil prices. The most divisive provision was the 8% petroleum and gas revenue tax, which provided the vehicle for the redistribution of oil wealth among the provinces. The National Energy Program ignited fierce backlash from the prairie provinces, led by Alberta and its formidable premier, Peter Loughheed. Trudeau was forced to back down on the most interventionist provisions of the policy, including the revenue tax, and within several years the entire program had been effectively dismantled (Doern and Toner 1985; James and Michelin 1989).

This perceived overreach of federal authority has cast a long shadow on the Canadian government and politics more generally. Electorally, the western alienation provoked by the NEP contributed to the demise of Pierre Trudeau's dominance of national politics.[3] It strengthened Alberta's sense of independent ownership of its oil resource and stoked hostility to federal intervention in the province. Alberta politicians have proven to have a long memory of the incident, which nearly four decades later is frequently revived as a boogeyman to discredit major federal energy-related initiatives, including climate action. According to Monica Gattinger, energy policy in Canada since the mid-1980s is best characterized as "third-rail federalism," in which any overassertion of federal authority could be politically deadly (Gattinger 2012).

The Role of Indigenous Groups in Natural Resource Decision-Making

As described in chapter 2, Indigenous groups in both Canada and the United States have played a pivotal role in the anti-pipeline movement. Much of this conflict has been about the appropriate role of Indigenous groups in making decisions about energy projects, such as pipelines, on their traditional territories. Indigenous land rights give them more power to veto unwanted energy projects than, for example, local governments. For most of North America, these rights were established in historical treaties, which have then been interpreted by courts in both countries, creating an evolving body of Aboriginal law. In British Columbia, however, the site of two of our pipelines cases, most of the province was settled without establishing treaties, which created a legal vacuum in provincial and federal law that courts have stepped in to fill.

American treaties tend to provide Native American tribes with the right to consent on decisions about projects directly involving their reservation's land and then a right to be consulted about Native American interests or values (such as heritage sites) off their reservations (Diotalevi and Burhoe 2016). In Canada, reserves tend to be much smaller geographically, and much of the legal conflict has been about the decision-making rights of Indigenous groups on resource projects outside their reserves but within their traditional territories. Over the past several decades, a legal doctrine of consultation and accommodation has emerged, with the obligation of the settler governments varying depending on the significance of their infringement of Aboriginal rights and the strength of the Aboriginal claim to rights in the area in question. The ambiguity of what constitutes sufficient consultation and accommodation, especially in cases where the Indigenous group is strongly opposed to the project, has led to continued conflict between settler and Indigenous governments. In the strongest case, where Indigenous groups have proven title, the duty amounts to a requirement for consent. But even in that context, settler governments can override Indigenous dissent in certain circumstances (Coates and Newman 2014; Wright 2018).

Many Indigenous groups have asserted the right to consent to projects in their traditional territories, and they point to the language of the United Nations Declaration on the Rights of Indigenous Peoples (UN General Assembly 2007). That declaration says: "States shall consult and cooperate in good faith with the Indigenous peoples concerned through their own representative institutions in order to obtain their free and informed consent prior to the approval of any project affecting their lands or territories and other resources, particularly in connection with the development, utilization or exploitation of mineral, water or other resources" (Article 32.1). This "free, prior, and informed consent" provision has become a rallying cry behind demands that projects not proceed without the approval of Indigenous groups.

While Canada has endorsed the UN declaration, it is not legally binding on signatories, so its consent provision only becomes Canadian law if it is adopted by either Canadian legislatures or courts. Thus far, the Supreme Court of Canada has made perfectly clear that its interpretation of consultation and accommodation *does not* constitute a veto power for Indigenous groups regarding resource projects (Newman 2017). While Canadian law has yet to adopt this standard of a right to consent, the idea has become enormously powerful within the anti-pipeline coalition. It articulates a standard that reinforces Indigenous sovereignty, and it has drawn in allies from the environmental community looking for more leverage over government and corporate decision-making (Hoberg 2018).

Oil Sands Policy and Governance through 2015: Alberta's Environmental Policies for the Oil Sands

The province of Alberta, with an economy so dependent on the oil and gas sector and a political system dominated by conservatives, has been extremely reluctant to impose any policies on the oil sands that might threaten the sector's growth and profitability (Carter 2016a; Adkin 2016). The government's strategy has been to resist regulation as long as that was politically palatable to the province, and when political pressure to act intensified, the tendency has been to adopt modest policies that did not significantly impact the bottom line. This pattern can be seen in what are arguably the three most important aspects of environmental policy in the oil sands: tailings regulations, land-use policy, and climate policy.

Regulatory Failure on Tailings

Tailings policy is a case of regulatory failure (Urquhart 2019, chapter 7). Prior to 2009, there was no regulation of tailings ponds. Regulations were finally enacted in 2009 and set to be enforced in 2013. By 2013, operators were required to capture and dry 50% of their tailings. Industry, however, failed to comply, and the Alberta government has failed to enforce the regulation. The government approved tailings management plans that did not meet the regulation's standards, and monitoring reports revealed that operators failed to comply even with those weaker plans (McNeill and Lothian 2017, 4).

In July 2016, the government changed the regulation, abandoning the 50% capture and dry requirement, and replaced binding standards with a planning-based approach requiring operators to submit plans showing how they would manage their tailings based on their reclamation plan within cumulative limits set in the broader land-use framework. The framework was designed to have tailings volumes peak in 2020 and then decline steeply thereafter, but analysis of the plans submitted shows that tailings volumes would not peak until 2037 (Pembina Institute 2017). Environmentalists have responded to the lack of action by Alberta by appealing to the NAFTA Commission on Environmental Cooperation in an effort to get the federal government to use the Fisheries Act to regulate tailings (Commission on Environmental Cooperation 2018).

Strategic Land-Use Planning in the Lower Athabasca Region

Strategic land-use planning for the oil sands was slow to emerge but can be considered somewhat more effective than the regulatory failure apparent in tailings management. Relatively early in the oil sands development process, there was a growing awareness of the need to move beyond a facility-by-facility assessment and regulation to consider the regionwide cumulative effects of oil sands development. In 2000, the Cumulative Environmental

Management Association (CEMA) was created as a voluntary stakeholder partnership among government, industry, environmentalists, and others (Urquhart 2018, 114–118). CEMA was slow to deliver, even as oil sands development was accelerating. Concerned by the pace of development in the absence of regional habitat protection plans, in 2008 the CEMA group working on habitat protection recommended a moratorium on new oil sands leases, but the government refused. Then the group recommended the protection of up to 40% of the region (Hoberg and Phillips 2011).

Rather than adopting that recommendation, the government of Alberta introduced a new Land Use Framework in 2008, creating a new planning process. Priority was given to the region where oil sands development had been the most intense, the Lower Athabasca Region, in the province's northeast. The Lower Athabasca Regional Plan (LARP) was finally completed in 2012. It provided regional standards for surface water quality and air quality, and most importantly created new conservation areas that increased protected areas from 6% to 22% of the region. It also committed to establishing specific targets for other regional ecosystem values (Government of Alberta 2012). The plan's commitment to addressing cumulative effects and its establishment of more protected areas are welcome improvements, but the plan only covers part of the area of oil sands development, and highly valued species, such as boreal caribou, remain in serious trouble. As a panel formed to review the plan's implementation concluded, "Despite the LARP's new conservation areas, the cumulative impacts on wildlife have exceeded or are reaching thresholds in significant adverse effects on biodiversity, some of which are likely permanent" (Lower Athabasca Regional Plan Review Panel 2015).

Alberta's Climate Policy Evasion

Given its desire to facilitate the rapid development of a carbon-intensive resource, prior to 2015 Alberta had adopted only modest climate policies that explicitly planned increases in greenhouse gas emissions and limited impacts on the costs of production to mere pennies per barrel (Urquhart 2018, chapter 8). While tailings and caribou have been significant issues regionally, it was the climate footprint of oil sands that put it on the map for the international environmental movement. The growing national, continental, and international interest in reducing emissions has long been considered a threat to the province of Alberta and its prosperity, particularly since environmental activists turned the spotlight on oil sands as a carbon threat in the early 2010s. Alberta was a strong opponent of the ratification of the Kyoto Protocol, and its recalcitrance contributed to the Liberal federal government's unwillingness to develop policy mechanisms to implement the treaty it had ratified (Harrison 2007; Carter, Fraser, and Zalik 2017).

Alberta preferred a "made in Alberta" climate plan, and its 2002 plan committed to "reducing greenhouse gas emissions." But, to Alberta, "reducing" has been defined as reductions in the rate of emissions per unit of GDP or barrel of oil produced, not in terms of actual reductions, so that as production grows, emissions would also continue to grow. The 2003 legislation implementing the plan committed to reducing emissions by 2020 "relative to Gross Domestic Product to an amount that is equal to or less than 50 percent of 1990 levels." It also authorized the government to enact sector-specific regulations (Bankes and Lucas 2004).

The province of Alberta introduced the innovative Specified Gas Emitters Regulation (SGER) in 2007 (Leach 2012; Urquhart 2018, 247–250). The regulation set *emission-intensity limits* for facilities producing more than 100,000 tonnes of GHGs per year. It required that the facilities reduce emissions per barrel produced to 12% below their baseline level within nine years. Facilities that didn't make those reductions were given the option to buy offsets for the equivalent amount or pay C$15 per tonne into a technology fund. Alberta has celebrated the introduction of SGER as the first carbon-pricing scheme in North America. That fact, in addition to its policy design, definitely makes it innovative, but its impact on emissions was quite limited by design. Because only the emissions above the 12% emission reduction target were charged, the cost to producers of the regulation's carbon price was small: equivalent to C$1.80 per tonne for a facility's emissions, or an inconsequential eight cents per barrel (Read 2014). As a consequence, the impact on emissions has been modest. After a brief dip in emissions as a result of the 2008 Great Recession, Alberta's emissions began growing again and in 2015 were 18% higher than in 2005 and 58% higher than in 1990 (Environment and Climate Change Canada 2017a).

Alberta undertook significant changes to its climate policy framework in 2015 after the NDP came into power. Those changes are described in chapter 8.

Canadian Energy and Climate Policy in the Harper Era

Whereas Alberta felt besieged by the climate change rhetoric (if not actual policy) of the Chretien and Martin Liberals, the province welcomed the ascension of Albertan Stephen Harper to dominance in national politics beginning in 2006. The Harper government was guided by four interrelated strategies in the energy-environment field: (1) relaxing environmental laws to facilitate project approval, (2) undermining capacity, (3) vacating intergovernmental relationships, and (4) evading meaningful action on climate change. Harper's core goal seemed to be to advance the interests of the resource sector as much as possible by minimizing regulatory costs and marginalizing opponents, but he pursued the strategy well past the time when it appeared to be fruitful. In the words of Monica Gattinger, Harper's strategy was tantamount to his government "shooting itself in the foot" (Gattinger 2016).

"Responsible" Resource Development

The Harper government responded to infrastructure approval delays it blamed on a process that was too participatory and open-ended by reforming review procedures to facilitate approvals. In spring 2012, it introduced Bill C-38, the Jobs, Growth and Long-term Prosperity Act, and its companion, Bill C-45. These mammoth omnibus budget bills rewrote much of Canadian environmental law in order to implement the government's Responsible Resource Development Plan. The stated purpose of the plan was to "create jobs, growth, and long-term prosperity" by "streamlin[ing] the review process for major resource projects" (Natural Resources Canada 2012). Bill C-38 amended the Canadian Environmental Assessment Act and amended the National Energy Board Act to streamline regulatory approvals, provide strict timelines for review, and limit participation in the process to those who are "directly affected" or have, in the regulators' judgment, "relevant information and expertise." These bills also moved the final decision on pipeline approvals from the National Energy Board to the cabinet, weakened the Fisheries Act to remove much of the protection afforded to fish habitats, and insulated pipelines from the provisions of the Navigable Waters Protection Act (Toner and McKee 2014; Olszynski 2015). These sweeping budget bills, unprecedented for the scope of their impact on authorizing legislation, provoked a massive outpouring of criticism and protest from the environmental community. They also outraged First Nations, some of which responded by mobilizing behind the indigenous-rights-oriented Idle No More movement (Wotherspoon and Hansen 2013).

Undermining Capacity

Harper's government actively worked to delegitimize environmentalists, the most extreme instance being the previously discussed open letter from his natural resources minister directly attacking pipeline opponents who "threaten to hijack our regulatory system to achieve their radical ideological agenda" with the help of "funding from foreign special interest groups" (Oliver 2012). Environmentalists were also alarmed when the Harper government's 2012 anti-terrorism strategy listed "eco-extremists" as threats. In 2014, a Royal Canadian Mounted Police Critical Infrastructure Intelligence Assessment listed the "anti-Canadian petroleum movement" in its document on "Criminal Threats to the Canadian Petroleum Industry" (Carter 2016b; McCarthy 2012; McCarthy 2015). Harper also used the tax code to discourage environmentalists and other critics of his administration from politically challenging the government by ordering audits of the charitable status of environmental groups under the tax law to ensure they were not exceeding the permissible amount of political advocacy.

One of the most striking features of Harper's record was his eagerness to undermine the capacity not just of his political opponents but of his own government as well (Carter 2016b). He eliminated the National Roundtable on the Environment and the Economy, a prestigious government forum for nonpartisan advice. Harper reduced funding for the environment compared to other programs; as a percentage of overall federal program funding, spending on environmental programs was halved between 2008 and 2014 (Hoberg 2016). The Harper government also restricted its own scientists from commenting publicly and became notorious for "muzzling" them. In the past, it was common practice for federal research scientists to be able to talk directly to the media or to attend conferences and openly discuss science-policy issues. The Harper government changed that by centralizing communications and prohibiting scientists from engaging with the media and others outside the government without getting approval from agency communication officials (Winfield 2013). Criticism from the international science community emerged, and the new practices became the target of political mobilization, giving birth to a new nongovernmental organization (NGO) called Evidence for Democracy and spawning headlines such as "Scientists March on Canadian Parliament" (Semeniuk 2012). Evidence-based decision-making became an issue in the 2015 election.[4]

Vacating Intergovernmental Relations

The fourth and final essential feature of Harper's energy and environmental record was his government's refusal to engage directly in intergovernmental discussions. Canada's constitution establishes a complex division of power between the federal government and the provinces. Since the 1980 National Energy Program debacle, the Canadian government has been reluctant to play a strong role in energy policy. But federal caution is unlikely to be effective at fostering the country's emergence as an "energy superpower," nor is it likely to be productive in resolving complex issues requiring difficult choices with strong regional distributional consequences. Harper's unwillingness to provide national leadership on these critical issues, driven in large part by his decentralist ideology, stymied effective policy and national conflict resolution.

The leadership vacuum resulting from Harper's unwillingness to engage with the provinces was evident in both climate and energy policy. On climate, Harper's goal was to evade meaningful policy, so the lack of engagement was at least consistent with his policy goals in the short term. His unwillingness to engage on energy issues is more puzzling. As early as 2007, initiatives began to emerge, both among the provinces and within civil society, to advance a national energy strategy of some kind. The initiatives began to take on momentum in 2012 when Alberta, anxious to create conditions for pipeline expansion, took on a leadership role. But Alberta's interest in oil sands expansion was not consistent with the interests of Ontario, Quebec, and British Columbia, which were keen to show leadership on fighting climate change. The premiers of British Columbia and Alberta ended up in a bitter conflict in 2012 over the Northern Gateway pipeline, yet Harper simply stood on the sidelines (Gattinger 2012). In July 2015, the provinces did succeed in producing a document called the *Canadian Energy Strategy* (Council of the Federation 2015), but it was too vague to resolve any of the enduring conflicts that characterized the Harper era and seems to have been quickly forgotten.

Harper's Climate Policy Evasion

Harper's record is one of evading meaningful policy actions to tackle climate change.[5] He came into office in 2006 criticizing the previous Liberal government for agreeing to targets without a plan to attain them, and he promised to develop a "made in Canada" policy. In May 2006, Environment Minister Rona Ambrose announced that it would be "impossible for Canada to reach its Kyoto targets" (CBC 2006), setting the stage for Canada's formal withdrawal from the Kyoto Protocol five years later.

The Harper administration replaced the nation's Kyoto Protocol commitment to reduce emissions by 6% below 1990 levels by 2012 with a new goal of a 20% reduction below 2006 levels by 2020 (equivalent to 2% below 1990 levels) (Harrison 2012). In April 2007, it proposed its "Turning the Corner" plan, which was a regulatory approach requiring that large emitters reduce their carbon pollution per unit of production by a specified percentage each year. That proposed plan was perhaps the most ambitious action on climate change ever proposed by the Harper government, under then environment minister John Baird. The plan was never implemented, however, and instead Harper abandoned any ambition to undertake meaningful climate action.

During the Copenhagen rounds of climate talks in 2009, Harper abandoned his "made in Canada" rhetoric for deference to the position of the United States. In January 2010, then environment minister Jim Prentice announced Canada's new national target—a 17% reduction below 2005 levels by 2020. Not only was that target weaker than the previously announced 6% reduction below 1990 levels (2020 emissions would be 6% higher), but the new position was striking for how it tied itself to US actions: the target was "to be aligned with the final economy-wide emissions target of the United States in enacted legislation." In his Calgary speech announcing the new targets, Prentice stated, "Our determination to harmonize our climate change policy with that of the United States also extends beyond greenhouse gas emission targets: we need to proceed even further in aligning our regulations. . . . [W]e will only adopt a cap-and-trade regime if the United States signals that it wants to do the same. Our position on harmonization applies equally to regulation. . . . Canada can go down either road—cap-and-trade or regulation—but we will go down neither road alone" (Fekete 2010).

In 2010, the legislative effort in the United States to develop a cap-and-trade program died in the US Senate (Lizza 2010). As Republican control of

Congress increased and environmental politics (like everything else American) became more polarized, President Barack Obama's only choice was to use his regulatory authority under the existing Clean Air Act. As president, his two most important areas of authority over greenhouse gases were regulation of coal-fired power plants and automobiles, where updating regulations was already authorized by the Clean Air Act. Recognizing that cap and trade was dead in the United States, the Harper government committed to pursuing regulations focusing on the three most substantial components of Canada's emissions: coal-fired power plants, the oil and gas industry, and automobiles.

Harper actually got out ahead of the US government in adopting regulations for coal-fired electricity. In 2012, Ottawa enacted regulations (to be brought into force in 2015) for new coal-fired power plants. The regulations effectively banned new coal-fired power plants that did not capture and sequester their carbon. The problem, however, was that the regulations would not affect existing power plants until they came to the end of their life, in 2030. Nothing in the regulations required or encouraged existing coal-fired plants to be shut down (Dion, Sawyer, and Gass 2014). When it came to automobiles, the Harper government—following a long tradition of Canadian governments—chose to adopt the new US vehicle standards adopted by the Obama administration in 2012. Given the integrated nature of the North American vehicle market, it would have been both costly and disruptive for Canada *not* to adopt the new standards. This action—following the US lead on auto regulations—was the single most consequential action by the Harper government in reducing greenhouse gas emissions.

In contrast to the coal and vehicle sectors, where Harper did at least enact some regulations, he never enacted regulations on Canada's fastest-growing GHG sector, the oil and gas industry. Harper's commitment to develop GHG standards for the oil and gas sector dates back to 2007. It was the third pillar in the government's regulatory approach and became salient as the Keystone XL controversy with the United States increasingly focused on the climate implications of oil sands expansion and Canada's lax climate policies.

While a regulatory proposal was never published, it was widely circulated that the Canadian government was considering a proposal to go beyond Alberta's oil sands GHG regulation. The Alberta regulation required emitters to reduce pollution per barrel 12% below certain emission targets, and if they couldn't meet that, pay C$15 a tonne for the amount in excess of that target.

The Harper government's proposal was to increase the emission reduction to 30% and increase the penalty for not meeting that to C$30 per tonne (Hoberg 2016), but it never enacted that proposal or any other GHG rule for the oil sands. Initially, Harper justified his reluctance by stating that the United States had yet to enact similar recommendations. Shortly after that justification had been shown to be false (Wingrove 2014), the price of oil began its spectacular decline. In December 2014, Harper essentially closed the book on the possibility of issuing a regulation, stating, "Under the current circumstances of the oil and gas sector, it would be crazy, it would be crazy economic policy to do unilateral penalties on that sector. We are clearly not going to do it" (McCarthy 2014b).

When the sector was booming and the price of oil soaring, Harper didn't regulate GHG emissions from the oil sands. When the price of oil collapsed, he didn't either. Harper never acted to address Canada's fastest-growing source of emissions and the single biggest symbol, domestically and internationally, of his government's climate evasions. He did enact regulations on future coal plants that would have no impact until 2030, and he did adopt US regulations for the auto sector. As a result, Canadian climate policy has lagged behind those of other countries and became a potent symbol of the Harper government's poor environmental record. In an international comparison of climate policy at the end of the Harper era, Canada ranked a very poor 56th of the 61 countries measured. Of OECD countries, only Korea, Japan, and Australia ranked lower (Burck, Marten, and Bals 2016).

Conclusion

Policy conflict is in part the conflict of ideas. The story of the battle over the oil sands reflects clashing worldviews between those who see the oil sands as an engine of prosperity and those who see them as destructive to the environment and Aboriginal cultures. Discourse on the pipeline conflicts reflects these same dynamics, as chapters 4–7 will demonstrate. Policy conflicts occur within a set of institutions that set the rules for decision-making, but strategic actors also struggle to redefine those rules to gain relative advantage over their adversaries. The following pipeline cases are filled with examples of this "politics of structure."

Through 2015, environmental policies in the oil sands were clearly inadequate. Tailings policies have been a failure, promising land-use policies have only been partly implemented, and the viability of boreal caribou populations in the region is at grave risk. Climate policies during the Harper era made Canada a laggard internationally, and Canada's poor climate record contributed to the opposition to pipelines at home and across the border in the United States. Alberta's heavy dependence on the oil sands for jobs, income, wealth, and government revenues contributed directly to its government's reluctance to pursue more-effective environmental policies. Canada's institutional culture of federal deference to provinces, especially on Albertan energy in the wake of the National Energy Program debacle, discouraged the federal government from compensating for Alberta's regulatory reluctance even when, prior to 2006, it was controlled by a party that might have had an inclination to do so. The elections of 2015—the Notley NDP in Alberta and the Trudeau Liberals nationally—created the promise of significant policy change. This shifting post-2015 policy landscape will be described in the chapters to come and assessed in chapter 8.

II Pipeline Resistance

4 Keystone XL and the Rise of the Anti-pipeline Movement

Tzeporah Berman is arguably the most influential Canadian environmental activist of her generation. From 2012 to 2018, she was coordinator of the Tar Sands Solution Network, the organization built to coordinate the multiple anti-pipeline campaigns. Prior to that, in 2011, she was working for Greenpeace International. A veteran protester who was one of the architects of the Clayoquot Sound civil disobedience campaign in the early 1990s (Pralle 2006a), she explains how she first learned of the idea of targeting Keystone XL with a massive protest campaign:

> I was in Amsterdam. One day I got a call and it was Bill McKibben. He said, look, there's this proposal, the Keystone XL pipeline. I think we should have hundreds of people come and get arrested in Washington. And I said, so you want to focus the US climate movement on a pipeline from Canada? At the point when there is all this backlash, with Sarah Palin out there shouting "drill baby drill"? I thought, well as a Canadian, thank you. But are you sure you want to do that? I know a lot of Canadians who would love that, because it's right in line with this strategy we developed several years ago. But I can't see that being successful with the US media. I was very, very wrong. That was the beginning. . . . That summer was the first protest on Keystone. (Berman 2016)

Keystone XL was the first major pipeline controversy of the 2010s and ultimately proved to be one of the most divisive environmental conflicts in recent American history. The 830,000 barrels per day project was proposed in September 2008 to connect the oil sands in northern Alberta to the Gulf Coast of the United States, promising the landlocked oil resource access to coveted "tidewater." A timeline for this extraordinary regulatory and political controversy is presented in figure 4.1. Pipeline opponents achieved an extraordinary victory, first in getting the project delayed numerous times and eventually having it rejected by President Obama in 2015. That decision

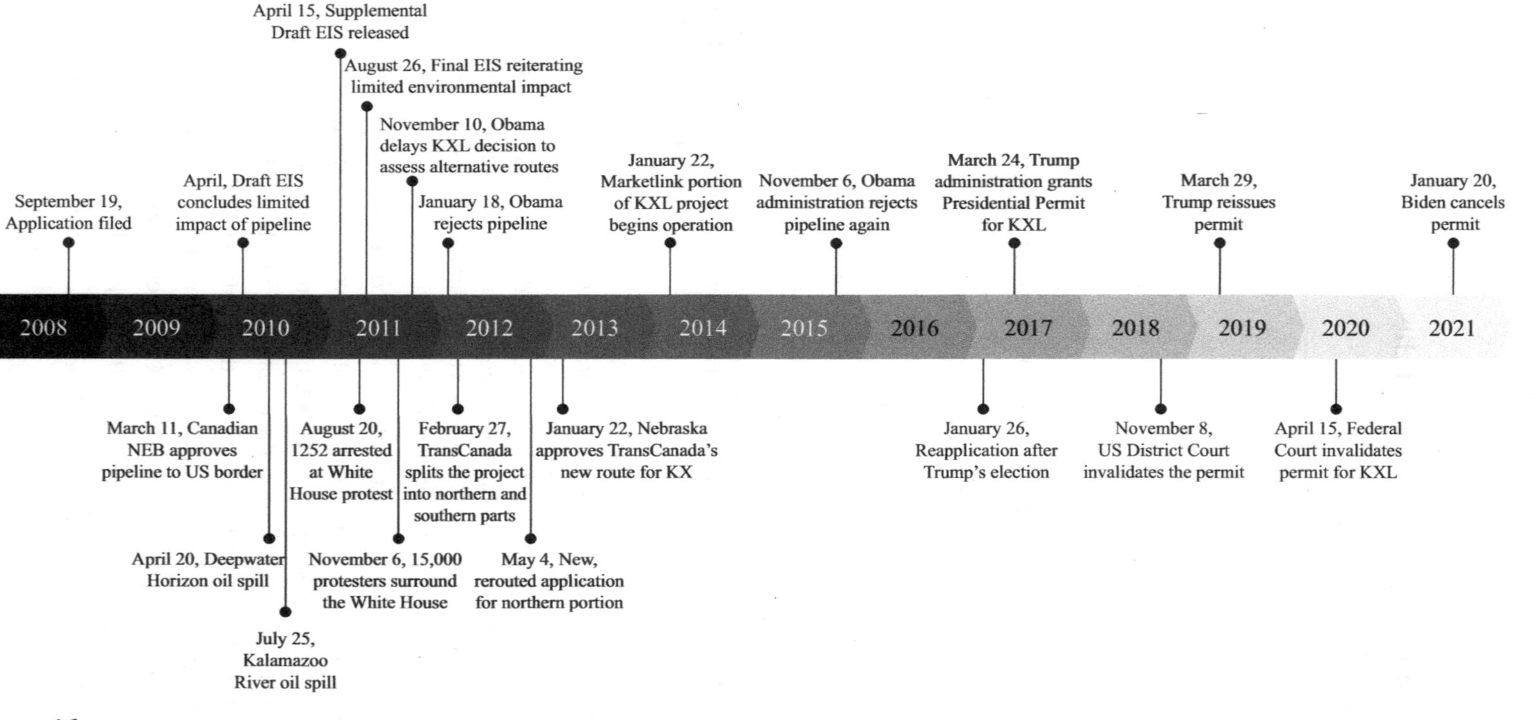

Figure 4.1
Keystone XL timeline.

was reversed by the Trump administration, but Trump's reversal was held up in the courts. Keystone XL was proposed in 2008, when George W. Bush was president, and ultimately terminated 13 years later by President Joe Biden on January 20, 2021. It is both a remarkable success for the pipeline resistance movement and an indication of the dysfunction of the US regulatory process (Zolden 2015; McConaghy 2017).

Applying the analytical framework developed in chapter 1, the Keystone XL pipeline certainly posed substantial political risks for pipeline proponents. The pipeline route was a new "greenfield" project through the prairie region. It crossed an international boundary, giving two national governments a veto. It crossed three US state boundaries, giving each of those states a veto. The Canadian project required approval by the US government, exposing the project to the notoriously fragmented American regulatory process, where vetoes were held by the president, the courts, and multiple actors within Congress. Given that its initial route crossed a precious aquifer, it activated intense place-based opposition. Risks and benefits were separated geographically and across national borders. Most of the benefits would accrue to the Canadian oil sector, whereas the risks were felt along the pipeline and by the climate more generally. Given these challenges, it's not surprising that the pipeline approval process has been a long, protracted, divisive affair.

Like the other pipeline controversies examined in this book, the Keystone XL story is complicated to tell because there were so many diverse actors, competing in overlapping venues and with different strategies. In an effort to bring order to the narrative, the first section of this chapter describes public and private actors, and their interests and strategies in the pipeline controversy. We'll focus in particular on the climate movement, because this is where the strategy of using place-based resistance to foster climate mobilization was first adopted. That discussion will allow us to explore the hypothesis introduced in chapter 1: that to strengthen their leverage, climate activists will ally themselves with groups representing place-based interests when possible.

The second section examines the ideas that have shaped the Keystone XL controversy by analyzing the discourses of the competing coalitions and how they become manifest in media reporting and public opinion polling. That section suggests some limitations to the prediction that opponents will focus on place-based risks. The third section describes how the controversy played out in separate but interdependent institutional arenas. That section will examine the hypothesis that strategic actors will focus their strategy on the institutional venue(s) most favorable to their interests. The fourth section explores the decision rationales used by Presidents Obama and Trump to evaluate the hypothesis that such rationales are more likely to emphasize place-based risks than climate risks. The concluding section puts this case in the context of the analytical framework and sets up the discussion of the Canadian pipelines to come.

Actors

Industry and Its Allies

The lead industry actor is unquestionably the pipeline company and project proponent, Calgary-based TransCanada Corporation (the company formally changed its name to TC Energy in early 2019). The company specializes in oil and gas pipelines and terminals and has substantial stakes in the US$8 billion Keystone XL project. In addition to the project's proponents, the oil sands companies that would use the pipeline also have a powerful interest in getting their expanded oil sands production to market. Pipeline capacity constraints have been a major concern for the landlocked Alberta oil industry, creating a significant discount for oil sands products, as described in chapter 2.

TransCanada and the oil companies have received substantial political support in this case from the government of Alberta and the government of Canada. Alberta is heavily dependent on the oil and gas sector, and government revenues are particularly sensitive to the price of oil. The government of Alberta has been a tireless champion of increasing oil sands access to tidewater, whether it was the Keystone XL pipeline or the alternatives linking the province to the Pacific or the Atlantic. This was the case when the pro-business Progressive Conservative Party ran the province but continued with the left-leaning New Democratic Party that came to power in 2015. In 2019, that NDP government was replaced by the more pro-industry United Conservative government of Jason Kenney. In March 2020, the Kenney government actually bought US$1.1 billion in equity in Keystone XL in an effort to rescue the ailing project (Government of Alberta 2020b).

The government of Canada has also been a strong advocate for the Keystone XL pipeline, under both Stephen Harper's Conservative government and Justin Trudeau's Liberal government.

Environmental Groups and Their Allies

Chapter 1 described the rationale for the climate movement's choice to shift from the frustrating politics of lobbying to blocking new fossil fuel infrastructure. The Keystone XL conflict was in fact the turning point for the movement, with 350.org's founder, Bill McKibben, being the catalyst. McKibben describes the evolution of his thinking in his 2013 memoir, *Oil and Honey: The Education of an Unlikely Activist.* He had already formed 350 .org as a global environmental group focused on climate. In spring 2011, he taught a course on social movement strategies, including a deep dive into the US civil rights movement, and it inspired a vision for a new strategy. He explained, "By the time I was done with the semester, I'd decided that 350.org should organize the first major civil disobedience action for the climate movement. I sensed, from the speeches I was giving and the e-mail that flowed in hourly, that people were ready for a deeper challenge—it was time to stop changing lightbulbs and start changing systems. If we were going to shake things up, we'd need to use the power [Martin Luther] King had tapped: the power of direct action and unearned suffering. We'd need to go to jail" (McKibben 2013c).

McKibben seized on the Keystone XL issue after reading a blog post by climate scientist turned activist James Hansen, arguing that fully exploiting Canada's oil sands would amount to "game over" for the climate:

> His calculations put a sudden spotlight on a previously little-known pipeline proposal called Keystone XL that was designed to carry almost a million barrels a day of that tar sands oil south from Canada to the Gulf of Mexico. Native leaders in Canada had been fighting tar sands mining for years, because it had wrecked their lands—only 3 percent of the oil had been pumped out. . . . And some ranchers in the United States had begun to rally along the planned route of the pipeline itself, particularly in Nebraska, where it was destined to run straight across the iconic Sandhills and atop the Ogallala Aquifer that irrigates the Great Plains. But these protests hadn't gained enough traction to stop the plan. Keystone XL awaited only a presidential permit. (McKibben 2013c)

The decision to launch a major resistance campaign against Keystone XL transformed the American wing of the climate movement and had profound repercussions not only for the project and the future of the oil sands but for American politics and US-Canada relations.

While the centerpiece of McKibben's message was about climate, the Keystone XL resistance movement was only able to be as effective as it was because it allied itself with place-based interests: farmers, landowners, and

Native Americans along the route, whose principal concern was the threat to local water quality. Much of the land on the proposed right-of-way was either privately owned or tribal land, so the positions of these groups were central to the conflict. The most influential voice of these place-based interests has been Bold Nebraska, a group created in 2010, originally to organize progressive groups in the state. But when the Keystone XL conflict emerged, the "small but mighty" group, as it calls itself, turned its attention to organizing the pipeline resistance movement (Ternes, Ordner, and Cooper 2020). The group's website describes itself this way: "Bold Nebraska is best known for our work with an unlikely alliance of farmers, ranchers, Tribal Nations and citizens to stop the risky Keystone XL pipeline. We work on issues including eminent domain, clean energy, small family farms and lifting up small businesses. . . . The typical national environmental model—where a campaign is run from DC and local folks brought in for only lobby days or press conferences—must be disrupted if we are going to win local and national fights against fossil fuel interests while advancing clean energy" (Bold Nebraska, n.d.).

While active in the campaign from the beginning, these local groups banded together in 2014 to form the "Cowboy-Indian Alliance" in a campaign labeled "resist and protect." The campaign website describes the initiative as follows: "Reject and Protect is led by the 'Cowboy-Indian Alliance,' a group of ranchers, farmers, and tribal communities from along the Keystone XL pipeline route" (Reject and Protect, n.d.). The addition of these groups diversified and strengthened the political coalition behind the resistance. This coalition was closely allied with 350.org and the climate movement, as the list of supporters on the "about" page of the group demonstrates.

In one of his efforts to articulate the philosophy of the movement, McKibben was quite direct about the central role played by making alliances with "front-line communities": "For all of us, it means standing with communities from the coal fields of Appalachia to the oil-soaked Niger Delta as they fight for their homes. They've fought longest and hardest and too often by themselves. Now that global warming is starting to pour seawater into subways, the front lines are expanding and the reinforcements are finally beginning to arrive" (McKibben 2013a). The "keep it in the ground" strategy represented by the anti-Keystone XL campaign clearly supports the hypothesis that, in the effort to increase their leverage, climate activists will ally themselves with groups representing place-based interests when possible.

Reframing the Oil Sands: The Battle over Ideas

Political conflict over public policy is not just a battle of strategic actors and institutions but also a battle of ideas (Baumgartner and Jones 2010). These battles are fought in both mainstream and social media and attempt to reshape public opinion to the advantage of strategic actors. On the pro-pipeline side, the core arguments were about job creation and energy security, particularly reducing dependence on Middle Eastern oil. On the anti-pipeline side, the core arguments were about risks to freshwater resources and, increasingly, the threat of climate change.

The core ideas motivating the "keep it in the ground" movement, most critically the carbon budget, were described in chapter 3. Environmental activists, led by 350.org's Bill McKibben, adapted the concept as the rhetorical centerpiece of the anti-Keystone XL campaign by highlighting the threat to a threshold for a safe climate of oil sands expansion enabled by the pipeline. The movement was given a framing metaphor by climate scientist James Hansen, then director of NASA's Goddard Institute for Space Studies. Hansen argued that, if fully exploited, Canada's oil sands would add so much carbon to the atmosphere that it would be "game over for the climate" (Mayer 2011; Hansen 2012). He labeled the Keystone XL pipeline the "fuse to the biggest carbon bomb on the planet" (McGowan 2011). This "game over" frame was picked up by many environmentalists, including McKibben (Mayer 2011). Many climate analysts took issue with Hansen's metaphor, showing that it was greatly exaggerated (Revkin 2011; Levi 2012; Leach 2014), but the frame helped mobilize urgency and opposition to Keystone XL and ultimately helped pave the way for Obama's rejection.

Issue Framing Analysis

Hodges and Stocking (2016) analyzed how this battle played out on Twitter, and Lawlor and Gravelle (2018) examined differences between national and local media reporting on Keystone XL through 2014. In this analysis, we examine a sample of mainstream media content. An analysis of mainstream media provides some indication of the relative importance given to different issues in the pipeline dispute. Figure 4.2 shows trends in US media mentions of Keystone XL from 2010 to the end of 2018, from Lexis Nexis. Mentions increased steadily through 2015, when Obama rejected the pipeline. It continued to receive media attention in 2016 because Trump made

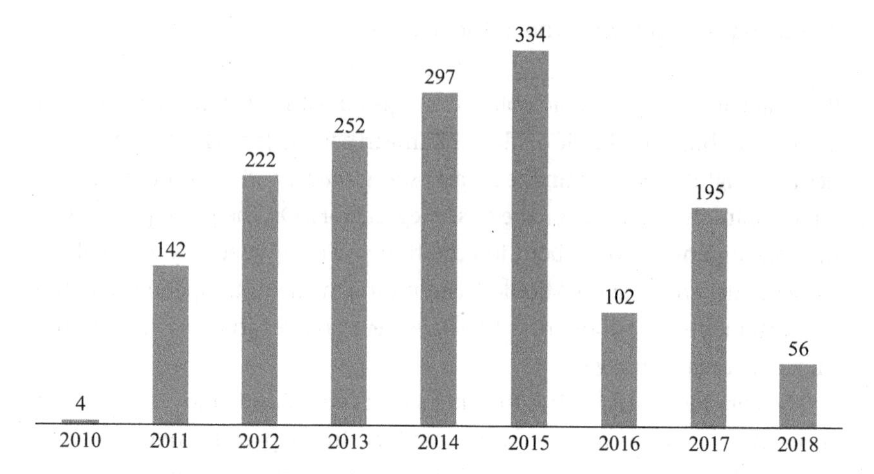

Figure 4.2
Total Keystone XL mentions in US media, 2010–2018.

it an election issue and in 2017 because his reversal of Obama's rejection was a prominent issue.

As a way to measure which values or issues were highlighted in discourses on the controversy, figure 4.3 compares the number of times news articles mentioning the Keystone XL pipeline also mention five particular issues: (1) climate change; (2) jobs and the economy; (3) energy security; (4) risks of pipeline accidents or tanker spills; and (5) rights of Indigenous groups. Data come from the Lexis-Nexis database and are for the *New York Times* and *Washington Post* from 2010 through August 31, 2019.[1] Analyzing mainstream media coverage provides only a proxy for the relative importance of issue frames in the discourse of the pipeline conflict. It is not a direct measure of the discourses of the competing coalitions, but it is a measure of how the media chose to report the conflict.[2] The data show that once media attention began in 2011, jobs and the economy were the dominant issues in this sample of mainstream media mentions. While jobs and the economy remained prominent in media discourses, mentions of energy security declined considerably, reflecting the remarkable surge in domestic oil production since 2010. Our hypothesis is about whether place-based environmental impacts (e.g., pipeline spills and accidents) are given more attention than the more diffuse issue of climate change. This analysis of media called that hypothesis into question. Climate mentions exceed spill and accident mentions from 2013 on. This shift no doubt reflects the TransCanada agreement to

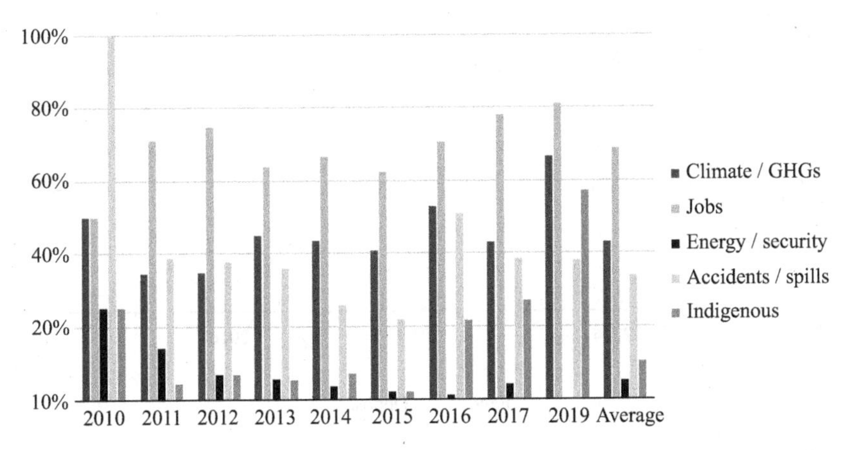

Figure 4.3
Mentions of the keywords climate change, jobs and the economy, energy security, risks of pipeline accidents or tanker spills, and rights of Indigenous groups as a proportion of total mentions of the Keystone XL pipeline in the *New York Times* and *Washington Post*, 2010–2019.

reroute the pipeline away from the Ogallala Aquifer in Nebraska. Mentions of Indigenous rights, another indicator of place-based concerns, increased as a fraction of total media attention from 2016 through 2019, probably as a result of the connections drawn to the Dakota Access Pipeline Standing Rock controversy, but Indigenous rights still received substantially less attention than climate.

In their Keystone XL campaign, environmentalists took care to ally themselves with place-based interests, but they clearly did not shy away from the climate message. Climate issues were the dominant environmental issue, far more so than local issues of spills or accidents, at least as reflected in this sample of mainstream media coverage of the Keystone XL controversy.

Public Opinion

Public opinion on Keystone XL has changed dramatically. Public sentiment was initially highly favorable toward the Keystone XL pipeline: in early 2013, 66% of respondents in a Pew poll favored the pipeline, with only 23% opposed (see figures 4.4 and 4.5). Opposition increased modestly over several years but then surged, apparently as it became an issue associated with Trump (Pew Research Center 2017). Gravelle and Lachapelle (2015) examined the factors influencing support or opposition to it before the precipitous decline.

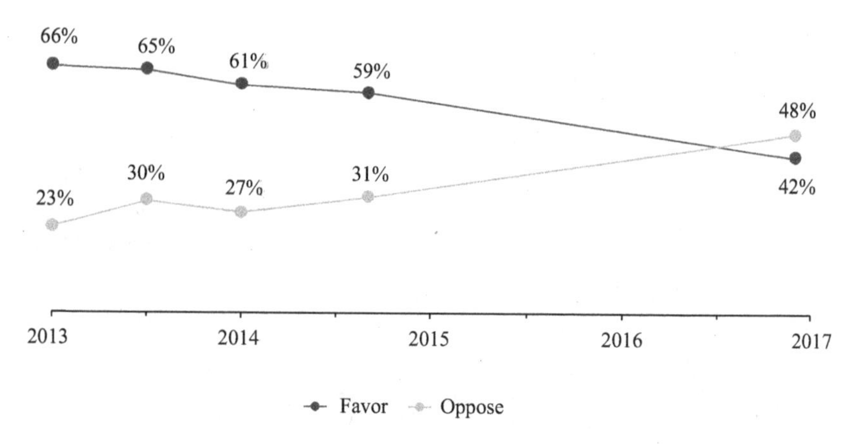

Figure 4.4
Poll results for Keystone XL pipeline support.
Note: "Don't know" responses are not shown.
Source: Pew Research Center (2017).

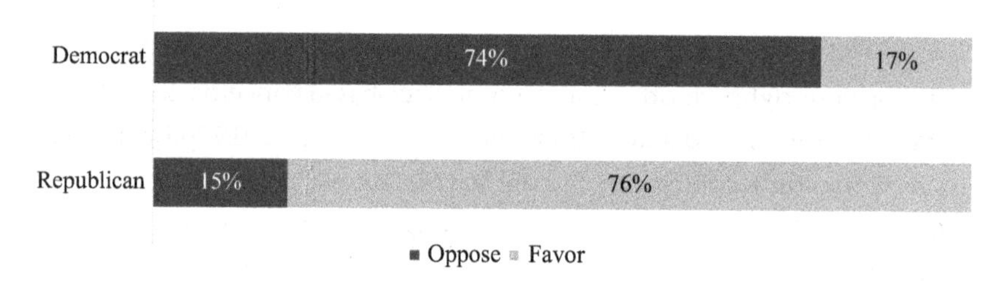

Figure 4.5
Partisan views of Keystone XL pipeline (February 2017).
Note: "Don't know" responses are not shown.
Source: Pew Research Center (2017).

They found, not surprisingly, that the level of opposition was strongly influenced by partisan affiliation and political ideology but also that the closer to the pipeline one lived, the more likely one was to support it, stating that "local framing of economic benefits appears to have outweighed counter-framing regarding risks" (Gravelle and Lachapelle 2015, 106). This finding certainly complicates the climate movement's strategy of allying itself with place-based interests and also reflects the fact that pipeline proponents focus heavily on jobs in part because they are a concrete, place-based phenomenon, but it does not undermine the logic of mobilizing resistance by appealing to concerns about risks to local environmental values.

Institutions and the Politics of Structure

Policy conflicts are struggles among strategic actors but also conflicts over institutions. Like many controversial issues, the struggle over the "rules of the game" has been a central component of this issue. Keystone XL has been caught in conflicts between Canada and the United States; between landowners, regulators, and corporations; between states and the US government; and between Congress and the president. The competing coalitions carefully chose the most strategic venues: opponents focused on the state of Nebraska, President Obama, and the courts. Proponents chose to focus on Congress and, after November 2016, President Trump. In this case, the political battle over who should play what role in decision-making is a clear demonstration that strategic actors will focus their strategy on the institutional venue(s) most favorable to their interests.

Because it crosses the Canada-US border, the Keystone XL pipeline required the approval of both federal governments. In Canada, the regulatory process was not controversial, and the project was approved by the National Energy Board in 2010 (National Energy Board 2010). The pipeline was strongly advocated by the Harper government (2006–2015), and even the more environmentally minded Prime Minister Justin Trudeau (2015–) has advocated for the pipeline's approval by the United States. In the United States, however, the regulatory process was anything but straightforward.

Conflict within the Executive Branch

Even within the executive branch of the US government, Keystone XL was the subject of division and conflict. On the US side of the border, the law requires that the transboundary pipeline receive a "national interest" determination issued by the secretary of state.[3] The State Department is responsible for the environmental impact assessment process under the National Environmental Policy Act (NEPA) in this case. TransCanada submitted its application for the Keystone XL pipeline process on September 19, 2008. The State Department began the process of environmental assessment, and it issued a draft environmental impact statement in April 2010 that suggested there were no significant environmental concerns. Even Secretary of State Hilary Clinton publicly stated, in October 2010, that the department was "inclined to approve" the project (McConaghy 2017, 50–53).

But a combination of accidents and political mobilization fundamentally altered the political environment of the entire Keystone XL project.

First, two major oil spills highlighted the risks of oil production and transport. The BP Deepwater Horizon accident in the Gulf of Mexico occurred in April 2010. More directly relevant to Keystone XL, an Enbridge pipeline carrying oil sands products ruptured and spilled into the Kalamazoo River in Michigan in late July 2010. Second, the environmental campaign against the project began in earnest, as described in the following section.

Within the Obama administration, environmental concerns about the project were pushed by the Environmental Protection Agency (EPA). Conflict within the administration began when the EPA, during the required period of interagency review, took issue with the State Department's finding of limited environmental impact. The EPA was particularly concerned that the draft impact statement did not adequately assess climate risks, declaring that the State Department draft was "inadequate" (Parfomak et al. 2013). This intervention delayed the process, and the State Department issued a draft supplemental environmental impact statement in April 2011, a full year after the draft had been issued. This time, the EPA revised its rating from inadequate to "Environmental Objections—Insufficient Information." The final environmental impact statement was released in August 2011. It examined risk to the Sand Hills and Ogallala Aquifer and considered alternative routes. It found that alternative routes were less desirable, either for reasons of cost or environmental impact, stating that "alternatives would be longer than the proposed route and would disturb more land and cross more water bodies than the proposed route" (US Department of State 2011a, ES-12). While it found that the greenhouse gas intensity of oil sands products was greater than for other sources of oil, the life cycle analysis it commissioned "reported that there would be no substantive change in global GHG emissions and . . . there would likely be no substantial change in oil sands imports to the gulf coast region" (US Department of State 2011a, 3.14–3.56).

The next stage in the process was the "national interest determination" made by the secretary of state. Even at this point, TransCanada was confident that the project would be expeditiously approved, anticipating that a certificate of approval would be received by the end of 2011 (McConaghy 2017, 53). But the company, like the entire oil sands coalition, dramatically underestimated the political power of the anti-pipeline coalition, in this case spearheaded by Bill McKibben of 350.org.

Spotlighting President Obama

Environmentalists seized on resistance to Keystone XL as a way to breathe new life into a climate movement devastated by defeats at Copenhagen and in Congress (Cheon and Urpelainen 2018; McConaghy 2017). Choosing a specific project allowed the movement to overcome some of the political challenges of climate mobilization. The choice to focus on Keystone XL was also based on an institutional opportunity: it provided an opportunity for a political win because the requirement for a determination of national interest focused the decision on the executive branch and the core veto player to whom they had access: President Barack Obama. Naomi Klein describes the choice this way: "Unlike so many other key climate policies, which either required approval from Congress or were made at the state level, the decision about whether to approve the Keystone XL pipeline was up to the State Department and, ultimately, the president himself, based on whether he determined the project to be in the 'national interest.' On this one, Obama would have to give his personal yes or no, and it seemed to us that there was value in extracting either answer" (Klein 2014, 140).

Bill McKibben explained the strategy in these words: "Congress didn't need to act, which was good since I knew there was no possible way to even think about convincing the Republican-controlled House of Representatives to block the pipeline. But this decision would be made by Barack Obama, and Barack Obama was fifteen months away from an election. Maybe we had an opening to apply some pressure—an opening to see if we'd nurtured a climate movement strong enough to make a difference" (McKibben 2013c).

Environmentalists were considered a core component of the Obama coalition, and the Keystone XL strategy was careful to seize the opportunity by, among other things, making Obama's own words the focus on protest signs at various events. If they were able to convince Obama to stop the pipeline, it would be a much-needed win. If they failed, that would also clarify that they needed to shift electoral strategies away from supporting him.

The campaign began in earnest in summer 2011, when McKibben, Klein, Hansen, and others called for a large-scale act of civil disobedience outside the White House. The open letter began: "We want you to consider doing something hard—coming to Washington in the hottest and stickiest weeks of the summer and engaging in civil disobedience that will likely get you arrested." It used evocative, even alarmist, language about the stakes: "The Keystone Pipeline would also be a 1,500-mile fuse to the biggest carbon bomb on the continent, a way to make it easier and faster to trigger the final overheating of our planet" (McKibben et al. 2011). The action was timed for the second half of August because that was when the State Department was expected to release the final environmental assessment.

The response was impressive: in August, 1,252 protesters, including McKibben, Hansen, and a number of celebrities, were arrested outside the White House. In a pivotal development for the US environmental movement, the civil disobedience campaign was endorsed by mainstream US environmental groups, including the NRDC, the Environmental Defense Fund, and the Sierra Club.[4] The letter couldn't have been more direct at threatening Obama with a political backlash from his base:

> Dear President Obama:
> Many of the organizations we head do not engage in civil disobedience; some do.
>
> Regardless, speaking as individuals, we want to let you know that there is not an inch of daylight between our policy position on the Keystone Pipeline and those of the very civil protesters being arrested daily outside the White House. This is a terrible project—many of the country's leading climate scientists have explained why in their letter last month to you. It risks many of our national treasures to leaks and spills. And it reduces incentives to make the transition to job-creating clean fuels.
>
> You have a clear shot to deny the permit, without any interference from Congress. It's perhaps the biggest climate test you face between now and the election. If you block it, you will trigger a surge of enthusiasm from the green base that supported you so strongly in the last election. We expect nothing less. (Henn and Kessler 2011)

The action was also endorsed by nine Nobel Peace Prize laureates, including Desmond Tutu and the Dalai Lama (Cheon and Urpelainen 2018, 71).

The campaign followed the August action with a larger call designed to pressure Obama on the national interest determination. On November 6, 2011, about 15,000 protesters surrounded the White House, demanding Obama reject the pipeline. Four days later, the Obama administration announced that it was delaying its decision on the pipeline for a year "to undertake an in-depth assessment of potential alternative routes in Nebraska" (US Department of State 2011b).

The environmental mobilization clearly had a profound policy impact, by obtaining first numerous delays and then, ultimately, Obama's decision to reject the pipeline outright. Obama's first rejection decision occurred when, as described in the next section, Congress tried to force his hand by requiring him to issue a decision within 60 days. But, in doing so, the administration also said that TransCanada was welcome to submit a new application with a new route to avoid the sensitive Sand Hills region. The company agreed and filed the new application in May 2012.

As the new environmental review proceeded, protests continued and the Keystone XL controversy became one of the most divisive political issues, and certainly the most divisive environmental issue, of the first half of the decade in the United States. In February 2013, what was billed as the largest climate rally in the United States brought 40,000 protesters to the Washington mall. In March 2014, a civil disobedience action by youths led to almost 400 arrests (Cheon and Urpelainen 2018, 72).

Obama had pledged action on climate since receiving the Democratic nomination for president, but in June 2013 he tied climate action directly to the Keystone XL pipeline for the first time, during a speech at Georgetown University:

> I know there's been . . . a lot of controversy surrounding the proposal to build a pipeline, the Keystone pipeline, that would carry oil from Canadian tar sands down to refineries in the Gulf. And the State Department is going through the final stages of evaluating the proposal. That's how it's always been done. But I do want to be clear: Allowing the Keystone pipeline to be built requires a finding that doing so would be in our nation's interest. And our national interest will be served only if this project does not significantly exacerbate the problem of carbon pollution. The net effects of the pipeline's impact on our climate will be absolutely critical to determining whether this project is allowed to go forward. It's relevant. (Obama 2013)

This new climate test for pipelines was a major victory for the US climate movement.

While the articulation of that climate test seemed to have set the stage for rejecting the pipeline, the State Department environmental assessment was unable to provide the rationale for a finding that the project would "significantly exacerbate" climate pollution. In January 2014, another 20 months of review after the formal draft, the State Department issued its second final environmental impact statement, again concluding that "the proposed Project is unlikely to significantly affect the rate of extraction in oil sands areas" (US Department of State 2014). Again, the ball was back in Obama's court to make a national interest determination, but there was no timeline for doing so. Developments in Nebraska, described later, provided political cover for further delays. Obama did ultimately reject the project in November 2015, for reasons described later.

Asserting Congressional Control

While pipeline opponents focused their campaign on President Obama, pipeline proponents focused much of their effort on Congress. Keystone XL became a major issue in the increasingly rancorous partisanship dividing Democratic president Barack Obama's White House from the Republican-dominated Congress. Republicans took a majority in the House in the 2010 midterm elections, and while they didn't gain a majority in the Senate until 2014, they had enough seats in the Senate to help thwart most of Obama's initiatives. When Obama decided to postpone a decision on the pipeline late in 2011, Congress tried to assert control by passing legislation requiring that a decision be made within 60 days. The president complied with the deadline but not the desired outcome: he rejected the pipeline because of insufficient information about risks to groundwater in Nebraska (Parfomak et al. 2015).

Republicans in Congress again tried to assert control over the decision in February 2015 by passing a bill requiring that Obama approve the pipeline. Obama vetoed the bill and Republicans were unable to get enough votes to override the veto. The Senate voted 62–36 in favor, enough to withstand a filibuster but not enough to override a presidential veto (Parfomak et al. 2015; Zoldan 2015). The veto decision did not reject Keystone XL. The text of the veto decision emphasized that its purpose was to protect presidential authority over cross-boundary pipeline decisions. As Obama explained, "Because this act of Congress conflicts with established executive branch procedures and cuts short thorough consideration of issues that could bear on our national interest—including our security, safety, and environment—it has earned my veto" (White House 2015). While the authorizing statute and the presidential veto allowed Obama to thwart the congressional assertion of control, his pipeline legacy was quickly reversed with Trump's election.

The Institutional Struggle within Nebraska

A further institutional complication for pipeline construction is that each individual state through which the pipeline would travel needed to approve the project. This multiplied the number of possible veto points in the American process. While approvals by Montana and South Dakota were relatively smooth, Nebraska became a major obstacle because of the concerns over the Ogallala Aquifer and the organizational activities of Bold Nebraska and other local groups. This battle over institutional rules was between landowners resisting the pipeline and TransCanada, with the issue being executive authority versus an independent regulatory commission.

As an indication of how effective Bold Nebraska and its allies were in turning public opinion against the project, when a TransCanada ad was shown at a University of Nebraska football game during the fall 2011 season, it was loudly booed by many of the 80,000 fans in attendance. The next day, the university announced that TransCanada would no longer be able to advertise at university events (Cheon and Urpelainen 2018, 81). To Greenpeace's Keith Stewart, this was a major blow to the project: "Where I think TransCanada lost in Nebraska was the day the football team gave them back their money and took down their ads from the stadium. As an indicator of social license: if your money isn't good to the Cornhuskers, that's a sign you've lost in Nebraska" (Stewart 2016).

In late August 2011, Nebraska governor Dave Heineman stunned TransCanada and its allies by writing to President Obama asking that the pipeline be rerouted to avoid the sensitive Sand Hills area. He also called a special session of the Nebraska legislature to pass a new law for pipeline reviews. Several months later, the Nebraska legislature passed the new pipeline siting law establishing an application process through the Nebraska Public Service Commission. In April 2012, the legislature amended the statute to give final decision authority to the governor. Relying on that authority, Governor Heineman approved the pipeline's route through the state in January 2013 (Omaha World Herald 2017).

Landowners challenged the legal basis for the governor's decision in court, and in February 2014 a Lancaster County district judge ruled that the action violated the state's constitution. In January 2015, the Nebraska Supreme Court overturned the decision (*Thomson v. Heineman*, Nebraska Supreme Court 289 Neb. 798 (2015)), so the governor's approval decision stood. But then legal battles turned to the "condemnation" process by which TransCanada was attempting to invoke eminent domain regarding landowners who refused to sign pipeline easements (Omaha World Herald 2017). After a setback in court, TransCanada decided to change course in November 2015 and submit its application to the Nebraska Public Service Commission. This move was considered a desperate "Hail Mary" to fend off what they believed was imminent rejection of the pipeline by Obama

(Morton 2015), but the Obama administration declined the company's request that it suspend its review and formally rejected the pipeline several days later.

In the wake of Trump's election and his January 2017 executive order to restart the Keystone XL review, TransCanada resubmitted its application to the Nebraska Public Service Commission in February 2017. After public hearings, the commission issued a ruling that cast further doubt on the pipeline's fate by rejecting the company's preferred route and approving instead an alternative route that followed the existing pipeline more closely (Hammel 2017).

The Battle in the Courts

While Nebraska-based opponents focused on that venue, other pipeline opponents challenged Trump's project approval in court.[5] Trump's decision has been successfully challenged in court by a coalition of environmental groups and Indigenous groups, including the Indigenous Environmental Network, the Sierra Club, the Natural Resources Defense Council, the Northern Plains Resource Council, and Bold Alliance (a coalition of citizens fighting pipelines). A federal district court in Montana first ruled, in August 2018, that the Trump administration was required to file a new environmental impact statement. Then, in November, the court struck down the presidential permit, ruling that the Trump administration had violated the National Environmental Policy Act and Administrative Procedures Act because it "disregarded prior factual findings related to climate change and reversed course" (*Indigenous Environmental Network et al. vs. Department of State et al.*, US District Court, Montana, Case 4:17-cv-00029-BMM, November 8, 2018).

After the US Court of Appeals for the Ninth Circuit refused to overturn the injunction, in March 2019 the Trump administration reissued a presidential permit for construction, ignoring the court's direction that it update the environmental assessment. Anti-pipeline groups challenged that action in court again, this time using the Clean Water Act's provisions as ammunition. In April 2020, the US district court in Montana invalided the generic permit the Army Corps of Engineers had implemented for violating the Endangered Species Act (*Northern Plains Resource Council et al. vs. Army Corps of Engineers et al.*, Case 4:19-cv-00044-BMM, April 15, 2020). This decision cast further doubt on the viability of the pipeline despite the Trump administration's enthusiasm for it.

This section supports the hypothesis that strategic actors will focus their strategies on the institutional venue(s) most favorable to their interests. Environmentalists chose the Keystone XL battle in large part to take advantage of the fact that the federal decision was controlled by their ally, President Obama. Pipeline advocates chose to focus their efforts on Congress, where environmental values were less well represented after the GOP takeover of the House. The 2016 presidential election gave pipeline advocates an ally in the White House, so pipeline opponents shifted their focus to the state of Nebraska and the federal courts. The Nebraska rerouting decision was a significant setback for the project, and the federal court decisions forced the company to postpone construction yet again. These delays thwarted progress on the pipeline long enough for Joe Biden to assume the presidency and immediately cancel the presidential permit.

The Reflection of Values in Presidential Determinations

The fourth behavioral hypothesis argues that decision rationales about pipelines will emphasize place-based risks far more than climate risks. The Keystone XL battle has witnessed four different presidential determinations, three by Obama and one by Trump (Trump's March 2019 presidential permit was not accompanied by a detailed rationale). Obama's first determination was the rationale for the fall 2011 decision to postpone action on the pipeline. It made scant mention of climate issues. In its four-paragraph statement explaining the decision, virtually all the focus was on the risks to a sensitive watershed along the pipeline route in the state of Nebraska, the Ogallala Aquifer, and the need to consider alternative routes to avoid jeopardizing that area. Only the final sentence gave a passing nod to climate change: "Among the relevant issues that would be considered are environmental concerns (including climate change), energy security, economic impacts, and foreign policy" (US Department of State 2011b).

The second determination was Obama's January 2012 response to congressional demands to make a decision. The brief statement from the White House does not mention climate at all. The decision rationale was based on the fact that more time was needed to conduct an "assessment of alternative pipeline routes that avoided the uniquely sensitive terrain of the Sand Hills in Nebraska" (US Department of State 2012). With the rerouting of the pipeline around the Sand Hills area and the resulting reduced threat to the

Ogallala Aquifer, it became much more difficult to justify decisions based on place-based environmental concerns.

In Obama's November 2015 decision to reject the pipeline, climate concerns did take center stage in the rationale. While the analysis suggested the pipeline was "unlikely to significantly impact the level of GHG-intensive extraction of oil sands," the principal rationale for the decision was reputational: "It is critical to the United States to prioritize actions that are not perceived as enabling further GHG emissions globally" (US Department of State 2015).

Revealingly, the decision was not based on the climate test from the Georgetown speech that Keystone XL "would significantly exacerbate" GHG emissions. The State Department analysis did not support that. The decision rationale directly states that "the proposed project by itself is unlikely to significantly impact the level of GHG-intensive extraction of oil sands" (US Department of State 2015). Instead, the decision cites the broader standard of the "net effects of the pipeline's impact on our climate." Despite finding little measurable impact on GHG emissions, the record of decision continued, "it is critical for the United States to prioritize actions that are not *perceived as enabling* further GHG emissions globally" (emphasis added).

In effect, the decision transformed Obama's climate test into a broader consideration of the impact of the decision on global climate politics: "A key consideration at this time is that granting a Presidential Permit for this proposed Project would undermine U.S. climate leadership and thereby have an adverse impact on encouraging other States to combat climate change and work to achieve and implement a robust and meaningful global climate agreement. Strong climate targets and an effective global climate agreement would lead to a reduction in global GHG emissions that would have a direct and beneficial impact on the national security and other interests of the United States" (US Department of State 2015).

In his public remarks, Obama expressed frustration over the exaggerated claims about the pipeline, not just from industry but also from the environmental movement: "Now, for years, the Keystone Pipeline has occupied what I, frankly, consider an overinflated role in our political discourse. It became a symbol too often used as a campaign cudgel by both parties rather than a serious policy matter. And all of this obscured the fact that this pipeline would neither be a silver bullet for the economy, as was promised by some, nor the express lane to climate disaster proclaimed by others" (Obama 2015). He stressed the role model justification for the decision, saying, "Today, we're continuing to lead by example, because ultimately, if we're gonna prevent large parts of this Earth from becoming not only inhospitable but uninhabitable in our lifetimes, we're going to have to keep some fossil fuels in the ground rather than burn them and release more dangerous pollution into the sky" (Obama 2015).

This decision was a remarkable rhetorical and political victory for the climate movement in two ways. First, Obama's adoption of the "keep it in the ground" framing was a major breakthrough for the movement. But more importantly, the Record of Decision essentially acknowledged that symbolism had triumphed over analysis. The power of the "carbon bomb/game over" framing is that it enabled Obama to reject the pipeline even without demonstrated climate impacts. The symbolic magnification of importance manufactured through the anti-pipeline coalition's framing had so damaged the global political interpretation of Keystone XL that the United States couldn't go to Paris with credibility on climate if it approved the project.

The climate movement, remarkably, transformed Keystone XL into a symbol of climate destruction, and it worked in getting the project killed at least until the White House was surrendered to an ally of the oil sands coalition.

Eighteen months later, Trump reversed Obama's decision and issued a presidential permit. The determination used the same GHG analysis but emphasized the absence of a significant impact on emission levels. While the Obama decision emphasized the project's "limited benefit for energy security," Trump's decision stated that the pipeline "will meaningfully support U.S. energy security." In the most important passage, Trump's decision replaced Obama's prioritization of the international reputation of the United States on climate with prioritization of energy security and economic growth. Given the commitments of many countries to act under the Paris agreement, Trump's determination concludes, "A decision to approve this Project at this time would not undermine U.S. objectives in this area." Rather, approving the project "would support U.S. priorities relating to energy security, economic development, and infrastructure" (US Department of State 2017).

The analysis of presidential decisions gives partial support to the fourth hypothesis. The decision rationales in the first round, when the route through Nebraska went through the sensitive Sand Hills region, gave clear priority to place-based risks over climate concerns. When the most salient place-based risks were eliminated by rerouting, Obama had no choice but to justify the decision on climate grounds. When Republicans took over the White House, economic values easily trumped climate concerns, and the pipeline was granted a presidential permit. When President Joe Biden reversed course and canceled the permit in January 2021, he stated: "The United States must be in a position to exercise vigorous climate leadership in order to achieve a significant increase in global climate action and put the world on a sustainable climate pathway" (White House 2021).

Conclusion

Keystone XL became one of the most controversial issues in American politics in the 2010s. Stung by losses in Congress and at the United Nations, US environmentalists deliberately reframed the pipeline as the poster child for the emerging "keep it in the ground" strategy to fight climate change. The requirement for an executive branch determination of national interest allowed them to deliberately target President Obama, who was ideologically sympathetic to the climate issue and anxious to maintain the committed electoral support of the environmental community. They also allied themselves with Native Americans, farmers, and landowners concerned about the risks pipeline accidents posed to their water supply. In doing so, they were able to ally themselves with place-based interests who had access to possible veto points at the state level and reduce the exceptional organizational challenges posed by the climate problem.

The analysis yields strong support for the first two hypotheses and mixed support for the third and fourth. Climate activists allied themselves with groups representing place-based interests in a conscious effort to strengthen their political leverage. The analysis of the institutional "politics of structure" clearly demonstrates that strategic actors will focus their strategies on the institutional venue(s) most favorable to their interests. The third hypothesis was that pipeline opponents would adopt framing that emphasized place-based risks. Unquestionably, climate activists did ally themselves with place-based groups and emphasized their concerns, but clearly not at the expense of climate issues. Climate issues were the dominant environmental issue, far more so than local issues, in mainstream media coverage of the Keystone XL controversy, at least as reflected in reporting by the *New York Times* and *Washington Post*. The analysis also found some support for the hypothesis that decision rationales about pipelines would emphasize place-based risks far more than climate risks. But once the most salient place-based risks were removed, climate issues took front and center.

The initial victories of pipeline opponents, first in having the decision postponed several times and then finally winning the rejection of the pipeline late in 2015, reveal the power of actors' strategic choices. Such significant policy changes are considered unlikely without "significant perturbations external to the subsystem" (Sabatier and Jenkins-Smith 1993), but in this case, it was the climate movement's shift to embracing the "keep it in the ground" strategy that seemed to be the most important force driving the policy change. Background economic conditions, most importantly the boom in US domestic energy production and the collapse of oil prices late in 2014, reduced the apparent economic benefits of the project. But those changes were not evident in 2011 and 2012, when aggressive campaigning by pipeline opponents won vital delays and the rejection of the initial proposal in 2012.

The case is also a clear reminder of the importance of presidential control and presidential power. On even relatively salient environmental controversies such as this one, the values of the president can be the most important determinant of policy, even in the face of a hostile Congress and public opinion. Recall that when Obama delayed and then rejected the pipeline, public opinion was in favor of the pipeline by a substantial margin. By the time Trump approved the pipeline in March 2017, public opinion had shifted to decisive opposition to it.

Led by McKibben, the environmental movement deliberately set out to elevate Keystone XL into a symbol of the climate crisis, and it worked. Thirteen years after it was initially proposed, President Joe Biden, as part of a package of climate initiatives announced on the first day of his presidency, terminated the project.

5 The Northern Gateway Pipeline: The Continental Divide in Energy Politics

Jess Housty is a member of the Heiltsuk First Nation from Bella Bella, a remote community on Campbell Island near the central coast of British Columbia, in the heart of the land that has become known as the Great Bear Rainforest. The community was strongly opposed to the proposed Northern Gateway Pipeline because of their concerns that a tanker spill would affect their way of life. Housty worked with other community members to help organize the response to the visit of the three-member Joint Review Panel, which was scheduled to hold hearings in the community on April 3, 2012. A formal delegation of hereditary leaders was poised on the tarmac to greet the panel members. Traditional dancers and drummers were outside the airport, supporting the chiefs. The road from the airport to town was lined by children from the community holding signs of opposition.

Rather than allow themselves to be greeted by the elders, panel members were whisked into the airport and to a waiting van. They drove straight to the town dock and boarded a water taxi to a nearby island, where they were staying. The panel abruptly canceled its planned hearing for that afternoon and the following day, saying they were concerned about their ability to "conduct the hearings in a safe and secure environment." Media reports give no reason for the panel members to fear for their safety, including assurances from the Royal Canadian Mounted Police that the protest was "very peaceful" (Hager 2012; CBC 2012a). But clearly, the vehemence of the opposition spooked the panel. In Housty's words, "Honestly, I was shocked to hear that they perceived that group of children standing at the roadside as a threat to their bodily safety. But there it is" (Housty 2017).

In many ways, the scene typified the Northern Gateway conflict. Touted by the oil industry and the governments of Alberta and Canada as a critical, even "nation-building," energy infrastructure project, it was met with well-organized and sustained opposition in British Columbia. While the review panel downplayed the depth and breadth of opposition and found the project to be in the national interest, by the time the Harper government approved the pipeline in June 2014, it was widely considered unviable, even by many of its strongest early supporters. Northern Gateway was defeated by sustained opposition that reverberated through provincial and federal elections and has proven to be a major setback for the oil sands coalition.

With respect to the analytical framework, Northern Gateway had a major advantage over Keystone XL in that it did not cross a national boundary and therefore was subject only to Canadian jurisdiction. It did, however, cross a provincial boundary. In this case, the border between British Columbia and Alberta proved to be much more than a separation of subnational jurisdictions within a federation. It became more like a Continental Divide of energy politics between the oil sands coalition and the anti-pipeline coalition.

When interest in the Northern Gateway Pipeline grew as Keystone XL stalled, what few members of the oil sands coalition knew is that a well-developed opposition movement had already been organizing in British Columbia, just waiting for the right project to elevate their combination of environmental, First Nations, and community control objectives in the political arena. Their focus on the risks of a tanker spill along British Columbia's rugged coast, as well as First Nations rights and title, ensured that place-based issues were front and center in the opposition campaign. The anti-pipeline coalition succeeded in mobilizing enough influence to induce the pro-business British Columbia government of Christy Clark to oppose the pipeline and eventually kill the project through political and legal opposition despite the Harper government's conditional approval.

The Northern Gateway Pipeline was the first of the Canadian oil sands pipelines to explode into the national media spotlight as the result of bitter political and regional conflict. The conflict intensified in early 2012, when Obama's decision to postpone acting on the Keystone XL pipeline alarmed the oil sands coalition, which was anxious to expand market access. Obama's reluctance strengthened the impulse to diversify away from the US market, and the Northern Gateway proposal was the most mature proposal to increase access to growing markets in Asia.

The timeline for the Northern Gateway case is depicted in figure 5.1. Enbridge, a Canadian pipeline company, proposed to build a 1,178-kilometer pipeline corridor from Bruderheim, Alberta, to Kitimat, a small community

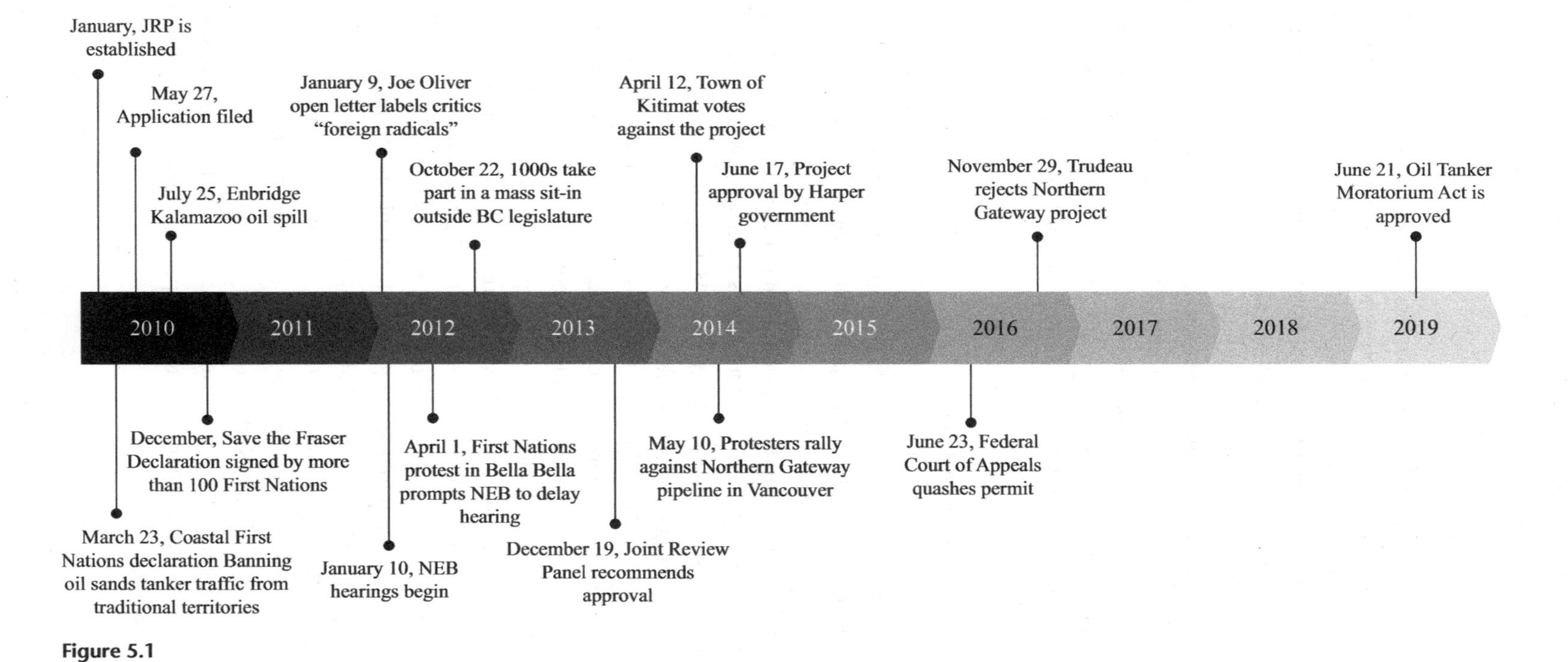

Figure 5.1
Northern Gateway Pipeline timeline.

on the northwest coast of British Columbia. The pipeline would carry diluted bitumen to Kitimat, and a parallel pipeline would transport condensate (a natural gas by-product used to dilute bitumen) from Kitimat back to the Edmonton area. The pipeline proposal was submitted for regulatory review in May 2010. After a protracted and adversarial hearing process, the Harper government ultimately approved the project, with 209 conditions, in June 2014. When the Federal Court of Appeal struck down the permit authorizing the pipeline in June 2016, the Northern Gateway Pipeline chapter finally seemed to be closed. Justin Trudeau nailed the coffin closed when he formally rejected the pipeline in November 2016 and promised to keep it closed by committing to a legislated moratorium on oil tanker traffic on British Columbia's North Pacific coast.

Actors—The Oil Sands Coalition

The core members of the oil sands coalition in support of the Northern Gateway Pipeline were the project proponent, Enbridge; the oil companies and their workers, who would benefit from improved market access; the government of Alberta; and the government of Canada under Prime Minister Stephen Harper (2006–2015). Enbridge is a Calgary-based company that is involved in a wide array of energy projects, from oil and gas pipelines to renewable energy generation and electricity transmission, but its core business is pipelines, and it is the largest oil pipeline company in the world. Its interest in the US$5.5 billion project is straightforward: increased revenues and profits from an expanded oil pipeline network.

Enbridge pursued a wide range of strategies toward this goal. It engaged in extensive consultations with First Nations and local communities in an effort to build support for, or at least reduce opposition to, the project. Those efforts included offering loans to First Nations so they could gain equity in the project. It invested a great deal in public relations, including a widespread print and video advertising campaign that, at times in 2013 and 2014, seemed to be everywhere in British Columbia—from Facebook, to movie theaters, to televised hockey games. It founded a community-based organization in 2009, the Northern Gateway Alliance, a coalition of northern British Columbia community leaders designed to build public support for the project. The company also lobbied political leaders and donated to political parties.

For the oil sands producers, Northern Gateway promised convenient access to growing Asian markets, an imperative that grew as the sector faced a pipeline capacity crunch. With American demand floundering and Keystone XL on the rocks, the oil companies' need for access to markets for growing oil sands production had become very intensely felt. The concern was greatly amplified when pipeline capacity constraints led to the emergence, in early 2011, of the widening difference between world prices (represented by Brent) and North American prices (West Texas Intermediate—WTI). Alberta academics, financial advisory firms, and the Alberta government began to produce estimates of the forgone revenues resulting from this price gap. One estimate suggested that oil companies could lose US$8 billion per year from 2017 to 2025 if the pipeline was not built (Wood MacKenzie Inc. 2011).

While companies such as Suncor, CNRL, and Cenovus represented themselves in regulatory proceedings, much of the political work for the sector was led by its trade association, the Canadian Association of Petroleum Producers (CAPP). CAPP played a very active role politically, including lobbying the Harper government to reform regulatory procedures, an initiative that bore fruit with the dramatic changes described in the institutions section later in the chapter. CAPP was also very involved in public relations and public education about the benefits of increased market access, focusing its message particularly on how widespread the economic benefits would be for Canada (described in more detail in the ideas section in this chapter).

The government of Alberta and Stephen Harper's government in Ottawa were also intensely interested in the Northern Gateway Pipeline's ability to expand market access and increase the oil sector's revenues and profits. The Alberta government had the most direct stake, with provincial revenues highly dependent on activity in the oil and gas sector. Until May 2015, the Alberta government was controlled by the Progressive Conservative Party, an aggressively pro-business, pro-oil party that unabashedly promoted increased market access for the landlocked province through new and expanded pipeline capacity. In 2012, Premier Alison Redford launched an effort to create a national energy strategy, in large part to smooth the way to increased access to tidewater through other provinces (Gattinger 2012).

In May 2015, as chapter 2 described, a political earthquake in Alberta brought Rachel Notley's leftist New Democratic Party to power. During the election campaign, Notley said she opposed Northern Gateway because of the extent of environmental concerns and First Nations opposition

(CBC 2015). She began to reconsider her opinion, however, when, as part of the broader discussions of national energy and climate policy, the issue of expanding British Columbia's electricity exports to Alberta emerged. Notley's government saw this as an opportunity to bargain Northern Gateway back into the picture (Bennett 2016). But as Trudeau's grand compromise, centered around getting British Columbia on board by approving the rival Trans Mountain expansion project, began to emerge, the deal linking Northern Gateway and British Columbia's electricity exports died.

The Harper government was unquestionably a part of the oil sands coalition, committed to using oil sands expansion as an engine of prosperity for the Canadian economy. Expanded access to tidewater with new pipelines was considered a critical ingredient in that strategy (Hoberg 2016). Once Obama's resistance to Keystone XL became apparent, Harper's focus shifted to championing Pacific coast access with the Northern Gateway Pipeline. While the government's talking points usually emphasized the importance of gaining access to the Pacific coast without specifying support for individual projects, at times it seemed like Harper's natural resources minister, Joe Oliver, could barely contain his enthusiasm for the project, saying in July 2001 that "Gateway, in our opinion, is in the national interest" (Vanderklippe 2011).

As opposition to the project escalated throughout 2011, the Harper government began a systematic effort to delegitimize Northern Gateway opponents. In a January 2012 interview, Prime Minister Harper characterized the pipeline resistance movement as foreign-funded groups seeking to hijack Canadian processes: "We have to have processes in Canada that come to our decision in a reasonable amount of time and processes that cannot be hijacked. In particular, growing concern has been expressed to me about the use of foreign money to really overload the public consultation phase of regulatory hearings, just for the purpose of slowing down the process" (Weber 2012).

These criticisms were ramped up by Natural Resources Minister Joe Oliver in an open letter to Canadians released the day before the commencement of the Northern Gateway Pipeline hearings. Oliver wrote:

> Environmental and other radical groups seek to block this opportunity to diversify our trade. . . . These groups threaten to hijack our regulatory system to achieve their radical ideological agenda. They seek to exploit any loophole they can find, stacking public hearings with bodies to ensure that delays kill good projects. They use funding from foreign special interest groups to undermine Canada's national economic interest. . . . We believe reviews for major projects can be accomplished in a quicker and more streamlined fashion. We do not want projects that are safe,

> generate thousands of new jobs and open up new export markets to die in the approval phase due to unnecessary delays. Unfortunately, the system seems to have lost sight of this balance over the past years. It is broken. It is time to take a look at it. (Oliver 2012)

The efforts to delegitimize opponents were augmented several months later when allegations of inappropriate use of charitable status by environmental groups were combined with concerns about inappropriate foreign influence. In a May 2012 CBC interview, Environment Minister Peter Kent accused environmental groups of "laundering" money: "There are allegations—and we have very strong suspicions—that some funds have come into the country improperly to obstruct, not to assist in, the environmental assessment process" (CBC 2012b).

As the pipeline's opponents grew in strength and the Northern Gateway deathwatch began, the Harper government started dialing back their project promotion in late 2012. In a speech to industry leaders, Joe Oliver acknowledged, "If we don't get people on side, we don't get the social licence—politics often follows opinion—and so we could well get a positive regulatory conclusion from the joint panel that is looking at the Northern Gateway, but if the population is not on side, there is a big problem" (Cattaneo 2012). Nonetheless, it was still clear which side the Harper government was on.

Actors—The Anti-pipeline Coalition

Environmental Groups

British Columbia's environmental community was deeply opposed to the pipeline and had strong support from environmental allies across Canada and in the United States. Organized opposition emerged from two converging sources. The first came from groups focused on environmental, First Nations, and community issues in British Columbia. Prior to around 2005, the primary focus of British Columbia's environmental community was forests, with a focus on opposition to clear-cut logging, preservation of old growth forests, and increasing the control of local communities. Environmentalists made significant progress in improving policy when the provincial NDP held power from 1991 to 2001 (Cashore et al. 2001), but the election of the "free market" government of BC Liberal Gordon Campbell threatened to roll back much of that progress, and environmental groups (and First Nations) organized to resist those changes (Hoberg 2010; Hoberg 2017).

A group of environmentalists and First Nations teamed up to form the Coalition for Sustainable Forest Solutions to oppose privatization of public lands and promote stronger environmental protection, reduced corporate control and increased community control, and respect for Aboriginal rights and title (BC Coalition for Sustainable Forest Solutions 2001). The issue of Aboriginal title became a meeting point between environmentalists and First Nations. Given the many unresolved title claims in the province, demands for consultation and the prospect of litigation served as potential veto points against industry-oriented forest policy reforms. British Columbia First Nations banded together to confront these changes with the Title and Rights Alliance, formed in 2003, and environmental groups, including the Dogwood Initiative and West Coast Environmental Law, provided strategy and legal support (Clogg 2017; Horter 2016).

One of the most prominent groups working against oil sands pipelines in British Columbia was the Dogwood Initiative, a grassroots organization founded in 1999 with the goal of helping "everyday British Columbians to reclaim decision-making power over the air, land and water they depend on" (Dogwood Initiative 2013). Its flagship campaign has been the "No Tankers" campaign, designed to keep oil tankers away from the British Columbia coast. The original focus of the campaign was the Northern Gateway Pipeline. Launched in spring 2007, over its history the No Tankers petition was signed by 290,111 British Columbians and about 90,000 other Canadians (personal communication, Dogwood Initiative).

In addition to a desire for greater environmental protection, what these groups had in common was a strong commitment to empowering local communities to make resource and environmental decisions. Several leaders in the coalition, including Dogwood's Will Horter, West Coast Environmental Law's Jessica Clogg, and Grand Chief of the Union of BC Indian Chiefs Stewart Phillip, began strategizing about how to build a strong movement within British Columbia. When the Enbridge Northern Gateway Pipeline proposal first appeared in 2005, they soon realized that pipeline resistance might become a powerful weapon in advancing their mutual objectives. In recounting discussions with First Nations, Will Horter explains the birth of the idea: "I said that they need a focal point that would get them out of the limitations of the forestry campaign, which was that one individual nation going against both Crowns. I said energy is the place where the fights of the future are going to happen, and the nature of energy infrastructure means that it will impact on multiple First Nations. So we should take on a pipeline" (Horter 2016).

While those groups based in British Columbia were mobilizing against Northern Gateway, a second source of opposition came from groups concerned about the impact of oil sands expansion. The pipeline resistance strategy began to take form when a group of environmentalists began meeting in 2005 under the name Upstream Strategy Working Group, whose activities and proposals were discussed in chapter 2.

When the Northern Gateway Pipeline proposal first emerged in 2005, the opposition to it was from the British Columbia–based coalition described earlier, but by the time the pipeline was formally proposed in 2010, the resistance movement had expanded dramatically. Opponents began to directly engage the pipeline in 2011. The broad coalition representing the resistance movement was illustrated by the authorship of one of the first major environmental group reports opposing Northern Gateway.[1] Published in November 2011 and titled *Pipeline and Tanker Trouble: The Impact to British Columbia's Communities, Rivers, and Pacific Coastline from Tar Sands Oil Transport*, the report was researched and written by members of Canada's Pembina Institute and Living Oceans Society, as well as the Natural Resources Defense Council, a highly influential US group. The report also listed numerous groups that had endorsed the report, essentially a who's who of the British Columbia environmental movement: the Dogwood Initiative, Douglas Channel Watch, ForestEthics, Friends of Wild Salmon, Headwaters Initiative, Pacific Wild, Raincoast Conservation Society, Sierra Club of BC, and West Coast Environmental Law. These groups, along with the Wilderness Committee, Ecojustice, and Toronto-based Environmental Defence, formed the environmental coalition in opposition to the Northern Gateway Pipeline (Swift et al. 2011).

The report focused on four issues that came to dominate the framing of the anti-pipeline coalition:

- the environmental impacts of oil sands production in Alberta on water quality and quantity, habitat health, air pollution, and climate change;
- the potential for a pipeline spill and the resulting risks to environmental values;
- the risks of a tanker spill along the coast; and
- First Nations concerns. (Swift et al. 2011)

These environmental risks were central to the argument against the pipeline and were fortified by strong opposition from First Nations.

For many of the environmental groups involved in the issue, most notably the Pembina Institute, West Coast Environmental Law, the Wilderness Committee, and Environmental Defence, climate concerns were a large part of the justification for mobilizing against the pipeline (as they told me in confidential interviews), but from the beginning, they understood the importance of allying with place-based concerns to broaden their coalition and strengthen their arguments. These links can be seen in the choice by environmentalists to tie their campaign so directly to advancing Indigenous rights and title, and their emphasis on place-based environmental risks. Climate risks were mentioned in reports, but the core framing for advocacy was built around spill risks and especially tanker accidents. This choice was driven by the belief, pushed strongly by Dogwood, that emphasizing the risk of tanker spills was critical to mobilizing opposition. As Horter put it:

> My gut feelings were pipelines are a difficult battle because no one has an opinion on them. But lots of people have opinions about tankers. . . . We put people on the streets with a petition. For pipelines, it would be a 15 minute conversation and it would be 55%–45% whether people would sign the thing. Asking about oil tankers we could get 100 signatures per hour with a team of two people. . . . Exxon Valdez was the only thing that was relevant really. It's an iconic idea that was locked in there. You didn't need to convince anybody. You just had to trigger it. So the frame was set, we just had to trigger it. (Horter 2016)

First Nations

From the beginning, the Northern Gateway Pipeline has faced resolute opposition from a large number of First Nations. The pipeline route would cross the territories of scores of First Nations. While in Alberta and the northeastern section of British Columbia First Nations are covered by treaty agreements, in British Columbia much of the route goes through unceded First Nations territory. First Nations opposition, especially west of Prince George, has been intense. The coalition between environmentalists and First Nations was critical to the defeat of Northern Gateway. In addition to the Coalition for Sustainable Forest Solutions mentioned earlier, the seeds for cooperation in the battle against the pipeline were also sown by the collaboration between environmental groups, First Nations, and US foundations over efforts to preserve the Great Bear Rainforest. The success of that forest conservation campaign, through the provincial land-use planning process and other initiatives

(Cullen et al. 2010), was an essential ingredient in the success of the campaign against Northern Gateway. It is noteworthy that several of the environmental leaders (Tzeporah Berman and Karen Mahon among them) who worked with First Nations leaders in the Great Bear Rainforest campaigns played leading roles in the Northern Gateway and other anti-pipeline battles (Berman 2011).

A core partner in that alliance was the Coastal First Nations, an alliance of nine First Nations on British Columbia's north and central coasts and Haida Gwaii, which was founded in 2003 (as the Turning Point Initiative Society) to represent regional interests (Smith and Sterritt 2016). In March 2010, the Coastal First Nations issued a declaration banning oil sands tankers and pipelines from their traditional territories and waters.[2] The declaration reads, in part:

> As nations of the Central and North Pacific Coast and Haida Gwaii, it is our custom to share our wealth and live in harmony with the broader human community. However, we will not bear the risk to these lands and waters caused by the proposed Enbridge Northern Gateway pipeline and crude oil tanker traffic. . . .
>
> Therefore, in upholding our ancestral laws, rights and responsibilities, we declare the oil tankers carrying crude oil from the Alberta tar sands will not be allowed to transit our lands and waters.
>
> To those who share our commitments to the well-being of the planet we invite you to join us in defending this magnificent coast, its creatures, cultures, and communities. (Coastal First Nations 2010)

In December 2010, the Yinka Dene Alliance joined a group of 61 First Nations in signing the Save the Fraser Declaration, stating that: "We will not allow the proposed Enbridge Northern Gateway Pipelines, or similar Tar Sands projects, to cross our lands, territories and watersheds, or the ocean migration routes of Fraser River salmon" (Save the Fraser Gathering of First Nations 2013). Both declarations state that they were made according to the authority of First Nations Ancestral Laws, Rights and Title.

Not all First Nations along the pipeline route are opposed to the project. Enbridge's engagement strategy was to offer First Nations an opportunity, collectively, to own up to 10% of the pipeline. Some First Nations accepted the offer and have become supporters of the pipeline. According to the *Gitxaala* decision, Enbridge has agreements with 26 Aboriginal equity partners. In January 2016, in the late stages of this pipeline controversy, a group calling itself Aboriginal Equity Partners published an op-ed in the *Vancouver Sun* and put up a website on behalf of 31 First Nations and Metis communities that support Northern Gateway (Aboriginal Equity Partners 2016).

First Nations opposed to the pipeline engaged in a wide variety of strategies in their efforts to block the project. They lobbied policymakers directly; engaged in discussions with Enbridge, the review panel, and governments; built coalitions formally with each other and informally with environmentalists and labor groups; conducted public information campaigns; protested extensively; and even sought to influence Enbridge shareholders. In the end, they also challenged government decisions in court, and it was the legal strategy that was ultimately the most effective in blocking the project.

Efforts to engage First Nations in the project review and decision process were conducted by the proponent, Enbridge, the Joint Review Panel, and the government of Canada. From the early days of the project consideration process in 2005, Northern Gateway was engaged in extensive consultations with First Nations groups. In 2005, the company began discussions and to offer protocol agreements to potentially affected First Nations. By the end of 2009, the company had entered 30 relationship agreements involving 36 First Nations (Joint Review Panel 2013, 28). Enbridge says that it engaged with 39 First Nations in Alberta and 41 in British Columbia (Joint Review Panel 2013, 31–33). Despite this level of activity, Enbridge was not successful at establishing effective relationships with a significant number of First Nations in British Columbia.

A number of First Nations also participated intensively as intervenors during the panel process, although the Coastal First Nations partially withdrew from the proceedings in February 2013. Information about the Canadian government's engagement with First Nations is contained in the discussion of the *Gitxaala* case later in the chapter.

Government of British Columbia

The British Columbia government had conflicting interests in the Northern Gateway proposal. On the one hand, British Columbia's Liberal Party, which dominated the province from 2001 to 2017, first under Gordon Campbell (2001–2011) and then under Christy Clark (2011–2017), was generally considered in favor of resource development, so under most circumstances it would have been open to this major resource development. But the government also needed to be sensitive to the intense opposition the pipeline had engendered and the political risks entailed in backing a project that risked becoming politically toxic.

During the early stages of the Northern Gateway controversy, Premier Christy Clark adopted a decidedly "wait and see" attitude toward the federal assessment of the pipeline (O'Neil 2011b). This stance is striking, given the enthusiastic support for the project by the previous premier and Clark's aggressive strategy of creating jobs through resource projects. But by summer 2012, the position of the government began to turn sharply against the pipeline. In addition to negative public opinion and the looming May 2013 election, pipeline politics were dramatically affected by a series of oil spills and their aftermath. The Deepwater Horizon accident on the Gulf Coast occurred in April 2010. Three months later, an Enbridge pipeline carrying diluted bitumen from the oil sands ruptured and created a major spill into Michigan's Kalamazoo River. In April 2011, just nine months later, there was another major pipeline accident in northern Alberta (Hoberg 2013).

Each of these spills heightened public concern over tanker and pipeline risks, but the development that broke the back of the Clark government's "wait and see" position was a US regulator's report on the Enbridge Kalamazoo spill. The National Transportation Safety Board's report, issued in July 2012, criticized Enbridge for a "culture of deviance" and "pervasive organizational failure" (National Transportation Safety Board 2012). Even more damaging were the remarks of the chair of the NTSB when releasing the report: "When we were examining Enbridge's poor handling of their response to this rupture you can't help but think of the Keystone Kops" (Shogren 2012). These phrases were a body blow to Enbridge's safety image and extremely damaging to elite and public perceptions of the Northern Gateway Pipeline. One columnist claimed the NTSB report was a "death knell" for the pipeline (Yaffe 2012). Another columnist pronounced the pipeline "dead and buried" (Palmer 2012). Just over two years after Enbridge submitted its application, and three and a half years before it was finally rejected, the Northern Gateway deathwatch began.

The Clark government announced its new position in July 2012. In a document titled *Requirements for British Columbia to Consider Support for Heavy Oil Pipelines*, the government outlined five conditions for its support:

1. Successful completion of the environmental review process
2. World-leading marine oil spill response, prevention and recovery systems

3. World-leading practices for land oil spill prevention, response and recovery systems
4. Legal requirements regarding Aboriginal and treaty rights are addressed
5. British Columbia receives a fair share of the fiscal and economic benefits of a proposed heavy oil project that reflects the level, degree and nature of the risk borne by the province, the environment and taxpayers. (Government of British Columbia 2012b)

This change in position reflected a far more adversarial approach by the province. The website linking to the formal document contained provocative pie charts highlighting the skewed distribution of risks and benefits between British Columbia and Alberta.

The fifth condition immediately became the most divisive. Clark demanded that the government of Alberta find a way to share economic benefits: "If Alberta doesn't decide they want to sit down and engage, the project stops. It's as simple as that." Alberta premier Alison Redford responded that, "We will not share royalties, and I see nothing else proposed and would not be prepared to consider anything else. . . . We will continue to protect the jurisdiction we have over our energy resources" (Fowlie 2012a). Redford accused Clark of trying to "renegotiate Confederation" (Bailey and Wingrove 2012). Clark theatrically left a Halifax meeting of premiers to develop a national energy strategy over the conflict. Clark characterized an October meeting between the two premiers as "short and frosty" and stated that, "As it stands right now, there is absolutely no way that British Columbia will support this proposal" (Fowlie 2012b). Clark also threatened to use sources of provincial leverage, such as control of electricity through BC Hydro and provincial permitting authorities, to thwart pipeline construction if the five conditions were not met (Fowlie 2012a).

Oil sands pipelines were a major issue in British Columbia's provincial election on May 14, 2013 (Hoberg 2013), but because Clark has taken a strong (if conditional) stance against Northern Gateway, the flashpoint was Kinder Morgan's Trans Mountain Expansion Project (see chapter 6), on which the two main political parties sharply differed. Clark's surprising come-from-behind victory in that election may have left the oil sands coalition feeling like they'd dodged a bullet, but the British Columbia Liberal Party's victory did not change the political logic on Northern Gateway.

On May 31, 2013, the provincial government submitted its final argument to the Joint Review Panel, and it took a very strong position against the Northern Gateway Pipeline. The submission focused most of its attention on whether Enbridge had demonstrated "world-leading spill response, prevention and recovery systems," the language of the second and third conditions for approval. The document was especially critical of the vagueness of Enbridge's spill response plans:

> The Project before the JRP is not a typical pipeline. For example, the behavior in water of the material to be transported is incompletely understood; the terrain the pipeline would cross is not only remote, it is in many places extremely difficult to access; the impact of the spills into pristine river environments would be profound. In these particular and unique circumstances, NG should not be granted a certificate on the basis of a promise to do more study and planning once the certificate is granted. The standard in this particular case must be higher. And yet, it is respectfully submitted, for the reasons set out below, NG has not met this standard. "Trust me" is not good enough in this case. (Government of British Columbia 2013)

Over the ensuing years, Clark sent signals that she was open to reconsidering this opposition, especially if emergency response readiness could be improved and if the province's share of economic benefits from the project could be increased, but the provincial government never formally backed away from this position of opposition.

Opposition to the pipeline also emerged among municipal governments. The Union of British Columbia Municipalities voted to oppose the pipeline at its annual meeting in fall 2010 (CBC 2010). Not all municipalities opposed the project, of course, but a majority of elected officials at that meeting voted for the resolution opposing the project, making it the official policy of the body representing municipalities in the province. The town of Kitimat, British Columbia, the Pacific coast port that would host the pipeline and tanker terminal, conducted a plebiscite on the pipeline proposal in April 2014. To the surprise of many, Kitimat residents voted against the project by a considerable margin, 58% to 42% (Rowland 2014; Bowles and MacPhail 2017). The result was a symbolic blow to Enbridge and the project and fed the political momentum against the project.

Nation Building versus Coastal Protection: Ideas and the Northern Gateway

Framing Arguments

The Northern Gateway battle was in part a battle over ideas framed and communicated by strategic actors within the dispute. The pro-pipeline oil sands coalition focused its framing around ideas of economic growth, jobs, and future prosperity. Enbridge, as well as both the Harper government and the Alberta government, stressed the economic benefits the pipeline would deliver to Canadians. Both the company and Harper government officials went beyond making the case that the project was in the "national interest" to the loftier claim that it was "nation-building." In a 2011 speech at the Empire Club in Toronto, Enbridge CEO Pat Daniel compared the project to the great leaps forward in national infrastructure in previous centuries:

> Northern Gateway will bring Canada's energy resources to the booming economies of the Pacific basin, while delivering sustainable local and regional prosperity to northern BC and Alberta and national economic advantage for all Canadians. . . . The benefits of this transformation will mean billions of dollars flowing into Canadian hands for decades. Reliable independent estimates of the project's impact over thirty years say it will deliver to all Canadians an additional $270 billion increase in Canada's GDP. . . . Where will the impact of that $270 Billion increase be felt? Not just in Alberta or BC. But here, in the industrial heartland of Canada, in the steel mills and manufacturing centres, and from heavy industry to high finance, for a long, long time. This isn't just another Alberta oil and gas project. *This is a nation-building project.* . . . Canada's earliest nation builders invested in massive infrastructure to make us a North American and European economic champion. Look at the St. Lawrence Seaway. It's an economic engine bringing maritime trade to the heart of the North American continent. It was and remains a massive and multi-generational undertaking that has cemented Canada's trade connections to the Atlantic nations. A gateway to Pacific markets will have the same advantage for Canada in the 21st Century that the St. Lawrence Seaway and key canals had for our country in the 19th and 20th Centuries. (Daniel 2011, emphasis added)

These comments were reiterated by Joe Oliver, Harper's natural resources minister, in December 2011. He called gaining access to new markets "nation-building, without exaggeration" (O'Neil 2011a).

To counter this frame of economic prosperity through a nation-building pipeline, Northern Gateway opponents focused on the risks of environmental damage and Aboriginal rights. Environmental groups focused on the relatively pristine environment through which the pipeline and tankers would travel and the threats to that environment posed by oil sands. A March 2012 report by Environmental Defence and ForestEthics characterized the case against Northern Gateway in the following terms:

> The project is premised on a rapid increase of the amount of tar sands oil being produced. The impacts of it would span from the tar sands region, which would

> deal with more habitat destruction, toxic tailings and air pollution, across pristine boreal forests and nearly 800 rivers and streams, to the coast. It would put at risk the survival of the threatened woodland caribou, the spawning grounds of all five species of wild salmon, and a unique and diverse marine ecosystem. The tankers would travel through the Great Bear Rainforest, where a spill would harm the iconic Spirit Bear, the animal that inspired one of the 2010 Olympic mascots. (Environmental Defence and ForestEthics 2012)

This frame of toxic oil threatening the pristine environment was central to the environmental campaign from the start.

First Nations also emphasized the risks of pipeline and tanker accidents, but they also focused on Aboriginal rights over traditional lands and waters. The Coastal First Nations declaration speaks of defending "this magnificent coast, its creatures, cultures and communities" (Coastal First Nations 2010). The Save the Fraser declaration refers specifically to the need to protect "the ocean migration routes of Fraser River salmon." In a media interview following the Harper government's approval decision, Grand Chief Stewart Phillip focused on Aboriginal decision rights: "The First Nations people of this province have made it abundantly clear that we have every right, based on our constitutional and judicially upheld rights, rights that are reflected in the UN Declaration on the Rights of Indigenous Peoples, to stand by our traditional laws and legal orders and defend our territories from the threats and predations of oil and gas. And we fully intend to do that" (Judd 2014).

Content of Media Coverage

Conflict over the Northern Gateway Pipeline proved to be quite newsworthy. Figure 5.2 shows how many times the pipeline was mentioned in Canadian news stories from 2010 and 2016. Media mentions of the pipeline peaked with 6,844 mentions in 2012—the first year of the hearings. While mentions tailed off before the pipeline was finally terminated in 2016, in 2013 and 2014 it continued to be mentioned over 3,500 times by Canadian media.

Figure 5.3 shows what issue the media focused on when it reported about the pipeline. The chart displays four categories of issues and depicts their relative priority in media coverage of Northern Gateway and how that has changed over time: climate change, jobs, pipeline or tanker accidents, and First Nations.[3] The analysis shows that First Nations concerns were the dominant issue overall, followed by concerns about accidents. Jobs ranked a close third, and climate change was a distant fourth. For this pipeline controversy,

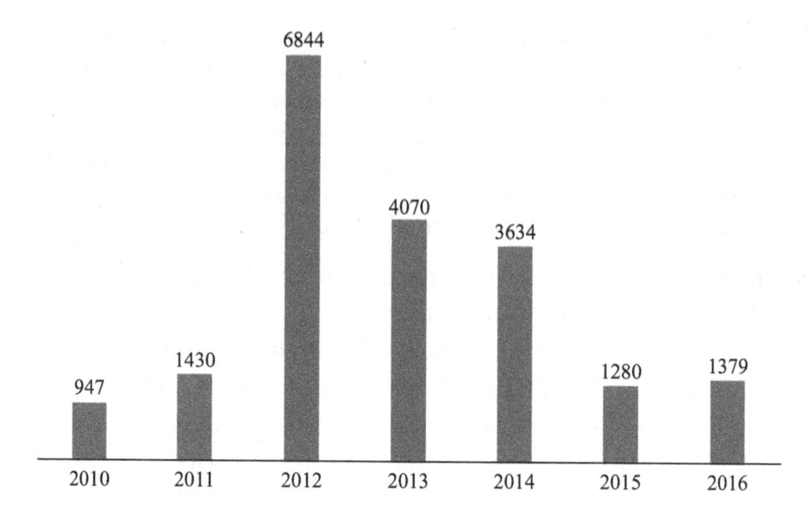

Figure 5.2
Northern Gateway Pipeline total media mentions, 2010–2016.

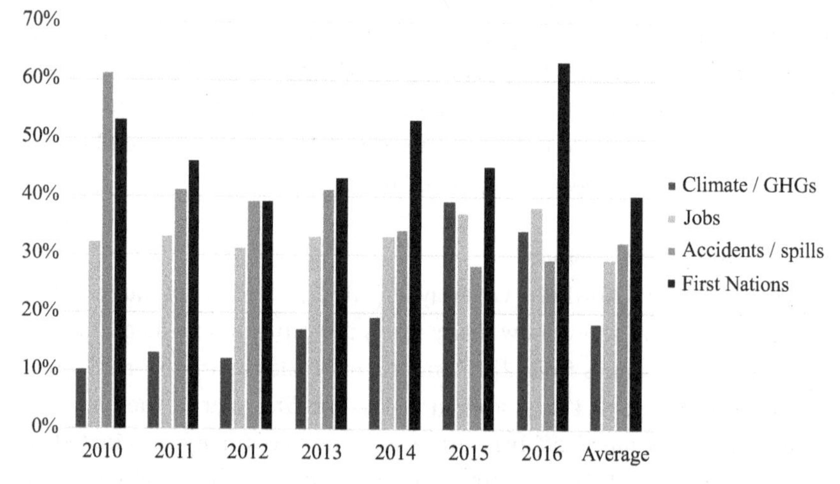

Figure 5.3
Northern Gateway Pipeline media focus, 2010–2016.

the issues given the greatest attention in opposition campaigns received more coverage than the central issue, jobs, featured in communications of the pro-pipeline coalition.

While overall these results support the behavioral hypothesis that opposition groups emphasize place-based risks, the trend over time shows that climate concerns became increasingly important in media coverage. While accidents and spills were mentioned more than three times as often as climate in 2010, 2011, and 2012, by 2015 and 2016, climate was mentioned more often than accidents and spills. The causes and implications of this shift are discussed further in chapter 6.

Public Opinion

In the early years of the Northern Gateway controversy, public opinion in Canada was generally opposed to the pipeline, but a big shift toward support appeared to occur by 2016. A March 2016 Forum Research poll showed that the Canadian public approved of the construction of the Northern Gateway Pipeline by a margin of 51% to 36%. However, between December 2011 and October 2013, Forum Research polls showed pipeline disapproval was greater than approval by between 12% and 16%. Even when national support for the pipeline surged in 2016, opposition in British Columbia was still strong: those opposed outnumbered those approving by a margin of 53% to 40%. In 2016, British Columbia was the only province other than Quebec that was opposed to the pipeline; Albertans approved of the pipeline by a 75% to 21% margin (Forum Research 2016).

An Ekos poll issued the same month showed less national support, but supporters still outnumbered opponents, 48% to 43% (Ekos 2016). While showing narrower national support, the Ekos poll showed even greater opposition in British Columbia. The pipeline was opposed by 60% and supported by 37% of respondents in British Columbia. An Insights West poll (figure 5.4) shows that British Columbia respondents consistently opposed the pipeline. In August 2016, the margin of opponents over supporters was 50% to 35% (Insights West 2016a).

Northern Gateway was not just a battle over interests but also a battle over values and ideas. Despite the structural advantages of economic issues over environmental concerns, appeals to jobs and prosperity through a nation-building pipeline were, in this case, trumped by concerns over environmental

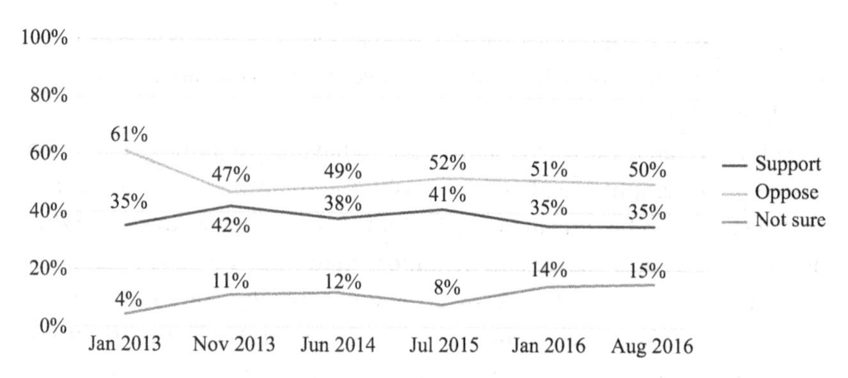

Figure 5.4
British Columbia poll results for Northern Gateway support when respondents were asked: "Given what you know about the proposed Enbridge Northern Gateway pipelines project, do you support or oppose the project?"
Source: Insights West (2016a).

damage to pristine environments and the rights of Aboriginal groups to make decisions affecting their own territory.

Institutions and the Politics of Structure

Intergovernmental Rules

One major institutional question with respect to pipeline review and approval is the allocation of decision-making authority between the federal and provincial governments (Olszynski 2018). Because the pipeline is interprovincial, the federal government has the clear lead responsibility for it, but there are also a number of areas within provincial jurisdiction, including the need for provincial permits to authorize various aspects of construction and operation (Bankes 2015). Canadian governments have dealt with this coordination challenge in a variety of ways in the past. At one end of the spectrum are cases where the provincial and federal governments conducted separate environmental reviews, as happened in the case of British Columbia's Prosperity Mine (Haddock 2010).

Over time, governments have made increased efforts to avoid duplication as much as possible. This could be done through conducting the process jointly, as occurred in the Site C Dam (see chapter 9). Alternatively, a formal agreement between the province and the federal government could be established to determine who would take the lead in performing the review and approval. This is how the governments proceeded in the Northern Gateway case. In June 2010, the British Columbia Environmental Assessment Office, under the authority of the British Columbia Environmental Assessment Act, entered an "equivalency agreement" with the National Energy Board (NEB) that essentially declared the federal review process equivalent to that which would be required under the act. The agreement states that "EAO accepts under the terms of this Agreement that any NEB assessment of a Project conducted . . . constitutes an equivalent assessment under Sections 27 and 28 of the BCEAA" (National Energy Board and British Columbia Environmental Assessment Office 2010).

The delegation of provincial authority to the federal government had the effect of transforming the British Columbia government from playing a direct role in pipeline approval to an intervenor in the regulatory proceedings. It was, in effect, a voluntary surrender by the province of a form of potential veto power it held. As will be described in the section on court cases, the British Columbia Supreme Court invalidated this equivalency agreement, ruling that the British Columbia Environmental Assessment Act clearly required the province to make a legal determination on the pipeline review.

Joint Review Panel Decision Rules

The regulatory review process for Northern Gateway was conducted in a changing legal landscape, with changes that in large part were precipitated by the conflict over the pipeline. When the Northern Gateway application was submitted, the process at that time provided for the establishment of a Joint Review Panel (JRP), constituting members of the National Energy Board and appointees of the Canadian Environmental Assessment Agency (CEAA) within the Ministry of Environment. The rules at that time, under the National Energy Board Act, provided that the panel itself was given the authority to make the final decision on the pipeline (Hoberg 2016). Once the hearings were completed, the JRP would issue an environmental assessment report as well as a recommended decision. The minister of environment would have an opportunity to provide a government response to the recommendation; the JRP would then consider the government's views and make a final decision.

In summer 2012, the Harper government introduced its controversial Bill C-38, an omnibus budget implementation bill, under the slogan of "responsible resource development." It followed several months later with

Bill C-45, a second massive omnibus bill. Together, the two bills rewrote much of the federal law designed to protect environmental values in the review, construction, and operation of major projects. The Fisheries Act was changed to significantly reduce the protection of fish habitat. The Navigable Waters Protection Act was altered so that far fewer projects triggered federal environmental reviews (Toner and McKee 2014; Olszynski 2015).

Of greatest concern here were the changes to the two core statutes most directly relevant to the review of pipelines and other major energy projects. The Canadian Environmental Assessment Act was repealed and replaced, and the National Energy Board Act was amended. The changes resulted in four significant implications for pipeline reviews. First, the regulatory approval process was "streamlined," including the number of agencies involved. Pipeline reviews would now be done exclusively by the NEB and not through Joint Review Panels with CEAA. Second, strict timelines were established for review. In the case of pipelines, the NEB hearing process was limited to 18 months. Third, participation in reviews was narrowed by changing eligibility requirements from "any person" to only those "directly affected" or who have, in the regulators' judgment, "relevant information and expertise" (Salomons and Hoberg 2014). Fourth, final approval authority for pipelines was moved from the National Energy Board to the federal cabinet. Under this new statutory scheme, NEB authority was restricted to evaluating proposals and making recommendations. The cabinet would make the authorizing decision by ordering the NEB to issue the appropriate permits.

These changes were potentially highly consequential because, in addition to reducing the number and variety of participants in the process, they resulted in the elimination of a veto point within the regulatory process. The final authority of a putatively expert independent panel was replaced by the elected politicians forming the government of the day. For the Harper government, this establishment of political authority over major resource projects was critical to its agenda of responsible resource development (Hoberg 2016). This change strongly supports the behavioral hypothesis that actors focus strategies on the institutional venues most favorable to their interests. In this case, Harper used his control over government to change the venue to one where his party had more control.

These new rules would end up playing a more important role in pipeline proposals after summer 2012, including the Trans Mountain Expansion Project and Energy East. Because the Northern Gateway process was already under way when the rule changes were made, the Joint Review Panel agreement needed to be amended to respect the ongoing process but also incorporate new timelines and the new allocation of decision-making authority. The agreement was amended in August 2012, bringing the process into compliance with the National Energy Board Act revisions by making the panel member appointed by the minister of environment a temporary member of the NEB and confirming that it would be the cabinet, not the Joint Review Panel, that would make the final decision. The amended agreement established a deadline for the completion of the assessment at the end of 2013 (Joint Review Panel 2013, appendix 4).

The Joint Review Panel Process

The project review stage began when Enbridge submitted its project application in May 2010. The environmental assessment and regulatory review were performed by a Joint Review Panel combining officials from the National Energy Board and the Canadian Environmental Assessment Agency. The agreement establishing the Joint Review Panel contained the Terms of Reference, including the scope of the assessment. The JRP was charged with conducting "a review of the Environmental Effects of the project and the appropriate mitigation measures based on the project description and consideration of the project application under the NEB Act" (Joint Review Panel 2013, appendix 4). The project scope included the oil sands pipeline and condensate pipeline, as well as the marine terminal in Kitimat. The scope of the review considered, in addition to the environmental effects, the commercial need and alternatives to the project, and comments from First Nations and the public. The potential impacts of the project-induced upstream operations in the oil sands on downstream use were explicitly excluded. The panel stated: "We do not consider that there is a sufficiently direct connection between the Project and any particular existing or proposed oil sands development, or other oil production activities, to warrant consideration of the environmental effects of such activities as part of our assessment of the Project under the CEA Act or the NEB Act" (Joint Review Panel 2011, 13). The JRP began hearings on January 10, 2012, and was originally scheduled to complete its work by fall 2013.

Procedures for the process were laid out by the JRP in a hearing order issued in May 2011. The process was designed to be open to a broad range of participants. In addition to formal intervenors and government participants, the process also invited oral statements and letters of comment. With the exception of the government category, no criteria were specified as to who could participate.

Environmentalists and other pipeline opponents took full advantage of these procedures by staging a "mob the mic" campaign to get as many opponents registered to participate as possible. The response was impressive. Over 4,000 signed up to participate, with the Dogwood Initiative claiming that its campaign was responsible for 1,600 of those successes (Dogwood Initiative 2011). The Joint Review Panel was clearly not anticipating this response, and it was forced to extend the timeframe for project review by a year in order to accommodate the large number of requests to be oral participants.

The hearings began in January 2012 and were completed in June 2013. The panel described the intense level of interest they witnessed in the 21 communities they visited: "Public hearings for the proposed project attracted a high level of public interest. There were 206 intervenors, 12 government participants, and 1,179 oral statements before the Panel. Over 9,000 letters of comment were received. The Panel held 180 days of hearings, of which 72 days were set aside for listening to oral statements and oral evidence. Most of the hearings were held in communities along the proposed pipeline corridor and shipping routes. The entire record of the proceeding is available on the National Energy Board website" (Joint Review Panel 2013, vol. 2, 2).

There was some controversy over public access to the hearings. Rather than conducting the hearings in public forums, oral statements and other hearings were held in a closed-room format with only directly participating individuals permitted. The hearings were webcast live. For proceedings involving intervenors, oral cross-examination was permitted. The overwhelming majority of public input expressed opposition to the project. Apparently referring to the oral statement submitted, an environmental group meme circulating in 2013 showed a football scoreboard giving the score "1159 against, 2 for." Ecojustice, an environmental law group representing a number of environmental groups in the hearing, states that "a staggering 96 per cent of written comments submitted to the joint review panel over the past two years. . . . oppose the proposed Northern Gateway" (Shearon 2013).

The panel, however, was apparently unmoved by the depth and breadth of opposition to the project. In its report, made public on December 19,

2013, it recommended the project be approved, subject to 209 conditions. Rather than provide any kind of quantitative summary of the tone or stance of the many submissions it received, the panel only referenced opposition obliquely in two passages:

> We acknowledged that different people placed different values on the burdens, benefits, and risks of the project. (Joint Review Panel 2013, vol. 1, 72)

> All parties did not agree on whether this project should proceed or not, and it was our job to weigh all aspects and deliver our recommendations to the Minister of Natural Resources for consideration by the Governor in Council. (Joint Review Panel 2013, vol. 1, 74)

The panel's core conclusions were as follows:

> We have taken a careful and precautionary approach in assessing the project. We are of the view that opening Pacific Basin markets is important to the Canadian economy and society. Societal and economic benefits can be expected from the project. We find that the environmental burdens associated with project construction and routine operation can generally be effectively mitigated.
>
> Some environmental burdens may not be fully mitigated in spite of reasonable best efforts and techniques. Continued monitoring, research, and adaptive management of these issues may lead to improved mitigation and further reduction of adverse effects. We acknowledge that this project may require some people and local communities to adapt to temporary disruptions during construction.
>
> The environmental, societal, and economic burdens of a large oil spill, while unlikely and not permanent, would be significant. Through our conditions we require Northern Gateway to implement appropriate and effective spill prevention measures and spill response capabilities, so that the likelihood and consequences of a large spill would be minimized. (Joint Review Panel 2013, vol. 1, 71–72)

The panel did not consider the risks of pipeline or tanker accidents significant: "It is our view that, after mitigation, the likelihood of significant adverse environmental effects resulting from project malfunctions or accidents is very low." It did find significant risks to two sensitive species—woodland caribou and grizzly bears—but it recommended that "the Governor in Council find these cases of significant adverse environmental effects are justified in the circumstances" (Joint Review Panel 2013, vol. 1, 72). With the completion of the regulatory review in December 2013, the Northern Gateway Pipeline completed the project review stage and entered the political stage of the pipeline controversy. The Harper government had six months to respond to the panel's recommendation (although the National Energy Board Act does allow the cabinet to grant itself extensions).

The Harper Government's Response

The Harper government took the entire six months to make its decision. While Harper's continued emphasis on the importance of access to tidewater to oil sands expansion and the prosperity of Canadians, it could not have been an easy or straightforward choice for the prime minister. With the 2015 election just over a year away, the electoral risk was considerable given the opposition to the project within British Columbia. The Harper government held 21 seats in British Columbia, some of which were highly vulnerable to the outcome of the pipeline decision. A Bloomberg-Nanos survey in early June 2014 had a very direct warning for Harper: "Forty seven percent of respondents said they would be less likely to support local Conservative candidates in BC if the Harper government approved the pipeline while only 11 percent said they would be more likely to support the local Conservative candidate if the project was approved by the Harper government" (Bloomberg-Nanos 2014). Moreover, while the Clark government sent signals that it could reconsider its position if the economic benefits to British Columbia were increased, the formal position of the provincial government remained a firm no.

Under the National Energy Board Act, the Harper cabinet had three options: approve the pipeline with the NEB's conditions, reject the pipeline, or delay the decision by directing the NEB to reconsider either its recommendation or its conditions. Ultimately, Harper chose to remain true to his vision and rolled the electoral dice by approving the pipeline. On June 17, 2014, the Harper government announced its agreement with the Joint Review Panel's report and directed the NEB to issue a certificate of public convenience and necessity with the 209 conditions. The NEB did so the following day. The June 17 cabinet decision statement addressed the requirements of the Canadian Environmental Assessment Act with the following terse statement:

> The Governor in Council has decided, after considering the Panel's report together with the conditions proposed in it, that the Designated Project is not likely to cause significant adverse environmental effects referred to in subsection 5(1) of CEAA 2012 but that it is likely to cause significant adverse environmental effects referred to in subsection 5(2) of CEAA 2012 to certain populations of woodland caribou and grizzly bear as described in the Panel's report. The Governor in Council has also decided that, pursuant to subsection 52(4) of CEAA 2012, the significant adverse environmental effects that the Designated Project is likely to cause to certain populations of woodland caribou and grizzly bear are justified in the circumstances. (National Energy Board 2014b)

Harper's approval decision brought an end to the political stage but hardly created a smooth path toward construction and operation of the Northern Gateway Pipeline. Five legal challenges had already been filed prior to the cabinet decision and another dozen quickly followed. The government of British Columbia remained formally opposed to the pipeline. Environmental groups and First Nations leaders vowed to stop the project through every legal means possible. The day Harper approved the pipeline, the price of WTI oil was US$107, but by the end of the year it had fallen below US$55. By that point, the youthful new leader of the Liberal Party of Canada had rejuvenated his party's electoral prospects and surged ahead of Harper's Tories in the polls.

Legal Challenges

There have been two major judicial decisions on Northern Gateway, both of which were major setbacks for pipeline proponents. The first decision was by the British Columbia Supreme Court about whether the equivalency agreement between British Columbia and the federal government was consistent with British Columbia's statutory framework for environmental assessment. The equivalency agreement between British Columbia and the federal government was challenged by the Coastal First Nations, who filed suit over the application of this agreement to the Northern Gateway Pipeline. The British Columbia Supreme Court ruled that the province had abdicated its decision-making authority under the British Columbia Environmental Assessment Act. The judge ruled that the act allows the province to defer to the federal government's review process but that it must still decide whether to issue an environmental assessment certificate (*Coastal First Nations v. British Columbia (Environment)*, 2016 BCSC 34). In a new twist on regulatory federalism in Canada, the judge in this case ruled that despite federal paramountcy over interprovincial pipeline approvals, it would be permissible for the provincial government to impose certain conditions on interprovincial pipeline approvals. The province could not use its regulatory authority to deny an approval to a pipeline that the federal government approved, but it could add conditions to the federal government's conditions.

The court decision (not appealed by the British Columbia government) shifts the intergovernmental politics of pipelines. For an equivalency agreement to pass muster, British Columbia can defer the assessment process to the federal government, but it would need to issue its own final decision. The approach the British Columbia government under Christy Clark used for Northern Gateway, where the province submits strenuous objections to the pipeline but then defers the final decision to the federal regulator, was no longer workable. The Trans Mountain case in chapter 6 will show how consequential that change can be.

The second major setback for the Northern Gateway Pipeline came in June 2016, when the Federal Court of Appeal struck an inevitably fatal blow to the project. In reviewing 18 appeals of the government's decision from First Nations and environmental groups consolidated into one decision, the court quashed the certificates of conditional approval provided by the government of Canada. The decision reflected a stunning victory for pipeline opponents, but the legal reasoning underlying the decision contained quite mixed ammunition for critics of pipelines and other large infrastructure projects.

The FCA decision was based on its conclusion that the Harper government engaged in a deeply flawed consultation process with First Nations that did not meet the government's obligations. Aboriginal engagement for the project was guided by a framework document issued by the federal government in February 2009. The process outlined five phases of the consultations: (1) a preliminary phase of consultation about the terms and conditions of the review process; (2) a prehearing phase to inform Aboriginal groups about the process and encourage their participation; (3) the hearing phase, where Aboriginal participation was encouraged and supported; (4) the posthearing phase to consult groups after the release of the Joint Review Panel's decision but before the cabinet's final decision; and (5) the permitting stage, where additional consultations on implementing the conditions and other legal requirements for authorization would be conducted (*Gitxaala Nation* 2016, 14–15). While laudatory about the federal government's consultations during the first three phases, it was the fourth phase, the posthearing stage, where the Federal Court of Appeal found major flaws in the government's performance. Two paragraphs from the decision effectively summarize the court's rationale:

> Based on our view of the totality of the evidence, we are satisfied that Canada failed in Phase IV to engage, dialogue and grapple with the concerns expressed to it in good faith by all of the applicant/appellant First Nations. Missing was any indication of an intention to amend or supplement the conditions imposed by the Joint Review Panel, to correct any errors or omissions in its Report, or to provide

> meaningful feedback in response to the material concerns raised. Missing was a real and sustained effort to pursue meaningful two-way dialogue. Missing was someone from Canada's side empowered to do more than take notes, someone able to respond meaningfully at some point. (*Gitxaala Nation* 2016, paragraph 279)

> We have applied the Supreme Court's authorities on the duty to consult to the uncontested evidence before us. We conclude that Canada offered only a brief, hurried and inadequate opportunity in Phase IV—a critical part of Canada's consultation framework—to exchange and discuss information and to dialogue. The inadequacies—more than just a handful and more than mere imperfections—left entire subjects of central interest to the affected First Nations, sometimes subjects affecting their subsistence and well-being, entirely ignored. Many impacts of the Project—some identified in the Report of the Joint Review Panel, some not—were left undisclosed, undiscussed and unconsidered. It would have taken Canada little time and little organizational effort to engage in meaningful dialogue on these and other subjects of prime importance to Aboriginal peoples. But this did not happen. (*Gitxaala Nation* 2016, paragraph 325)

While these passages show that the court was quite critical of the Harper government's consultation approach, the court emphasized that it was merely applying existing law: "In reaching this conclusion, we have not extended any existing legal principles or fashioned new ones. Our conclusion follows from the application of legal principles previously settled by the Supreme Court of Canada to the undisputed facts of this case" (*Gitxaala Nation* 2016, paragraph 9). The court did not see itself as advancing the duty of Crown governments any closer to the "free, prior, and informed consent" advocated by many First Nations.

While the effect of the decision was very positive for environmental opponents to the project, the Federal Court of Appeal's ruling was actually a major setback for the ability of environmentalists to challenge environmental reviews in court. The court emphasized the distinction between three aspects of the decision process: the report of the Joint Review Panel, the federal cabinet's order in council, and the certificates issued by the NEB. The court ruled that it was only the order in council that was reviewable, not the process or report of the JRP. The court ruled that unless the cabinet asked for reconsideration, the JRP report is considered "final and conclusive." "Any deficiency," the court wrote, "in the Report of the Joint Review Panel was to be considered only by the Governor in Council, not this Court" (*Gitxaala Nation* 2016, paragraph 125). If this interpretation is applied to other cases, it will dramatically reduce the capacity of environmental groups to challenge energy projects in the future. Law scholar Martin Olszynski argues that the court erred by applying the wrong section of the Canadian Environmental Assessment Act in the ruling (Olszynski 2016). Nonetheless, the Supreme Court of Canada declined to hear the appeal by environmental groups of that part of the ruling. The Trans Mountain pipeline judicial rulings (discussed in the next chapter) continued to find that environmental assessment reports in cases like these are not subject to review.

The decision had the effect of putting the Trudeau government in the position of either simply canceling the pipeline certificates or restarting the post-JRP consultation proceedings with First Nations. Given his commitments in the 2015 election campaign and the lack of reasons to believe the positions of any First Nations had changed since the Harper government's process, it really was not much of a decision at all.

The Final Rejection

On November 30, 2016, Justin Trudeau announced that the government was dismissing the application for the Northern Gateway Pipeline. "It has become clear," Trudeau stated, "that this project is not in the best interests of the local affected communities, including Indigenous peoples." Repeating the line he used during the election campaign and since then, he reiterated that "The Great Bear Rainforest is no place for a pipeline, and the Douglas Channel is no place for oil tanker traffic." To bring finality to the issues, Trudeau also announced, "We will keep our commitment to implement a moratorium on crude oil tanker shipping on British Columbia's north coast" (Trudeau 2016). Trudeau's focus on place-based concerns in his decision to reject the pipeline is consistent with the behavioral hypothesis about decision rationales.

Trudeau's attachment to the Great Bear Rainforest and his choice to use it to inform his stance against the pipeline were influenced by several factors. First, Trudeau's longtime friend and political adviser Gerald Butts had a background as the leader of an environmental group active in the area. From 2008 to 2012, Butts was executive director of the World Wildlife Fund (WWF) for Canada. WWF had been active in efforts to protect the Great Bear Rainforest. While historically the organization had not taken positions on specific projects, the group did come out to formally oppose the Northern Gateway Pipeline while Butts was executive director. Butts explicitly linked the group's pipeline opposition to its support for rainforest protection in a 2012 op-ed, saying, "The hearings to decide the future of the Great Bear Sea and Rainforest got off to quite a start this week. Big oil, foreign intrigue, a grassroots uprising, duelling polls, angry ministers—this one has all the makings of a blockbuster. But the fervour obscures the heart of the matter: whether and under what conditions we should permit supertankers and a bitumen pipeline in one of the last intact temperate coastal rain forests on Earth" (Butts 2012).

Trudeau himself also got a personal tour of the region during the campaign. Art Sterritt, then executive director of the Coastal First Nations, was approached by Jody Wilson-Reybould, a longtime friend and newly designated Liberal Party candidate, with a request that he lead Trudeau on a tour of the region. According to Sterritt, "We took a seaplane to Digby Island and Prince Rupert and flew down to Hartley Bay and let him meet the elders and the kids and everybody else. It was soon after that he came out and said that he was opposed to Northern Gateway" (Sterritt 2017).

The decision statement issued by the cabinet reflected how fundamentally the political calculus on the pipeline had changed since Harper's approval decision as a result of the Federal Court of Appeal's decision and the election of a government with different values:

> In a 23 June 2016 decision, the Federal Court of Appeal quashed Order in Council P.C. 2014–809 dated 17 June 2014, which was the order directing the National Energy Board (Board) to issue Certificates OC-060 and OC-061 for the Project. The Court also quashed the certificates at that time and remitted the matter to the Governor in Council for redetermination.
>
> Through Order in Council P.C. 2016–1047 dated 25 November 2016, the Governor in Council does not accept the finding of the joint review panel (Panel) that the Project, if constructed and operated in full compliance with the conditions set out in the Panel's report, is and will be required by the present and future convenience and necessity. The Governor in Council does not accept the Panel's recommendation and is of the view that the Project is not in the public interest. (National Energy Board 2016d)

Pipeline opponents used their access to the judicial veto point to alter the government's decision calculus. When the court quashed the license, "they had no pathway out" (Clogg 2017). With a new federal government in power that was more supportive of protecting environmental values, the long-running conflict over the Northern Gateway Pipeline was brought to a close. While the announcement was made at the same time as the Trudeau government approved Kinder Morgan's Trans Mountain Expansion Project and Enbridge's Line 3 Replacement Project, there is no denying that the final rejection of Northern Gateway—once the oil sector and Harper government's first choice to improve oil sands access to tidewater and Asian markets—was an extraordinary victory for the anti-pipeline coalition.

Conclusion

Stopping the Northern Gateway Pipeline is a remarkable accomplishment for the anti-pipeline coalition. Up against Canada's most powerful economic sector and a determined federal government, environmentalists and First Nations were enormously successful mobilizing opposition to the pipeline. While their political clout was not enough to stop the Harper government from approving the pipeline, it was enough to prevent it from being built and, ultimately, to get Harper's decision reversed. That political clout was attained, for the most part, through a sustained focus on place-based concerns: the risk of pipeline and tanker accidents damaging precious environmental values, and Aboriginal rights to decision-making on their traditional lands. Climate concerns were a part of the opposition argument but not as central as they were to the three other pipeline cases.

Opponents gained access to the two critical authoritative decision bodies in this case: the federal courts and the federal government. Aboriginal rights were used by the Federal Court of Appeal to negate the pipeline's approval, forcing the federal government to either reverse its decision or undertake a new, more sincere process of Aboriginal engagement. The federal election of October 2015 swept the Harper government out of power, replacing it with a prime minister already determined to stop the pipeline and impose a tanker ban on the north coast of British Columbia. It would be too much to say that the 2015 election shifted the federal government from the oil sands coalition to the anti-pipeline coalition. After all, the same day Trudeau formally rejected Northern Gateway he approved the Trans Mountain Expansion Project and Enbridge's Line 3 pipeline. But there is no question that the Trudeau government is greener than the Harper government.

The potential veto point of the Alberta-British Columbia border also proved to be very important. Sustained public opposition in British Columbia made advocating the pipeline undesirable even for the pro-business government of Premier Christy Clark. The British Columbia Supreme Court nullified the British Columbia government's effort to abdicate responsibility for approving or rejecting the pipeline. While the position of British Columbia did not constitute a legal veto, it did act as a political veto of sorts. The coalition supporting oil sands was powerful enough to get a conditional approval over the objections of the British Columbia government but not powerful enough to get the pipeline constructed and into operation.

As Keystone XL stumbled in the pro-environment Obama administration, the oil sands coalition's hope for tidewater access quickly shifted to Northern Gateway. But the advocates of nation-building pipelines ran straight into a well-organized, well-resourced coalition of pipeline opponents that proved sufficiently formidable to put the pipeline into a death spiral as early as 2012 and eventually get the project formally terminated. Northern Gateway's demise was not, however, the defeat of the oil sands coalition's vision of access to Pacific tidewater. Waiting in the wings was Enbridge's competitor Kinder Morgan, whose proposal to triple the capacity of the Trans Mountain pipeline was the next major oil sands pipeline to go through regulatory review and the surrounding political turmoil.

6 The Trans Mountain Expansion Project: The Politics of Structure

Overview

Pipeline resistance in British Columbia was initially focused on the Northern Gateway Pipeline. Ben West, an activist with the Wilderness Committee, based in Vancouver, first got wind of another plan to get increased oil sands production to the West Coast when he got a call from Rex Weyler, a cofounder of Greenpeace, in early 2011. Weyler had noticed an increase in tanker traffic through the Burrard Inlet and was interested in taking action to stop it. Weyler and West knew effective action would require the support of area First Nations, and they were soon introduced to Reuben George of the Tsleil-Waututh Nation in spring 2011. George invited West to join him in spiritual ceremonies before they got down to discussing political strategies. Before long, West found himself a "fire keeper" for George's sweat lodge, and George was pressing his chief and council to take a strong stand against the nascent Trans Mountain Expansion Project. From that meeting, a formidable resistance coalition of environmentalists and First Nations was born to fight the project (West 2016).

Kinder Morgan`s Trans Mountain Expansion Project would "twin" an existing pipeline from the Edmonton area to Burnaby, British Columbia, and Vancouver's harbor. The pipeline began operation in 1953 (Kheraj 2013). The expansion project would virtually triple the capacity of the pipeline, from 300,000 to 890,000 barrels per day. One potential advantage of the project was that 74% of the new pipeline would be along the existing pipeline right-of-way, 16% would follow other rights-of-way established by utilities, and only 11% would be new rights-of-way. The project was originally expected to cost $6.8 billion (Trans Mountain, n.d.). When it was formally proposed, Kinder Morgan was expecting to start construction in 2015, with an in-service date of 2017 (Hoekstra 2013).

The project would significantly increase tanker traffic in Vancouver's harbor. Three new terminal berths would be constructed at the Westridge Marine Terminal in Burnaby, British Columbia, designed to handle Aframax class vessels (245 meters in length, with a capacity of 750,000 barrels). According to the Trans Mountain application, tanker traffic would increase sevenfold, from 5 per month to 34 per month (Trans Mountain 2013, vol. 2, 2–27).

The timeline for the Trans Mountain Expansion Project is presented in figure 6.1. The project was formally proposed in December 2013 (National Energy Board [NEB] 2016a) and became engulfed in protests in late 2014. Despite considerable controversy, on November 29, 2016, Prime Minister Justin Trudeau approved it, subject to 157 conditions. As part of the same announcement, his government rejected the Northern Gateway Pipeline.

Like Northern Gateway, Kinder Morgan's project has also been opposed by environmentalists and many First Nations. Municipal governments in the Greater Vancouver area also strongly opposed the project. That anti-pipeline coalition confronted a similar oil sands coalition, but the proponent was Texas-based Kinder Morgan rather than Enbridge. This meant that a similar geographical distribution of risks and benefits across the Continental Divide between Alberta and British Columbia was in play. But there were two critical differences with Trans Mountain. Unlike the greenfield Northern Gateway, the bulk of the Trans Mountain route would follow the existing pipeline's right-of-way. While the terminus of Northern Gateway would have been in the remote northern coastal town of Kitimat, Trans Mountain's terminus was in Burnaby, adjacent to Vancouver, a city that is aspiring to be "the greenest city in the world."

Trudeau's decision hardly brought finality to the case. The election of an anti-pipeline NDP government in British Columbia in summer 2017 led to a chain of events that heightened modern Canada's sustainability dilemma and challenged the core of the Canadian federation. British Columbia proposed to restrict increases in the shipment of oil sands through the province. Alberta retaliated by temporarily banning wine imports from British Columbia and then by enacting legislation authorizing the province to restrict oil and gas shipments to British Columbia. The Trudeau government then nationalized the pipeline project, altering the structure of the pipeline policy regime. Shortly thereafter, the Federal Court of Appeal struck down the

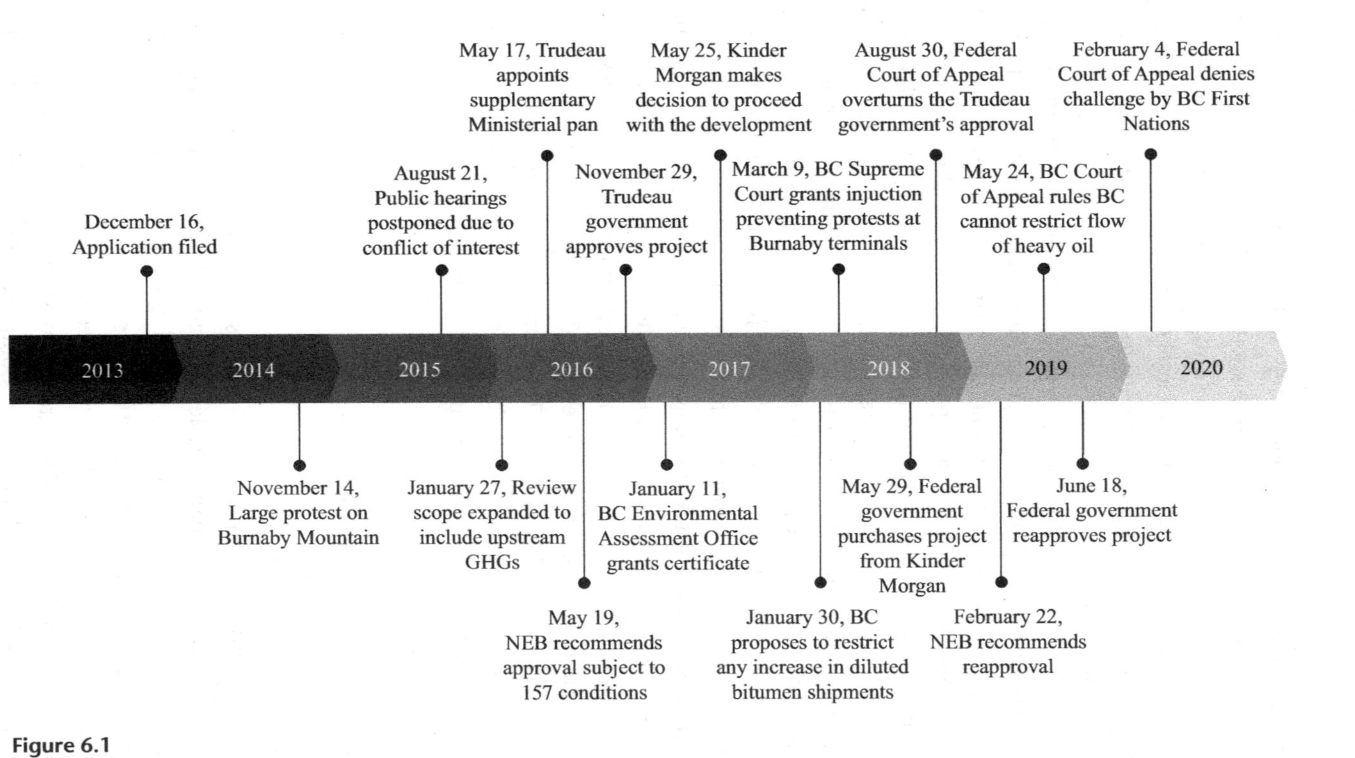

Figure 6.1
Trans Mountain Expansion Project timeline.

approval decision for failing to adequately address tanker risks and engage in meaningful consultations with First Nations. This led to a reconsideration by the National Energy Board and a second approval by the Trudeau government. That second decision survived legal challenges in the federal courts. Three years after it was originally supposed to come into service, the pipeline was in the early stages of construction.

The conflict has pitted the most powerful economic interests in Canada against a surprisingly formidable resistance movement. A prominent theme that emerges from the analysis is that the institutional rules of the game have been highly contested. Major political, legal, and physical conflicts have erupted over the scope of issues that should be considered and where decision-making authority lies. As a result, the politics of this pipeline has in large part been "the politics of structure," or the struggle over defining the rules of the game (Moe and Wilson 1994; Hoberg 2018).

The next section describes the public and private actors and their interests and strategies in the pipeline controversy, focusing on how they differ from those of the Northern Gateway case described in chapter 5. The second section looks at public opinion on the issue and presents a media analysis. The third section describes the institutional rules at work and conflicts that erupted over the rules of the game. These sections will allow us to test our four behavioral hypotheses. The fourth section builds on chapter 5 to elaborate on the aftermath of Trudeau's approval, including the bitter jurisdictional conflict unleashed after the May 2017 election in British Columbia.

Actors—The Oil Sands Coalition

The Private Oil Industry

The principal industry supporters of the Trans Mountain Expansion Project have been the parent company, Kinder Morgan, a Texas-based energy pipeline company, their shippers among the oil sands companies in Alberta, and the refinery companies in export markets.

These oil sector interests have been supported through campaigns by other business groups and new advocacy groups favoring resource development. For example, the BC-based Independent Contractors and Businesses Association, a group whose members would benefit enormously from construction associated with the pipeline, produced a 30-second ad for the Super Bowl in Canada. The ad featured a young family man whose ability to work is thwarted by protesters. The voiceover states, "Our province is being held hostage by a loud few with too much free time and very few facts. We're tired of a promising future constantly being blocked by no" (Independent Contractors and Businesses Association of 2016). Resource Works is a new group designed to counter the resistance movement by "communicat[ing] with British Columbians about the importance of the province's resource sectors to their personal well-being" (Resource Works, n.d.).

Governments

Governments have expressed a variety of interests in the pipeline. The Harper government was a resolute supporter of increasing oil sands access to global markets. Trudeau has taken a more nuanced approach but has also consistently expressed the importance of increasing access to global markets. In his February 2015 speech in Calgary, he emphasized the importance of regaining public trust in the regulatory process in order to improve market access: "Getting our resources to market is a priority for Canada, and we know that our economic success depends on keeping our word on the environment" (Trudeau 2015). Once it approved the Trans Mountain project in late 2016, the Trudeau government vigorously defended it. On numerous occasions, Trudeau himself declared resolutely that "this pipeline will be built" (Hall 2018). The strongest indication of that commitment was the Trudeau government's nationalization of the project in May 2018.

The Alberta government has been a major champion of pipeline expansion, given the importance of the oil sector to the province's economy and the government's dependence on oil revenues. Once she became premier in 2015, Rachel Notley also became an enthusiastic champion of pipeline expansion. In her address to Albertans in April 2016, Notley emphasized the importance of pipelines to the provincial economy: "Every Canadian benefits from a strong energy sector. But we can't continue to support Canada's economy, unless Canada supports us. That means one thing: building a modern and carefully-regulated pipeline to tidewater. We now have a balanced framework to develop our industry and every government in Canada understands this issue must be dealt with. But I can promise you this: I won't let up. We must get to 'yes' on a pipeline" (Notley 2016).

Actors—The Anti-pipeline Coalition

Environmental Group Resistance

Along with the Northern Gateway Pipeline, the Trans Mountain Expansion Project has been one of the core issues for the vast and diverse environmental movement in British Columbia. Again, much of the concern driving larger environmental groups was climate change, but for the same reasons described in the previous chapter, resistance to this project focused largely on place-based risks, particularly spills along the pipeline route but especially the risks of tanker spills in the coastal waters around Vancouver, particularly early in the campaign. As described in the section on media coverage, greater attention was given to climate issues after 2014, when 350.org became more involved in this conflict.

Table 6.1 lists 20 organized environmental groups that actively opposed the project. They range from large binational groups such as Stand (formerly ForestEthics), Canadian sections of large international groups (e.g., Greenpeace and 350.org), multi-issue groups that have chosen Trans Mountain as one of their campaigns (such as LeadNow), mainstay British Columbia environmental groups (such as Wilderness Committee and Sierra Club of BC), and local groups organized specifically to fight the project (such as Tanker Free BC and Burnaby Residents against Kinder Morgan Expansion, or BROKE).

Focused environmental opposition to the project began in summer 2012, when the Wilderness Committee, under the leadership of Ben West, launched a campaign in collaboration with Tanker Free BC. The campaign initially focused on the risks to Vancouver's Stanley Park, a cherished urban park in the region (Wilderness Committee and Tanker Free BC 2012). The Wilderness Committee's 2012 annual report and its previous four annual reports make no mention of Kinder Morgan or Trans Mountain, but the 2013 report describes a series of town halls on the pipeline proposal as one of the year's major accomplishments (Wilderness Committee 2013). The centrality of place-based risks in the campaign is clearly signaled in the headline of the Wilderness Committee's first major publication on this pipeline: "Want to Help Stop an Oil Spill?" The early focus on Stanley Park—an iconic urban park on a peninsula directly between the tanker terminal and more open Pacific Ocean shipping lanes—was critical in galvanizing concerns about the project among the local populace and municipal politicians (West 2016).

Table 6.1
Environmental groups active on the Trans Mountain Expansion Project (July 2018)

Group name	Facebook likes	Twitter followers	Intervenor?
350 Vancouver	467	309	
BROKE	506	184	
Georgia Strait Alliance	1,736	5,105	Yes
350 Canada	2,510	2,252	
Tanker Free BC	2,529	2,230	
Living Oceans Society	4,954	4,395	Yes
West Coast Environmental Law	5,714	11,800	
Wilderness Committee	6,708	8,367	
Sierra Club of BC	7,354	6,490	
Raincoast Conservation Foundation	10,567	7,952	Yes
Nature Canada	12,186	49,800	Yes
Dogwood BC	24,144	10,788	
Ecojustice	52,460	22,700	
LeadNow	55,474	17,100	
Greenpeace Canada	176,935	38,000	
Stand	183,601	18,100	
BC Nature	NA	273	Yes
Force of Nature	NA	NA	
PIPEUP Network	NA	807	

Dogwood BC's (previously the Dogwood Initiative) "No Tankers" campaign was originally focused on the Northern Gateway Pipeline, but in 2011 the group expanded its coverage to include the Trans Mountain project. This change is reflected in the group's annual reports. The 2009–10 and 2010–11 annual reports contain no mention of the Kinder Morgan proposal. It is first mentioned in the 2012–13 annual report, where the group states: "We spent the summer of 2011 re-designing the No Tankers campaign. . . . to include opposition to Kinder Morgan's oil tanker proposal on B.C.'s south coast, and began focusing on the role of the government of British Columbia in the debate" (Dogwood Initiative 2013). In 2014, Dogwood focused on forcing the British Columbia government to assert jurisdiction over the project with its "Let BC Decide" campaign. The campaign was designed to mobilize support for a citizen initiative against tankers if "politicians try to force these projects on B.C." (Dogwood BC, n.d.).

First Nations Resistance

Oil sands pipelines have been a major issue for British Columbia's First Nations. While some First Nations are open to oil sands pipelines and have signed impact benefit agreements with pipeline companies, a large number of First Nations have taken a principled stance in opposition to them. Much of this opposition originally focused on the Northern Gateway Pipeline, and in 2010 two large coalitions of First Nations, the Coastal First Nations and the Yinka Dene Alliance, appealed to their ancestral laws in banning the pipeline and tankers from their territories. While the Trans Mountain route does not directly affect the territories of the Coastal First Nations, the geographical scope of the Yinka Dene Alliance`s Save the Fraser Declaration is much larger and has been signed by many First Nations along the Trans Mountain route. The declaration makes it clear that the Kinder Morgan project is also covered: "We will not allow the proposed Enbridge Northern Gateway Pipelines, or similar Tar Sands projects, to cross our lands, territories and watersheds, or the ocean migration routes of Fraser River salmon" (Save the Fraser Gathering of Nations 2013).

Some First Nations along the pipeline route have been very active in opposing the pipeline, most notably the Tsleil-Waututh Nation, whose traditional territory encompasses the terminus for the pipeline in Burnaby (see the section on legal action by First Nations). The Union of British Columbia Indian Chiefs (UBCIC) has also been a vocal opponent of Trans Mountain. Referring to a bill that would bring Canadian law into harmony with the UN Declaration on the Rights of Indigenous Peoples, former UBCIC president Grand Chief Stewart Phillip argued that "Bill C-262 further validates what we already know: Kinder Morgan cannot proceed without the consent of the First Nations along its path, so many of which oppose it" (UBCIC 2018).

On the other hand, according to Kinder Morgan, 43 Aboriginal groups signed mutual benefit agreements with Kinder Morgan, and approximately 100 agreements were made in total, including memoranda of understanding, capacity funding agreements, integrated cultural assessments, and relationship agreements. The financial value of the mutual benefit agreements was over C$400 million (Trans Mountain 2018). A database put together by Discourse, APTN, and HuffPost Canada lists 41 groups that signed agreements with Kinder Morgan. They also noted that the company's own information about affected First Nations listed a total of 140 groups, of which 41 have agreements, 14 initially challenged the project in court, and an additional 85 groups do not have agreements (Owen 2018).

From the start of the campaign against Trans Mountain, First Nations worked closely with environmentalists. As the project moved into the on-the-ground stage in spring 2017, new initiatives and coalitions were created to coordinate resistance. Coast Protectors was an initiative of the Union of BC Indian Chiefs. As of July 2018, the group's pledge, "Whatever it takes, we will stop the Kinder Morgan pipeline and tanker project," had 25,630 signatures online (Coast Protectors, n.d.). A related group, Protect the Inlet, or Kwekwecnewtxw, describes itself as an Indigenous-led initiative, supported by allied organizations (Protect the Inlet, n.d.).

Municipal Governments

Municipal governments have been highly politicized by recent pipeline controversies. In British Columbia, 21 municipal governments, including virtually all those on the lower mainland, expressed formal opposition to the project. The most active opponent has been the city of Burnaby, which challenged the NEB and Kinder Morgan in court over their plans to perform seismic drilling on Burnaby Mountain (to be discussed). Burnaby mayor Derek Corrigan pledged to get arrested if the pipeline was approved (Morneau 2015). The city of Vancouver, under the leadership of Mayor Gregor Robertson, has been strongly opposed to the pipeline. (When Robertson declined to run for reelection in the 2018 municipal election, he was replaced by Kennedy Stewart, who, as a sitting member of Parliament, was actually arrested for civil disobedience against the Trans Mountain project (Larsen 2018). The city's 23-page submission to the NEB focused on the risks of a tanker accident and climate change. The day the NEB announced its recommendation to approve the pipeline, Robertson called the decision a "call to action" and launched a campaign to convince his friend the prime minister to reject it (Hunter and Hume 2016). The following day, Mayor Corrigan committed to a "mass citizen campaign" (Sinoski 2016). More details on Burnaby's anti-pipeline efforts will be given later in the chapter.

Government of British Columbia

The British Columbia government, under BC Liberal premier Christy Clark, first adopted a position of conditional opposition. Relying on the five conditions introduced during the Northern Gateway proceedings (Government of British Columbia 2012b), the province emphasized the lack of emergency response preparedness in its rationale, stating, "During the course of the NEB review the company has not provided enough information around its proposed spill prevention and response for the Province to determine if it would use a world leading spills regime. Because of this the Province is unable to support the project at this time, based on the evidence submitted" (Government of British Columbia 2016). Changes in emergency preparedness policy, particularly those supported by Trudeau's Oceans Protection Plan, allowed Clark to reconsider the government's position and issue the required environmental assessment certificate.

But British Columbia's position changed with the May 2017 election, which resulted in the anti-pipeline British Columbia New Democratic Party forming a minority government, supported by three Green Party Members of the Legislative Assembly (Shaw and Zussman 2018). The defeat of the pro-development BC Liberal Party changed the formal position of the government of British Columbia, triggering the intensified political conflict described here.

The anti-pipeline coalition arrayed against Trans Mountain was broad and deep. Resistance strategies support the first two behavioral hypotheses. Environmentalists allied themselves with local environmental groups, Indigenous groups, and municipal and provincial governments. They focused their communications on place-based concerns, most importantly the risk of a tanker spill. As the next section shows, these concerns strongly influenced media reporting.

Ideas—Public Opinion and Media Issue Framing

Nationwide public opinion polls have shown that support for the pipeline generally exceeded opposition. In a nationwide sample by Ekos in February 2016, supporters outweighed opponents 47% to 42% (Ekos 2016). Many polls showed sharp differences among provinces, with support being highest in Alberta (Angus Reid Institute 2016; IPSOS 2018c).

Public sentiment in British Columbia was initially opposed to the pipeline. Insights West has been surveying British Columbians about the pipeline since January 2013, as shown in figure 6.2. Opposition was greater than

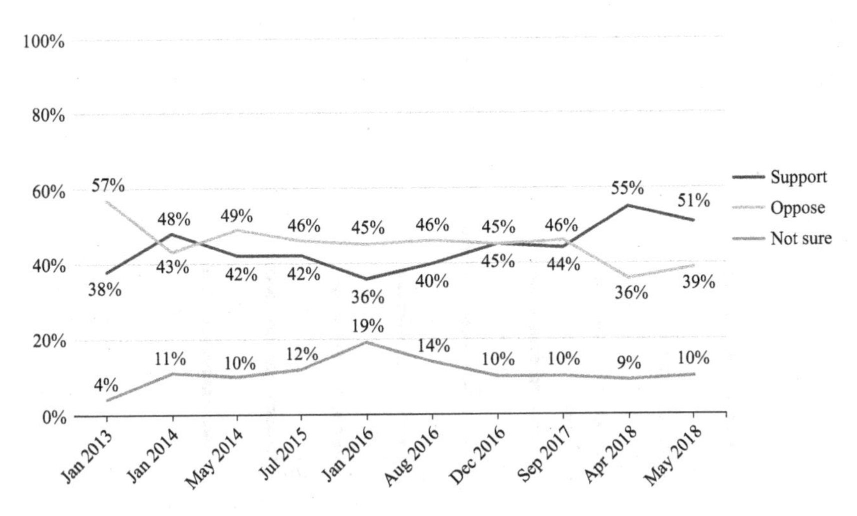

Figure 6.2
British Columbia poll results for the following question regarding support for the Trans Mountain Expansion Project: "Given what you know about the proposed expansion of Kinder Morgan's Trans Mountain pipeline, do you support or oppose the project?"
Source: Insights West (2018).

support in all but one of the polls prior to April 2018, but that month's poll showed a sharp spike in support to 55% before dropping down to 51% in the May poll following the federal government's purchase of the pipeline. An April 2018 Angus Reid poll showed that 54% supported the pipeline, while 38% opposed it (Angus Reid Institute 2018b). A July 2018 IPSOS poll showed that 59% of British Columbians supported (strongly or somewhat) the pipeline, while 35% opposed it (IPSOS 2018c).

An analysis of media coverage provides some indication of the relative importance given to different issues in the pipeline dispute. This analysis compares the number of times news articles mentioning the pipeline also mention four particular issues: climate change, jobs and the economy, risks of pipeline or tanker spills, and First Nations.[1] Figure 6.3 shows the dominance of the economy-jobs frame. It also provides insight into the hypothesis that pipeline opponents will adopt framing that emphasizes place-based risks. While the anti-pipeline coalition's communications have not been measured directly, the figure shows that through 2015, placed-based environmental risks outweighed climate risks, but that changed in 2016. Over the entire 2012–2019 period, climate risks received more attention than place-based risks.

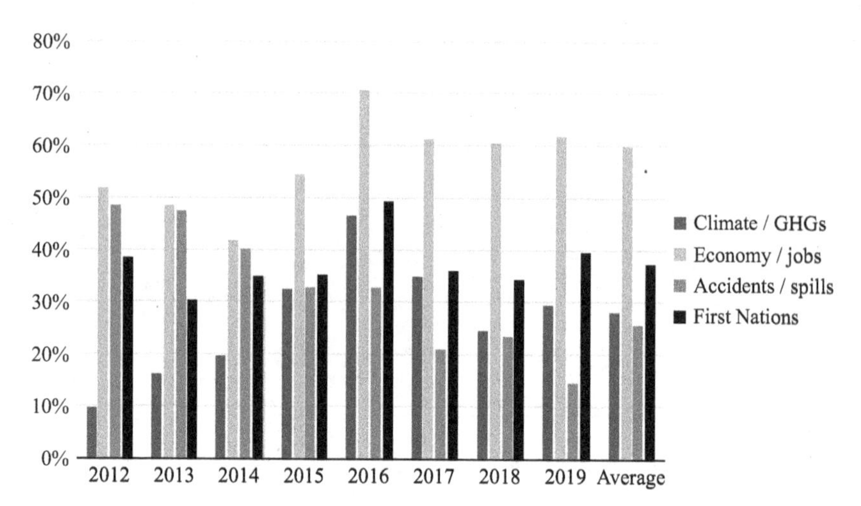

Figure 6.3
Trans Mountain pipeline keyword mentions.

This shift was clearly related to a change in the tone of environmental discourse. Campaigns initially focused on a "no tankers" message, but the relative attention to climate shifted as time went on. This change can be credited, at least in part, to the emergence of 350.org in the pipeline controversy in 2014, as well as the emergence of nationwide strategies to use climate concerns to undermine the credibility of the regulatory process in the leadup to the 2015 federal election (Fenton 2017a).

Institutions and the Politics of Structure

Kinder Morgan submitted its project description in May 2013 and its full application to the NEB in December 2013 after two years of advanced consultations through public meetings. After the NEB determined that the application was complete and initiated the process, Kinder Morgan notified the NEB that it wanted to change the route through Burnaby. This led to a seven-month "suspension" of the hearing time clock so that the company could do additional testing of the proposed route. Hearings began in August 2014 and continued, with an additional delay because of a conflict of interest created when Prime Minister Stephen Harper appointed a Kinder Morgan witness to the NEB, until February 2016. On May 19, 2016, the NEB released its final report. It found "that the Project is not likely to cause significant adverse environmental effects" and that it was in the public interest. It recommended that the federal cabinet approve the pipeline, subject to 157 conditions (National Energy Board 2016b, 2-3). The Trudeau government, after three additional reviews described later in this chapter, approved the project with conditions in November 2016.

The third behavioral hypothesis is that actors will focus their strategies on the institutional venues most favorable to their interests. These strategic incentives frequently create conflict among competing interests about the allocation of decision-making authority across different institutions. The rules of the game governing the Trans Mountain project have been contested along three dimensions: horizontally at the federal level, over whether final decision-making authority rests with an independent regulatory agency or elected officials in the cabinet; vertically among federal, provincial, and municipal governments; and finally between colonial governments and First Nations.[2] The horizontal conflict emerged in response to efforts by environmental groups to overwhelm the regulatory review process with demands for participation, but the even bigger political-institutional conflicts have been about federalism and Aboriginal rights.

Harper's Assertion of Political Control

Part of the institutional conflict over Trans Mountain is spillover from the Northern Gateway case, as described in chapter 3, which precipitated four important changes:

1. Regulatory review authority for pipelines was consolidated in a single organization, the NEB, eliminating the need for a Joint Review Panel with the Canadian Environmental Assessment Agency as was the case for the Northern Gateway Pipeline.
2. More stringent timelines were imposed; for pipelines, the NEB hearing process was limited to 18 months.
3. Participation was narrowed from the original language of "interested parties" to those who are "directly affected" or have, in the review panel's judgment, "relevant information and expertise."
4. Decision-making authority was shifted from the NEB to the federal cabinet, relegating the NEB's role to one of project review and recommendation.

These changes were designed to streamline the decision-making process and give the elected political arm of government more direct control. The change that became the most controversial in the Trans Mountain case was the new restrictions on participation, which are discussed in the following section.

The Struggle over Scope

One critical aspect of political structure is the issues that are determined to be within the scope of the regulatory review. From the start, the NEB determined in its list of issues that it would consider only the greenhouse gas emissions resulting from construction and operation of the pipeline and not the upstream emissions from the oil sands or downstream emissions when the products were refined and combusted. It is worth noting that the US State Department considered upstream and downstream impacts in its review of the Keystone XL pipeline.

Pipeline opponents tried to combat the restrictions on participation and scope by shifting the venue to the courts. A number of individuals, including a group of academics, applied to participate for the express purpose of discussing climate impacts, with the expectation that they would be rejected by the NEB. And indeed they were (National Energy Board [NEB]) 2014a).

A group of those who were denied their application to participate, led by Simon Fraser University (SFU) professor Lynne Quarmby, renowned Canadian environmentalist Tzeporah Berman, and the group ForestEthics Advocacy, challenged the NEB's action in the Federal Court of Appeal with the novel claim that their charter right to freedom of expression had been violated. The Federal Court of Appeal dismissed the appeal without giving reasons, and the Supreme Court of Canada took the same action when that dismissal was appealed. In October 2014, three months before the Federal Court of Appeal dismissed the case, it ruled on a very similar charter claim made about the NEB's decision to deny applicants who sought to talk about climate change with respect to Enbridge's Line 9 application. In that case, it did issue a written decision that dismissed the application for judicial review because the plaintiffs had not brought their charter claim to the NEB before seeking judicial review. In doing so, the court went so far as to denounce ForestEthics as a "busybody" (*ForestEthics Advocacy Association v. Canada (National Energy Board*, 2014 FCA 245).

The city of Vancouver also filed suit challenging the NEB's decision to exclude from consideration the upstream and downstream environmental and socioeconomic impacts of the pipeline. Its suit was also dismissed by the Federal Court of Appeal without reasons.

The NEB says it received 2,118 applications to participate and denied 22% of those applications, or 468 participants. While the pre-2012 rules would have allowed the additional 468 to participate, the number of intervenors (400) and commenters (1,250) who were permitted to participate in the process was still quite substantial. For comparison, the total of 1,650 was just shy of the 1,790 parties that participated in the more accessible Northern Gateway joint review process (Northern Gateway Joint Review Panel 2013, 14–15).

Strategic actors are always searching for better approaches to advance their interests. The Harper government, frustrated with delays from mass participation in the Northern Gateway case, narrowed the range of eligible participants. For environmentalists, it was very important to force a climate lens onto pipeline decision-making. When frozen out of NEB hearings by scoping rules and the new limits on participation, they attempted to shift the venue to the courts, but that effort failed. Clearly, not every strategic choice will be successful. With the charter challenge rejected by the courts on procedural grounds, pipeline opponents shifted to other strategies.

Forcing Jurisdiction on British Columbia

Traditionally in Canada, the decision-making authority to approve interprovincial pipelines has rested with the federal government and the NEB (Bankes 2015; Olszynski 2018). However, since pipeline and terminal construction and operation affect many areas under provincial jurisdiction, provinces also have a role to play. The British Columbia government, however, was willing to cede authority to the federal government through an equivalency agreement whereby British Columbia agreed to accept the NEB review process as its own (National Energy Board and British Columbia Environmental Assessment Office 2010).

The equivalency agreement under which the British Columbia government deferred to the review and decision-making by the federal government was the result of an intergovernmental compromise. The industry and the Harper government were particularly concerned about reducing jurisdictional overlap and conflict, and they promoted a one-project, one-process approach to regulatory reviews, where feasible. The Harper government, with its strong pro-development orientation, was understandably reluctant to devolve regulatory authority to provinces with strong environmental sentiments.

Normally, a provincial government would be expected to be reluctant to give up its authority over regulatory processes, but the pro-development BC Liberal government, although keenly aware of the strong environmental movement in the province, was happy to use the equivalency agreement to "pass the buck" to the federal government and avoid blame for contentious decisions (Harrison 1996). The agreement shifted the role of British Columbia's government from sharing regulatory authority to being an intervenor in the federal regulatory proceedings.

Environmentalists and First Nations lobbied forcefully to have British Columbia reassert its jurisdiction over the project, and in the 2013 election campaign, the provincial NDP had a "made-in-BC" environmental assessment process as a core part of its election platform (British Columbia New Democratic Party 2013). Despite a formidable NDP lead going into the election, Christy Clark's BC Liberals defeated the NDP. The midcampaign decision by NDP leader Adrian Dix to come out in opposition to the Trans Mountain project is credited with contributing to Clark's comeback (Hoberg 2013).

Having lost that political battle, the anti-pipeline coalition turned to fighting the agreement in court. The Coastal First Nations challenged the equivalency agreement in the context of Northern Gateway. As described in chapter 4, the British Columbia Supreme Court ruled that the province could defer to the federal government review process but that it must still decide whether to issue an environmental assessment certificate. The province could not use its regulatory authority to deny an approval to a pipeline that the federal government approved, but it could add conditions to the federal government's condition (*Coastal First Nations v. British Columbia (Environment)*, 2016 BCSC 34).

The political implications of the ruling were formidable because they shifted the intergovernmental politics of pipelines. For an equivalency agreement to pass muster, British Columbia could allow the federal government the lead in conducting the assessment, but it would still need to make its own final decision on the basis of that assessment. This forced the provincial government to share accountability for the final decision. The preexisting process, where British Columbia submitted strenuous objections to the pipeline but then deferred the final decision to the federal regulator, was unlawful. This ruling gave pipeline opponents another opportunity to question the legitimacy of the decision and another potential veto point to access.

In response, British Columbia launched its own environmental assessment process on the Trans Mountain project (British Columbia Environmental Assessment Office 2016). In January 2017, Clark announced that Kinder Morgan had fulfilled the province's five conditions. Therefore, she issued an environmental certificate for the British Columbia portion of the expansion, with 37 conditions (British Columbia Office of the Premier 2017). Days earlier, she had announced a deal with Kinder Morgan for the province to receive C$1 billion over 20 years. This secured British Columbia's fifth "fair share" condition in an unprecedented way (Cryderman 2017). The role of the British Columbia government following the 2017 election will be discussed in more detail.

Asserting First Nations Control

Aboriginal rights and title have become increasingly important to Canadian resource development, particularly in British Columbia. Since the 1970s, courts have increasingly acknowledged that First Nations in untreatied areas must be consulted and, in some cases, accommodated about resource developments proposed for their lands, even if their title claims have not been recognized by Crown governments (Christie 2006; Wright 2018). The latest advance for Aboriginal rights was the 2014 Supreme Court decision *Tsilhqot'in Nation v. British Columbia*, which granted title for the first time in British Columbia's history. While much attention has been given to the court's declaration that once title has been granted, First Nations should be accorded the right to consent to development on their title lands, the decision still permits the Crown to infringe on First Nations title lands so long as the Crown goes through a careful justification process (Coates and Newman 2014). As a result, current Canadian law still falls short of the "free, prior, and informed consent" doctrine of the UN Declaration on the Rights of Indigenous Peoples (Hoberg 2018; Wright 2018).

Among First Nations, the Tsleil-Waututh and, more recently, the Squamish and Coldwater have played the most active roles in opposing the project. The Tsleil-Waututh carefully established their position to challenge an eventual government approval of the pipeline. They filed a lawsuit in the Federal Court of Canada to appeal the NEB`s hearing order establishing the terms of the review process, charging that the government had not sufficiently consulted and accommodated the First Nations on the procedures. They performed their own independent assessment of the project and in May 2015 rejected the pipeline according to their own laws (Tsleil-Waututh First Nation 2015). The Tsleil-Waututh also formed an alliance with Indigenous groups on the Salish Sea from across the forty-ninth parallel, whose waters may also be affected by increased tanker traffic (Connolly 2016). More details about asserting First Nations control are provided in the discussion of the Federal Court of Appeal's August 2018 decision.

Trudeau Asserting Liberal Control

Market access and environmental implications associated with pipelines played a significant role in the 2015 federal election campaign when Justin Trudeau's Liberals ended nine years of Conservative Party rule under Stephen Harper. Throughout the campaign, Trudeau sent mixed signals about his support for pipelines, arguing both that oil sands needed greater market access and that the NEB's process for regulating pipelines was flawed and needed to be reformed to regain the trust of Canadians: "Canadians must be able to trust that government will engage in appropriate regulatory oversight, including credible environmental assessments, and that it will respect the rights of those most affected, such as Indigenous communities. While governments grant permits for resource development, only communities can grant permission" (Liberal Party of Canada 2015a).

This statement that "only communities can grant permission" became a core part of the Liberal campaign of 2015. Trudeau first used the slogan in public in October 2013 in a speech on energy policy to the Calgary Petroleum Club. Criticizing Prime Minister Stephen Harper for his inability to get new pipelines approved and built, Trudeau argued that Harper "needlessly antagonized" both the Obama administration and the Canadian public: "Times have changed, my friends. Social license is more important than ever. Governments may be able to issue permits, but only communities can grant permission" (Liberal Party of Canada 2013).

Despite its prominent role in the campaign, the slogan disappeared from the Trudeau government's communications as soon as it was elected. In fact, since the election, there has been only one instance in the public record where Trudeau seems to have used a version of the phrase in public.[3] The phrase cannot be found using the search function on the prime minister of Canada's news page.[4] The phrase is also absent from the government of Canada's website, according to the search function. Searching Hansard for the 42nd Parliament beginning with the first Speech from the Throne of Trudeau's government, the phrase has not been used in Parliament by any member of Trudeau's government.[5] Given that the Trudeau government has taken a number of actions that are inconsistent with the slogan, it is a perfect case study of how rhetorical incentives when political parties are in campaign mode differ from those when they are in governing mode.

In January 2016, the Trudeau government announced "Interim Measures for Pipeline Reviews" as it geared up to initiate the promised review of environmental assessment and regulatory processes. The interim measures were to be guided by the following five principles:

1. No project proponent will be asked to return to the starting line—project reviews will continue within the current legislative framework and in accordance with treaty provisions, under the auspices of relevant responsible authorities and Northern regulatory boards.
2. Decisions will be based on science, traditional knowledge of Indigenous peoples and other relevant evidence.
3. The views of the public and affected communities will be sought and considered.
4. Indigenous peoples will be meaningfully consulted, and where appropriate, impacts on their rights and interests will be accommodated.
5. Direct and upstream greenhouse gas emissions linked to the projects under review will be assessed. (Natural Resources Canada 2016a)

It's revealing that their commitment to addressing First Nations concerns is a rather tepid restatement of obligations under Canadian law and a far cry from the platform's apparent commitment to accord the right to consent: "This will ensure that on project reviews and assessments, the Crown is fully executing its consultation, accommodation, and consent obligations, in accordance with its constitutional and international human rights obligations, including Aboriginal and Treaty rights and the United Nations Declaration on the Rights of Indigenous Peoples" (Liberal Party of Canada 2015).

For the Trans Mountain project, they committed to an assessment of the "upstream greenhouse gas emissions associated with this project" and to additional consultations with Indigenous groups and other affected communities. Environment and Climate Change Canada's assessment of upstream greenhouse gas emissions found that "the upstream GHG emissions could range from 13 to 15 megatonnes of carbon dioxide equivalent per year" (Environment and Climate Change Canada 2016).

The government also created a three-person ministerial panel "to create additional opportunities for communities close to the proposed pipeline and shipping route to share views on the project" (Major Projects Management Office 2016). The ministerial panel's process was not intended to repeat the NEB hearing process or make recommendations. The government recognized that the NEB hearing process was surrounded by a lack of public confidence and that the conditions under which it gave its recommendation in 2016 were very different from those when the project was proposed in 2013. With the mandate to address gaps that may have been left out of the NEB process, the panel held a series of 44 public meetings and considered 20,000 email submissions and 35,000 survey responses. Its November 2016 report represented the positions and concerns of the public on marine impacts, earthquake risk, pipeline routing, rail transport, diluted bitumen characteristics and behavior, aging infrastructure, economic arguments, climate change, and public confidence in the regulatory process. In lieu of specific findings or recommendations, the panel concluded with six pointed questions that illustrated key controversies for further consideration:

1. Can construction of a new Trans Mountain Pipeline be reconciled with Canada's climate change commitments?
2. In the absence of a comprehensive national energy strategy, how can policymakers effectively assess projects such as the Trans Mountain Pipeline?
3. How might Cabinet square approval of the Trans Mountain Pipeline with its commitment to reconciliation with First Nations and to the UNDRIP principles of "free, prior, and informed consent?"
4. Given the changed economic and political circumstances, the perceived flaws in the NEB process, and also the criticism of the Ministerial Panel's own review, how can Canada be confident in its assessment of the project's economic rewards and risks?
5. If approved, what route would best serve aquifer, municipal, aquatic and marine safety?
6. How does federal policy define the terms "social licence" and "Canadian public interest" and their interrelationships? (Ministerial Panel 2016)

The federal government also conducted a separate consultation process with First Nations. It was deliberately designed to avoid the procedural errors the Harper government made that led the Federal Court of Appeal to strike down the permit for Northern Gateway (see chapter 5).

These supplementary processes, while far more modest than pipeline opponents had hoped for, were the mechanisms the Trudeau government chose as a way to put its stamp on the review process and build sufficient social acceptance.

Asserting Local Control

In the Trans Mountain case, the authority of municipalities to influence pipeline regulation through zoning or permitting authority became a major issue. While a number of lower mainland British Columbia municipalities took positions against the project, the cities of Vancouver and Burnaby have been the most active opponents. Vancouver created an elaborate website that hosts 12 research reports supporting its position, acted as a formal intervenor, and challenged several federal decisions in court (City of Vancouver, n.d.). For the most part, it has acted like other interested parties in the sense that the project's physical location is not within the city's boundaries and thus the city was not involved in any permitting decisions.

The role of Burnaby has been the most controversial and has involved the most jurisprudence. Controversy erupted when Kinder Morgan decided to reroute the pipeline through Burnaby Mountain rather than a more residential area. The change led the NEB to request more information about route design, which required that the company perform seismic testing by drilling in the Burnaby Mountain Conservation Area. When Burnaby sought to block the drilling by enforcing its bylaws against that type of disruption without a permit, conflict erupted in the regulatory tribunal, in the courts, and on the ground.

Kinder Morgan appealed to the NEB, and the NEB, referring to the doctrines of federal paramountcy and interjurisdictional immunity, ruled that the National Energy Board Act clearly gave Kinder Morgan the authority to perform the testing without the consent of the local government. Burnaby appealed that ruling to the Federal Court of Appeal, but that court refused to grant leave to appeal several times. In response, Burnaby also appealed to the British Columbia Supreme Court. In December 2015, that court rejected Burnaby's argument. The court was clearly of the view that the case did not belong before it and called Burnaby's application "an abuse of process." Nevertheless, it gave reasons for its rejection, concluding that the doctrine of federal paramountcy was properly interpreted and applied by the NEB, stating: "Where valid provincial laws conflict with valid federal laws in addressing interprovincial undertakings, paramountcy dictates that the federal legal regime will govern. The provincial law remains valid, but becomes inoperative where its application would frustrate the federal undertaking" (*Burnaby (City) v. Trans Mountain Pipeline ULC*, 2015 BCSC 2140).

As these cases were winding their way through the courts, resistance on the ground emerged once Kinder Morgan sought to begin the seismic testing. Protesters disrupted the activities as Kinder Morgan employees began work, and they established an encampment around Bore Hole 2, including a "sacred fire" being nurtured by local First Nations. For nearly a month, Burnaby Mountain became the site of daily protests against the pipeline and, eventually, the arrest of over 100 protesters, including several prominent SFU academics (Prystupa 2014). Kinder Morgan went to court to get an injunction preventing protesters from disrupting its testing activities and took the additional step of filing a civil suit against some of the protesters for damages resulting from project delays and harassment of workers.

While the threat of having significant damages leveled against them was alarming, the protesters also had a bit of fun with what they viewed as overreaction by the company. Part of the claim for damages was based on allegations of assault by protesters on Kinder Morgan workers, including any intimidating facial expressions. This provoked a social media campaign where numerous anti-pipeline advocates posted their own "#KMface" to express their anger at the prospects of pipeline construction (Burgman 2014). When it was revealed in court that the GPS coordinates for the work site given to the court and used as the basis for the injunction were incorrect, the judge threw out the charges against the protesters (Keller 2014). At that point, Kinder Morgan decided that it would not continue its case against them. Despite protests and the arrest of over 100 demonstrators in November and December 2014, the conflict quieted for over a year until Trans Mountain was approved in November 2016 and the company began preparing for construction around the terminal in mid-2017.

The Decision and Its Aftermath

Trudeau's Announcement

In Trudeau's approval speech of November 29, 2016, he emphasized his core rhetoric about the compatibility of environmental protection and economic growth. He pointed to the fact that his government "created a policy to put a price on pollution, and an Oceans Protection Plan to preserve our coasts." In emphasizing the economic benefits, he said the project "will give much needed new hope to thousands of hard-working people in Alberta's conventional energy sector, who have suffered a great deal over the past few years." But he also emphasized environmental protection, saying, "We approved this project because it meets the strictest of environmental standards, and fits within our national climate plan" (Trudeau 2016). He went out of his way to link the decision to Alberta's new climate plan: "And let me say this definitively: We could not have approved this project without the leadership of Premier Notley, and Alberta's *Climate Leadership Plan*—a plan that commits to pricing carbon and capping oilsands emissions at 100 megatonnes per year" (Trudeau 2016). He also spoke directly to those concerned with environmental values in British Columbia: "But to them—and to all Canadians—I want to say this: if I thought this project was unsafe for the BC coast, I would reject it. This is a decision based on rigorous debate, on science and on evidence. We have not been and will not be swayed by political arguments—be they local, regional or national" (Trudeau 2016).

More Conflict with Burnaby

One condition on the approval was that the company be required "to apply for, or seek variance from, provincial and municipal permits and authorizations that apply to the Project" (Bankes and Olszynski 2018). Conflict quickly developed over whether Burnaby was deliberately delaying the issuance of necessary permits. Kinder Morgan applied to the NEB to be exempted from the requirement to obtain certain permits, and requested the establishment of a "process for Trans Mountain to bring similar future matters to the Board for its determination in cases where municipal or provincial permitting agencies unreasonably delay or fail to issue permits or authorizations in relation to the Project" (National Energy Board 2017f).

In another major blow to municipal powers, the NEB ruled that, despite no evidence of "political interference or improper motives," Burnaby's processes "were not reasonable, resulting in unreasonable delay." That delay "constitutes a sufficiently serious entrenchment on a protected federal power." As a result, the NEB declared the Burnaby bylaws in question "inapplicable" (City of Burnaby 2018). Burnaby and the government of British Columbia applied for leave to appeal to the Federal Court of Appeal, but its applications were dismissed, again without any reasons given. In responding to this decision, Mayor Derek Corrigan took issue with the decision and announced that they would appeal to the Supreme Court of Canada:

> The federal court has refused to review the decisions made by the National Energy Board. They're not giving consideration to the arguments being made by the City and the provincial government that oppose the NEB ruling. The Court System should be the body that decides whether or not this is fair and just, but they dismissed our application without reasons. Very clearly, it's something the court should have dealt with and given reasons why it's not allowing the provincial government to exert its authority to protect the environmental interests of the province. We will, therefore, now ask the Supreme Court of Canada to perform this function. (City of Burnaby 2018)

The city's news release emphasized that the NEB "found that there was no evidence of political interference or deliberate obstruction" (City of Burnaby 2018).

While the "only communities grant permission" slogan disappeared from the Trudeau government's discourse once it was in power, it became a staple of opposition discourse. It not only clearly articulates a standard requiring community support but also punctuates the hypocrisy of the Trudeau government. In response to the Trudeau government's approval of the Trans Mountain pipeline in November 2016, Burnaby mayor Derek Corrigan employed the slogan directly: "Prime Minister Trudeau said 'Governments grant permits; ultimately only communities grant permission.' We agree. He does not, however, have our permission and we will continue to make that clear" (City of Burnaby 2016).

Pipeline opponents have worked hard to mobilize affected communities against pipelines. For the most part, that opposition has been expressed politically, taking advantage of the ethic of community consent as well as the influence of local political leaders in swaying votes in elections in senior jurisdictions. The legal powers of municipalities are limited to local zoning and permitting authority. Burnaby's efforts to use those powers to throw a wrench in the gears of the Trans Mountain project have been resoundingly rejected by the NEB and the courts (Olszynski 2018), but they have contributed to delays and cost increases for the project and contributed to the political risks that forced Kinder Morgan to sell the project to the government of Canada.

The 2017 Election in British Columbia

Trudeau's decision marked what should have been the formal end to the political stage of this pipeline conflict. While at first it looked like Trudeau's announcement had ended the political stage, everything changed with the chaotic May 2017 provincial election in British Columbia. With Clark's issuance of the environmental assessment certification in January 2017, the BC Liberals became enthusiastic champions of the pipeline project and denounced their opponents as "the parties of no." The provincial NDP opposed the pipeline, promising in its platform to "use every tool in our toolbox to stop the project from going ahead" (British Columbia New Democratic Party 2017). The province's Green Party also opposed the project.

The 2017 election failed to produce a majority government. The BC Liberals won 43 seats, one short of the 44 needed to form a majority. The NDP won 41, and the Greens won 3. Clark was given the opportunity to form a government, but the NDP and Greens teamed up to bring her party down in a nonconfidence motion. The two parties had agreed to a Confidence and Supply Agreement that committed them to cooperation so the NDP could govern with a minority (Shaw and Zussman 2018). The agreement committed the two parties to working together to "immediately employ every tool available to the new government to stop the expansion of the Kinder Morgan pipeline, the seven-fold increase in tanker traffic on our coast, and the transportation of raw bitumen through our province" (BC Green Caucus and the BC New Democrat Caucus 2017).

Once the NDP, with the support of the three members of the Green Party caucus, replaced the BC Liberals as the government of British Columbia, consultations with government lawyers convinced them that a commitment to "stop the pipeline" created legal risks for the province.[6] Thus, when Premier Horgan sent mandate letters to his cabinet, the phrasing changed from "stopping the pipeline" to the much vaguer "defend BC's interest" and "Employ every tool available to defend B.C.'s interests in the face of the expansion of the Kinder Morgan pipeline, and the threat of a seven-fold increase in tanker traffic on our coast" (Horgan 2017).

Constitutional Conflict between British Columbia and Alberta

Once they were in power, the BC NDP's actions appeared tentative at first. As they unveiled their "tools," they stuck to the rhetoric of either "defending BC's interests" or "protecting the coast." In August 2017, the government took the obvious step of seeking intervenor status in legal challenges against the project's approval in the Federal Court of Appeal (British Columbia Ministry of Environment and Climate Change Strategy 2017).

However, the politics of structure escalated dramatically in January 2018, when British Columbia proposed a regulation to place "restrictions on the increase of diluted bitumen ('dilbit') transportation until the behaviour of spilled bitumen can be better understood and there is certainty regarding the ability to adequately mitigate spills." The press release and background information were careful not to mention the Trans Mountain project and instead emphasized areas of concern within provincial jurisdiction: "The potential for a diluted bitumen spill already poses significant risk to our inland and coastal environment and the thousands of existing tourism and marine harvesting jobs. British Columbians rightfully expect their government to defend B.C.'s coastline and our inland waterways, and the economic and environmental interests that are so important to the people in our province, and we are working hard to do just that" (British Columbia Ministry of Environment and Climate Change Strategy 2018).

Within a week of this announcement, Alberta premier Rachel Notley, calling British Columbia's action an "unprovoked and unconstitutional attack," retaliated by banning British Columbia wines from the province. Three days later, Notley stated, "This is not a fight between Alberta and B.C. This is B.C. trying to usurp the authority of the federal government and undermine the basis of our Confederation" (Notley 2018a). A bit later, her criticism intensified: "That is completely unconstitutional, it's a made-up authority, it's a made-up law, it's ridiculous" (Rabson and the Canadian Press 2018).

After several weeks of heated rhetoric and threats of escalation, Premier Horgan decided to change course and refer the question of whether British Columbia had constitutional jurisdiction to regulate diluted bitumen to the courts. He stated, "We believe it is our right to take appropriate measures to protect our environment, economy and our coast from the drastic consequence of a diluted bitumen spill. And we are prepared to confirm that right in the courts" (British Columbia Office of the Premier 2018). Alberta responded by dropping its wine boycott. It took British Columbia two months to prepare the reference question to the British Columbia Court of Appeal, which it announced in April. In making the case for the reference question, Attorney General David Eby stated, "We believe B.C. has the ability to regulate movement of these substances through the province. This reference question seeks to confirm the scope and extent of provincial powers to regulate environmental and economic risks related to heavy oils like diluted bitumen" (Boothby 2018).

Earlier in April, in the midst of this constitutional sparring between British Columbia and Alberta, Kinder Morgan sent shock waves through the

Canadian political system by announcing it would cease all nonessential spending on the Trans Mountain Expansion Project. It issued an ultimatum giving governments in Canada until May 31 to resolve their differences in a way "that may allow the Project to proceed." Their media release stated: "'[We] have determined that in the current environment, we will not put KML shareholders at risk on the remaining project spend,' said KML Chairman and Chief Executive Officer Steve Kean. 'The Project has the support of the Federal Government and the Provinces of Alberta and Saskatchewan but faces continued active opposition from the government of British Columbia. A company cannot resolve differences between governments. While we have succeeded in all legal challenges to date, a company cannot litigate its way to an in-service pipeline amidst jurisdictional differences between governments,' added Kean" (Kinder Morgan Canada 2018).

The company put the blame squarely on the government of British Columbia, saying, "Unfortunately BC has now been asserting broad jurisdiction and reiterating its intention to use that jurisdiction to stop the Project. BC's intention in that regard has been neither validated nor quashed, and the Province has continued to threaten unspecified additional actions to prevent Project success. Those actions have created even greater, and growing, uncertainty with respect to the regulatory landscape facing the Project" (Kinder Morgan Canada 2018).

While the government under the BC NDP has been careful to modify its rhetoric somewhat since coming into power, pipeline proponents continue to refer back to the NDP's preelection statement of intent. In its release announcing the ultimatum, Kinder Morgan stated, "Since the change in government in June 2017, that government has been clear and public in its intention to use 'every tool in the toolbox' to stop the Project" (Kinder Morgan Canada 2018).

In response to the ultimatum, Notley promised that "Alberta is prepared to do whatever it takes to get this pipeline built—including taking a public position on the pipeline. Alberta is prepared to be an investor in the pipeline" (Notley 2018b). On Twitter, she promised retaliation, stating, "We will be bringing forward legislation giving our gov't the powers it needs to impose serious economic consequences on British Columbia if its government continues on its present course. Let me be absolutely clear, they cannot mess with Alberta" (Baldrey 2018). She also suggested that the conflict could amount to a constitutional crisis, saying, "There are those out there who are, at this point, calling this . . . a constitutional crisis for the country. And I don't know really if that's too far off. If the federal government allows its authority to be challenged in this way, if the national interest is given to the extremes on the left or the right, and if the voices of the moderate majority of Canadians are forgotten, the reverberations of that will tear at the fabric of Confederation for many many years to come" (Hall 2018).

On April 16, 2018, Notley introduced Bill 12 (Preserving Canada's Economic Prosperity Act), which created an export license requirement for crude oil, natural gas, and refined fuels. It gave the minister of energy authority to deny the issuance of a license if "it is in the public interest of Alberta to do so" (Alberta Ministry of Energy 2018). In announcing her intention to introduce the legislation, Notley stated, "Alberta must have the ability to respond. This is not an action that anyone wants to take. And it is one that I hope we never have to take. And it's not how Canada should work. And it's not how neighbours, frankly, should treat one another" (Cryderman, Tait, and Hager 2018). Sarah Hoffman, Alberta's deputy premier, stated, "Their government has caused pain to Alberta families. We can certainly do the same, and we've put a bill on the order paper that enables us to do that" (Braid 2018). In a letter to David Eby, Alberta's minister of justice, Kathleen Ganley, declined to refer the legislation to the courts and stated, "Given B.C.'s transparent attempt to sow legal confusion by claiming constitutional authority it does not have in order to harass the pipeline investors into abandoning the project, the government of Alberta has a responsibility to its citizens to protect the interests of its citizens" (Attorney General of British Columbia 2018; Hunter 2018).[7]

In responding to the announcement, British Columbia's environment minister, George Heyman, expressed his dismay, saying, "I see no reason for the government of Alberta to take any action when all BC has been doing is standing up for our interests in proposing some regulations that are well within our jurisdiction. We are determined to defend our environment, our economy and our coastline. We have tried to be the adults in the room here" (Zussman 2018). On May 22, 2018, British Columbia launched a constitutional challenge to the Alberta legislation. In justifying the move, Eby decried the Alberta legislation as "blatantly unconstitutional" (Judd and Zussman 2018).

The legal confrontation between the provinces deescalated somewhat in May 2019, when the British Columbia Court of Appeal forcefully rejected British Columbia's reference question (Reference re Environmental Management Act [British Columbia]). The decision essentially ripped away the legal foundation of British Columbia's capacity to block the pipeline, but before that legal question was settled, other events shifted the fate of the pipeline back to the federal government and federal courts.

The Government of Canada Buys Out Kinder Morgan Canada

Shortly after Kinder Morgan announced its ultimatum, Canada's finance minister, Bill Morneau, entered into negotiations with the company. After a month of apparently limited progress, he stated publicly that the Canadian government was prepared to offer Kinder Morgan, and any future owner of the project, indemnity for any financial losses resulting from political opposition by British Columbia's government.

Then, on May 29, 2018, the entire political economy of the oil sands policy regime underwent a seismic shift. Morneau made the stunning announcement that the government of Canada was purchasing Kinder Morgan Canada's Trans Mountain assets for $4.5 billion. Alberta would also contribute up to $2 billion to cover costs resulting from "unforeseen circumstances" (Department of Finance Canada 2018).

In her comments, Notley referred to the project as nation building three times and emphasized its pan-Canadian support and benefits, saying, "I believe in Canada, not just as a concept, but as a country" (Notley 2018c). With the project now government owned, the government of Canada's stakes in its success increased, which could bolster the political image of the project. But it didn't change the constitutional conflicts or how they were being framed by competing interests in the pipeline dispute.

In response to the federal government's buyout, Premier Horgan made it clear that this did not change British Columbia's position: "It's not about politics. It's not about trade. It is about British Columbians' right to have their voices heard. To do so is squarely within our rights as a province, and our duty as a government. Ottawa has acted to take over the project. . . . At the end of the day, it doesn't matter who owns the pipeline. What matters is protecting B.C.'s coast—and our lands, rivers and streams—from the catastrophic effects of an oil spill" (Horgan 2018).

Courts Grant Anti-pipeline Coalition Major Victory

The judicial stage was formally initiated with a series of legal challenges from First Nations, environmentalists, and municipalities. In total, there were 15 challenges to the adequacy of the NEB review process and the

Trudeau government's order in council. The challenges were filed by 10 separate plaintiffs: seven First Nations,[8] Raincoast Conservation Foundation and Living Oceans Society, and the cities of Burnaby and Vancouver. The Federal Court of Appeal chose to consolidate the challenges and consider them together. Hearings were held in October 2017. Reflecting the complexity of the case, the hearings were the longest in the court's history (West Coast Environmental Law 2017). The previous record was held by the case involving Northern Gateway, *Gitxaala v. Canada*.

The oil sands coalition received another fundamental shock in *Tsleil-Waututh Nation v. Canada (Attorney General)*. In a decision released August 30, 2018, the Federal Court of Appeal again quashed the certificate of a pipeline to the West Coast. This outcome surprised many because the Trudeau government claimed to have learned from, and be applying the principles of, the *Gitxaala* case involving the Northern Gateway Pipeline. In its decision, the court noted that the federal government had taken some specific steps to ensure "that the flaws identified by the Court in Gitxaala were remedied and not repeated," and the court agreed that there were "significant improvements in the consultation process" (*Tsleil-Waututh Nation v. Canada (Attorney General) 2018)*.

Nonetheless, the court found that the consultation was "unacceptably flawed and fell short of the standard prescribed by the jurisprudence of the Supreme Court." In making the finding, the court emphasized the importance of "meaningful two-way dialogue":

> I begin the analysis by underscoring the need for meaningful two-way dialogue in the context of this Project and then move to describe in more detail the three significant impediments to meaningful consultation: the Crown consultation team's implementation of their mandate essentially as note-takers, Canada's reluctance to consider any departure from the Board's findings and recommended conditions, and Canada's erroneous view that it lacked the ability to impose additional conditions on Trans Mountain. I then discuss Canada's late disclosure of its assessment of the Project's impact on the Indigenous applicants. Finally, I review instances that show that as a result of these impediments the opportunity for meaningful dialogue was frustrated (Tsleil-Waututh Nation v. Canada (Attorney General) 2018).

In addition to the flaws in the First Nations consultation, the court also ruled on the NEB's choice to exclude marine shipping issues in its definition of the project: "The Board unjustifiably defined the scope of the Project under review not to include Project-related tanker traffic. The unjustified exclusion of marine shipping from the scope of the Project led to successive, unacceptable deficiencies in the Board's report and recommendations. As a result, the Governor in Council could not rely on the Board's report and recommendations when assessing the Project's environmental effects and the overall public interest" (Tsleil-Waututh Nation v. Canada (Attorney General) 2018).

The year 2018 was a tumultuous one for the Trans Mountain project. The political obstacles created by the anti-pipeline coalition forced Kinder Morgan, the original proponent, to abandon the project. The federal government stepped in to take it over. Now the project's approval had been quashed by the Federal Court of Appeal. Having doubled down on the project by nationalizing, the Trudeau government did not believe it could afford to walk away.

Reconsideration and Reapproval

The federal government acted quickly to start a new process. On September 20, 2018, the government referred the project back to the NEB for reconsideration of project-related marine shipping, including its effects on southern resident killer whales. The reconsideration was an accelerated review, and the NEB issued a new recommendation on February 22, 2019. The NEB concluded that the project "is likely to cause significant adverse environmental effects on the Southern resident killer whale" but recommended the project be approved, again, because it believes those effects "can be justified in the circumstances, in light of the considerable benefits of the Project and measures to minimize the effects" (National Energy Board 2019). While the NEB reviewed the marine impacts, Trudeau appointed a former Supreme Court justice, Frank Iacobucci, to oversee the consultation process with Indigenous groups.

As expected, the Trudeau government reapproved the project on June 18, 2019. The announcement of the decision came with an interesting political twist: the approval announcement contained the commitment "that every dollar the federal government earns from this project will be invested in Canada's clean energy transition" (Trudeau 2019). As expected, Indigenous groups and environmentalists challenged the decision in court, so the project was sent back into another legal stage.

The reapproval decision was challenged again in the courts. The Federal Court of Appeal refused to hear the challenge from environmental groups, but it did hear the challenge from four First Nations about the adequacy of consultation. In a major victory for the oil sands coalition, the court rejected the First Nations' complaints and upheld the reapproval decision. The court ruled that the government's purchase of the pipeline project did not create any conflict of interest that undermined its capacity to deliberate fairly, saying that "there is no evidence that the Governor in Council's decision was reached by reason of Canada's ownership interest rather than the Governor in Council's genuine belief that the Project was in the public interest" (*Coldwater et al. v. Canada (Attorney General) et al.*, 2020 FCA 34, paragraph 23). The court held that the government had clearly met its obligations: "In this case, the Governor in Council's key justifications for deciding as it did are fully supported by evidence in the record. The evidentiary record shows a genuine effort in ascertaining and taking into account the key concerns of the applicants, considering them, engaging in two-way communication, and considering and sometimes agreeing to accommodations, all very much consistent with the concepts of reconciliation and the honour of the Crown" (*Coldwater et al. v. Canada (Attorney General) et al.*, 2020 FCA 34, paragraph 76).

The court reiterated that the Crown's consultational obligation was not equivalent to the right to consent: "Canada was under no obligation to obtain consent prior to approving the Project. That would, again, amount to giving Indigenous groups a veto" (*Coldwater et al. v. Canada (Attorney General) et al.*, 2020 FCA 34, paragraph 194). The decision has been appealed to the Supreme Court of Canada, but the Supreme Court declined to hear the case, bringing apparent legal finality to the dispute (Supreme Court of Canada 2020).

Conclusion

The Trans Mountain Expansion Project was the third, and ultimately most explosive, of the oil sands pipeline controversies that emerged in the United States and Canada in the 2010s. Of the four, it was the one that faced the fewest apparent political risks. There was no international border in question, as with Keystone XL, and most of the project followed an existing right-of-way, unlike Northern Gateway. Yet a stark separation of risks and benefits across a provincial boundary, and significant place-based concerns, helped mobilize opposition.

The controversy has altered the policy regime and created some unusual bedfellows, given historical federal-provincial conflict over energy policy in Canada. The core antagonists in the conflict over the 1980 National Energy

Program were the federal government of Pierre Trudeau and the government of Alberta. The curious politics of Trans Mountain turned the government of Alberta into an enthusiastic champion of the national interest and federal authority and, for a time at least, a pivotal ally with the federal government of the elder Trudeau's son.

Justin Trudeau's nationalization of the Trans Mountain project was more than just a victory for the oil sands coalition. It was a transformation of the political economy of the oil sands policy regime. The government of Canada moved from being a regulator to an owner of a major oil sands asset. Because the federal government now has a much more direct financial interest in the profitability not just of the project but of the Albertan oil sector, it has shifted its core incentives with respect to regulation. The Canadian government's financial interest will likely make it even more reluctant to take regulatory actions that increase costs of production. This includes environmental protection measures needed to improve the sustainability of the oil sands, such as actions to reduce GHGs, protect caribou habitat, or reduce and clean up tailings ponds. With its purchase of the pipeline, the federal government of Canada is now at the core of the pro-pipeline oil sands coalition.

Even more so than the other cases, Trans Mountain provoked a contentious politics of structure, as predicted by the third behavioral hypothesis. Multiple strategic actors competed for institutional authority, with courts frequently acting as the ultimate arbiter. Harper sought to advance his pro-industry agenda by restructuring regulatory reviews, tightening deadlines, and narrowing participation. Environmentalists sought to challenge these restrictions in court, but to no avail. While Harper sought to exert more direct political control, the government of British Columbia worked to evade direct responsibility by deferring to the federal process. A lawsuit by First Nations provoked a court to strike down that equivalency agreement and hand some decision-making authority back to the provincial government. Local governments in the Vancouver area tried to use their zoning and bylaw powers to assert jurisdiction, but that effort was also rebuffed by the courts. First Nations seem intent on leveraging pipeline resistance to acquire greater control over decision-making on their traditional territories.

The 2017 election of a BC NDP government opposed to the pipeline challenged the extent of federal authority over pipeline decision-making and provoked a political and constitutional conflict with the government of Alberta. Interprovincial tensions have calmed, legally at least, with the British Columbia Court of Appeal's forceful rejection of British Columbia's authority to block increases in shipments of diluted bitumen. The Federal Court of Appeal's rejection of challenges to the reapproval decision, and the Supreme Court's apparent validation of that decision, was a major breakthrough for the oil sands coalition. As of the end of 2020, the pipeline is in the early stages of construction. More of the policy and political consequences of the Trans Mountain Expansion Project will be described in chapter 8, after the fourth and final pipeline case study is presented.

7 After Careful Review of Changed Circumstances: The Demise of Energy East

with Xavier Deschênes-Philion

Overview

When the oil sands coalition became more serious about building a pipeline from the oil sands across Canada to the east, environmentalists knew that their strongest cards were in Quebec. Greenpeace Canada's Keith Stewart was quite direct: "From the very beginning, I said we're going to kill this pipeline in Quebec, and in French" (Stewart 2016). Quebec had a well-established environmental movement with strong grassroots connections to local citizens groups. According to Équiterre's Steven Guilbeault, to foster awareness of the project and its risks, his group began conducting community meetings even before the official announcement of the project: "A bit like a rock band on tour, we travelled across southern Quebec and even northern New Brunswick, not to tell people what to think or do, but to share our views on the matter. We were often invited by local groups or organizations, and quite often by elected representatives" (Guilbeault 2017).

Resistance in Quebec was the most formidable obstacle to the Energy East Pipeline. Proposed by TransCanada, Energy East was the most ambitious of the four major proposed oil sands projects, covering the most territory and promising to move the most product. The 4,500-kilometer pipeline would have carried up to 1.1 million barrels of oil per day from Alberta, Saskatchewan, and North Dakota to refineries and a marine terminal at tidewater in eastern Canada. It would have connected the oil sands to refineries in Quebec and New Brunswick and given the oil sands access to Atlantic tidewater in New Brunswick. In doing so, it would have avoided the formidable pipeline resistance in British Columbia, but it would have to run through Quebec, which has a long tradition of environmental activism and resistance to external imposition of costs. Energy East was the most recently proposed of the four cases but the second to come to finality, with TransCanada's decision to withdraw its application and cancel the project in October 2017.

That a west to east pipeline promoted by TransCanada would become a deeply divisive regional and partisan issue should come as no surprise to anyone familiar with Canadian history. The converted TransCanada Mainline natural gas pipeline from Alberta to Montreal was to be the backbone of the Energy East proposal. That very same natural gas pipeline, proposed, approved, and constructed in the 1950s, was the focus of Canada's first "great pipeline debate." Partisan conflict erupted during deliberations over the pipeline in the House of Commons with regard to government subsidies to an American company and the government's procedural conduct. The conflict is credited with leading to the defeat of the Liberal government of Louis St. Laurent by John Diefenbaker's Conservatives in 1957 (Kilbourn 1970).

The timeline for the project is presented in figure 7.1. Energy East was first announced on August 1, 2013, a few weeks after a train containing crude oil from North Dakota's Bakken shale formation derailed and exploded in the heart of Quebec's municipality of Lac-Mégantic, causing the tragic deaths of 47 people and raising major debates about oil transportation safety across the country. TransCanada first expected to start operating the pipeline by the end of 2017 (Krugel 2013). Initially considered a promising alternative to the lightning-rod West Coast pipelines, Energy East quickly became mired in a similar conflict over environmental values and Aboriginal rights, but this time the claims to the legitimacy of provincial authority were accentuated by Quebec's distinct culture and long-standing grievances within the Canadian federation. The regulatory review process became mired in a crisis as conflict of interest accusations emerged against National Energy Board panel members, the hearing panel was disbanded, and a new panel had to be established. Hearings were scheduled to begin in fall 2017, but before they could get going, TransCanada terminated the project. Worn down by sustained and intense public opposition, market uncertainties, and self-inflicted wounds, the company used new analytical requirements imposed by the reconfigured NEB panel as a justification to cancel the project.

This chapter applies the regime framework to analyze the Energy East controversy but focuses in particular on competing images of the pipeline in discourses across three communities. For the pipeline's supporters across Canada, the pipeline was a nation-building project, but the very concept

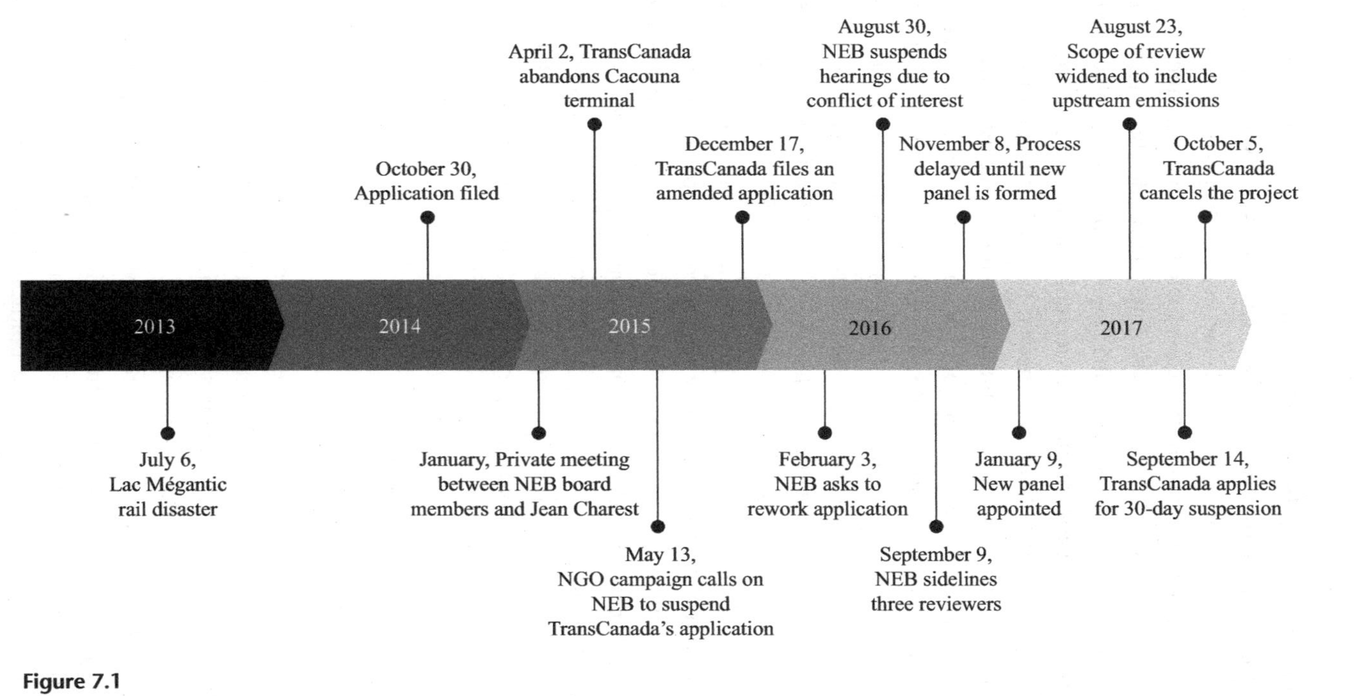

Figure 7.1
Energy East timeline.

of "nation" is contested, both by First Nations, who see themselves as sovereign within their territories, and by Quebec, which has its own form of nationalism. Both the Aboriginal and Quebecois concepts of "nation" challenge the Canadian one and, in this case, fueled pipeline resistance.

The oil sands coalition was initially quite optimistic about Energy East because, when first introduced, it seemed less risky politically. Its all-Canadian route avoided the vagaries of the American political process endured by Keystone XL. Like the Trans Mountain project, Energy East could capitalize on preexisting infrastructure. For two-thirds of the route, an existing natural gas pipeline (TransCanada's Mainline) would be converted and upgraded to suit oil transportation (TransCanada 2015), but Energy East had the added benefit of avoiding the entrenched West Coast environmental opposition and legal uncertainties of working in unceded Aboriginal territory. Moreover, by delivering bitumen to refineries in Quebec and New Brunswick, it promised a larger share of economic benefits to provinces through which it traveled. As a result, unlike British Columbia, the province at the end of the Energy East line, New Brunswick, enthusiastically supported the project. But that initial optimism was not sustained for long. The pipeline ran through six provinces (Alberta, Saskatchewan, Manitoba, Ontario, Quebec, and New Brunswick), but it was opposition in Quebec that proved to be the most frustrating and eventually fatal to the project. An increasingly influential anti-pipeline coalition harnessed Quebec's pro-environment, nationalist political culture into an unbeatable wall of resistance.

Actors—The Oil Sands Coalition

Oil Industry

The same company behind Keystone XL, TransCanada Corporation (later renamed TC Energy) is the main stakeholder in the Energy East project. The Calgary-based company operates oil and natural gas pipelines as well as energy production facilities (including natural gas, nuclear, wind, and solar power) across Canada, the United States, and Mexico. The company's interest in the project was magnified when their Keystone XL project became ensnarled in American political conflicts. In an effort to be proactive and more effective at gaining social and political acceptance for the Energy East proposal, TransCanada hired the world's largest public relations firm, Edelman. Edelman had committed 40 employees out of its Washington,

D.C., office, along with nine TransCanada employees, to collaborate on the campaign (Goldenberg 2014). The public relations firm's proposed tactics included distracting opposing groups and "causing them to redirect their resources" by working through "supportive third parties who can in turn put the pressure on, particularly when TransCanada can't," and a budget for the mobilization of 35,000 pipeline support activists. The strategy also incorporated digging into any negative aspects of funding sources or litigation issues of opposition groups, such as the Council of Canadians and the David Suzuki Foundation (McCarthy 2014a).

TransCanada had a major setback when the Edelman documents were leaked to, and subsequently released by, Greenpeace Canada in November 2014. Given the explicitness of the bold, aggressive tactics proposed by Edelman, Greenpeace denounced the strategy as "dirty tricks" (Greenpeace Canada 2014), and the incident exacerbated tensions in TransCanada's relations with opponents and reluctant communities.

From west to east, different oil-related companies also had strong interests in the project. Upstream of the pipeline, oil companies had essentially the same interests they did with the other pipelines: they were hoping for new ways to access tidewater and increase oil transportation capacity to export Alberta's oil sands and shale oil from Saskatchewan and North Dakota. Increasing revenues by reducing or eliminating the oil price differential was also a core interest of oil companies. This argument was critical when Energy East was first announced because the gap between the international Brent price and continental West Texas Intermediate price reached over $20 per barrel at different times between 2011 and 2014 (Ycharts 2017). Accessing international markets was therefore perceived as a key element of oil companies' business strategy.

On the other end of the proposed pipeline, eastern refineries wanted to diversify their feedstocks and improve their supply security with Canadian oil. Two refineries in Quebec (Suncor in Montreal, Valero in Levis) and one in New Brunswick (Irving Oil in Saint John) would have benefited from the project. Irving Oil would also have operated the marine terminal through a joint venture with TransCanada (TransCanada 2015).

Governments

When Energy East was announced in 2013, Stephen Harper's Conservative government endorsed the project with no hesitation. "We think it's a good idea in principle," Harper said, "It is, I think, a good idea that we find pan-Canadian solutions so that all of this country benefits from our energy products and that we enhance our own energy security" (Whittington 2013). In contrast, Justin Trudeau's Liberal government has taken a more prudent approach toward the project since its election in 2015, arguing that the assessment process must run its course before it takes a position on the project. Trudeau's approach attracted criticism both from the Conservatives, who wanted the government to show public support for the project, and from the NDP, who believed the NEB review process was skewed in favor of the oil companies.

The government of Alberta strongly supported the Energy East Pipeline. Previous Conservative premiers Alison Redford and Jim Prentice both advocated for the pipeline. Redford praised Energy East because it would reduce the oil price differential and create new markets for Alberta's oil sands, saying, "I am very pleased with today's announcement that Energy East is moving forward. My government made a commitment to the project as part of our efforts to build new markets and get a fairer price for the oil resources Albertans own" (Government of Alberta 2013). After she resigned in March 2014, her successor, Jim Prentice, continued the Alberta government's efforts to promote Energy East across the country.

Rachel Notley's New Democratic government, elected in 2015, has also lobbied in favor of the pipeline while promoting more stringent climate actions through a provincial carbon tax and a cap on provincial greenhouse gas emissions from the oil sands sector (described in chapter 2). Despite federal approval of Kinder Morgan's Trans Mountain Expansion Project and Enbridge's Line 3 Replacement Project in 2016, and Donald Trump's approval of the Keystone XL Pipeline in January 2017, Notley continued to press the case for Energy East. Notley stressed the Canada-wide benefits of the project, saying, "This is not just about the strength of Alberta's economy, even though we are feeling the drop in oil price struck more here than anywhere else. This is about the long-term sustainability of the Canadian economy" (Notley quoted in Global News 2016).

In fact, according to TransCanada, a larger share of the benefits stemming from the Energy East Pipeline would have gone not to Alberta but to other Canadian provinces, in terms of job creation, but also with regard to tax revenues and GDP growth. Table 7.1 presents TransCanada's expected benefits by province for the project's construction and operational phases.

Table 7.1
Energy East Pipeline benefits by province

	Job creation (full-time employment/year)						Tax revenues (2013 C$millions)			GDP impact (2013 C$millions)		
	Construction		Operations		Combined		Constr.	Oper.	Comb.	Constr.	Oper.	Comb.
Provinces	Direct	Indir.	Direct	Indir.	Direct	Indir.						
Alberta	754	524	499	196	1,253	720	350	777	1,127	1,926	5,465	7,389
British Columbia	–	179	–	73	–	252	240	412	652	327	289	616
Manitoba	485	149	49	157	534	306	157	309	466	621	2,354	2,975
New Brunswick	3,123	648	132	129	3,255	777	482	371	853	3,248	3,322	6,570
Ontario	2,148	1,741	114	1,409	2,262	3,150	1,320	2,935	4,255	5,440	18,520	23,960
Other Atlantic	–	123	–	26	–	149	97	149	246	249	127	376
Quebec	1,922	1,206	33	291	1,955	1,497	972	1,151	2,123	3,942	5,315	9,257
Saskatchewan	773	190	49	157	822	347	185	354	539	1,012	3,283	4,295
Territories	–	2	–	1	–	3	9	17	26	10	9	19
Total	9,246	4,762	891	2,447	10,081	7,201	3,813	6,475	10,288	16,776	38,683	55,459

Source: Adapted from TransCanada (2017b). The construction phase is projected over a nine-year period, while the operational phase is projected over a 20-year period.

TransCanada expected the Energy East project to create more than ten thousand full-time jobs per year during the construction phase and over seven thousand full-time jobs during the operational phase. Table 7.1 demonstrates that the project's economic benefits were much more evenly spread across provinces along the route than for the two other Canadian pipelines. New Brunswick, Quebec, and Ontario would all have seen job benefits, especially in construction. These jobs were especially coveted in New Brunswick, which in 2014–2015 faced significantly higher unemployment than Ontario and Quebec. New Brunswick's unemployment rate in 2015 was 9.8%, compared to 7.6% for Quebec and 6.8% for Ontario (Newfoundland and Labrador Statistics Agency 2017). After the construction was completed, however, Alberta would have received the lion's share of operating jobs.[1]

Apart from Alberta, two other provinces explicitly supported the project: Saskatchewan and New Brunswick. The Legislative Assembly of Saskatchewan unanimously adopted a motion supporting Energy East in November 2014 (Government of Saskatchewan 2014). New Brunswick endorsed a similar motion in December 2016 (with MLA and Green Party leader David Coone being the only voice against the motion) (Government of New Brunswick 2016). Saskatchewan premier Brad Wall and New Brunswick premier Brian Gallant were among the strongest public advocates for the project. Both premiers attempted to publicize the benefits of the project and downplay the risks around which a majority of Quebecers had mobilized. Premier Brad Wall took a somewhat combative stance against Montreal mayor Denis Coderre for his official opposition to the project. Two weeks after being elected in 2014, Gallant went to Calgary and visited TransCanada's headquarters, where he showed strong support for the project (CBC 2014). Premier Gallant also went to Quebec and Ottawa to publicly defend Energy East's economic benefits, and he agreed to debate Energy East and oil transportation issues with Denis Coderre on *Tout le monde en parle*, one of the most popular shows on French Canadian television (CBC 2016a).

Actors—Anti-pipeline Coalition

Environmental Groups

As with the other oil sands pipelines analyzed in this book, a wide range of environmental groups joined the battle against the Energy East Pipeline. This environmental opposition included well-known Canadian organizations or their local chapters, such as Greenpeace, the David Suzuki Foundation, Environmental Defence, the Council of Canadians, and Équiterre. As in the other cases, environmentalists carefully cultivated relationships with place-based groups, and the risks to precious places posed by oil spills or tanker activity were the foundation for resistance. But perhaps more so than other pipelines, climate played a very important role in environmental strategies and discourse.

The foundation for resistance to Energy East was the alliance of environmental groups formed with local actors in Quebec. In Quebec, more than 60 environmental groups, Indigenous communities, and civic associations formed the Common Front for the Energy Transition to campaign against further development of the oil sector or new oil transportation projects and to promote further transition toward clean and renewable energy (Front commun pour la transition énergétique 2019).

As part of this coalition, the Don't Spill in My Yard (Coule pas chez nous) campaign received greater attention when a former leader of Quebec's 2012 student strike and now MLA for Québec Solidaire, Gabriel Nadeau-Dubois, decided in November 2014 to personally give C$25,000 to the campaign (Poitras 2018, 208–209). Following this highly publicized move, the campaign raised $80,000 in 24 hours and $400,000 in one week from more than 14,000 donors (Coule pas chez nous 2016a). Coule pas chez nous is very active online (its Facebook page has over 18,000 likes) and on the ground; it has supported and financed more than 30 projects and campaigns in opposition to Energy East, including the Climate Action March with more than 25,000 citizens in Quebec City in April 2015, information sessions across the province, and judicial actions against the hydrocarbon sector (Coule pas chez nous 2016b). Opposition to Energy East in Quebec had also stretched to groups that rarely take positions on energy issues, such as the Union of Agricultural Producers and the Federation of Quebec Workers (Bergeron 2016; Shields 2016a).

Another strategic asset for the anti-pipeline coalition was the Quebec anti-fracking movement, which after several years of intense mobilization won a de facto moratorium halting fracking in the province (Montpetit, Lachapelle, and Harvey 2016). A substantial network of local groups emerged in opposition to fracking in the province. Rivard et al. (2014) report that in 2012 there were "over 30 local (municipal to regional levels) opposition groups, 3 provincial protest groups and 63 municipalities" opposed to fracking. The anti-pipeline movement saw these groups as ripe for mobilization. According to Stewart, "The anti-fracking movement had won a moratorium, and was kind of looking around for what it would do next. And this is a very similar thing. It's really easy to transition from that to 'oh, new threat, from oil companies to water'" (Stewart 2016). As a result, a formidable environmental resistance movement emerged soon after Energy East was announced. As will be described, a series of missteps by TransCanada played right into the hands of environmentalists.

Indigenous Groups

As with the other pipeline proposals, indigenous communities had become highly mobilized against the Energy East Pipeline. Given the length of the project, addressing Indigenous concerns was sure to be quite challenging; in its application, TransCanada stated that it was "engaging with a total of 166 First Nations and Métis communities and organizations across the length of the Project" (TransCanada 2016b, 5). By the time the Energy East regulatory review got into full swing, the level of organization and cooperation among Indigenous groups in North America had increased dramatically. In September 2016, First Nations across North America signed the Treaty Alliance against Tar Sands Expansion. The Treaty Alliance was initially formed by 50 First Nations but as of August 2017 had expanded to more than 120 First Nations in Canada and the United States. The Treaty Alliance aims "to prevent a pipeline/train/tanker spill from poisoning their water and to stop the Tar Sands from increasing its output and becoming an even bigger obstacle to solving the climate crisis" (Treaty Alliance 2016a).

The Treaty Alliance was inspired by the Save the Fraser Declaration and the Yinka Dene Alliance (YDA) described in chapters 5 and 6. YDA's representatives organized the West Meets East Tour and traveled across Canada to meet with First Nations communities living along the Energy East Pipeline route and raise awareness over oil sands' environmental and social issues. In turn, Grand Chief Serge Simon of the Mohawk Council of Kanesatake, Grand Chief Derek Nepinak of the Assembly of Manitoba Chiefs, and Chief Arnold Gardner of Eagle Lake First Nation were invited to address the 47th Annual Chiefs-in-Assembly of the Union of BC Indian Chiefs, a key event in the creation and solidification of a coast-to-coast alliance against tar sands expansion and transportation (Treaty Alliance 2016b). The Treaty Alliance brought together Indigenous communities from each of the six provinces Energy East was expecting to cross.

Governments

The Liberal government of Quebec has taken a wait-and-see approach toward Energy East. Former premier from Quebec Philippe Couillard decided to wait until the end of Quebec's environmental assessment process in 2018 before taking a position. Still, as will be discussed in the discussion of institutions, Quebec's government took a somewhat combative approach toward TransCanada, perhaps as a result of the mounting opposition in the province. Ontario chose to conduct its own review process, but former premier Kathleen Wynne seemed somewhat more open to the project.

For instance, in November 2014, Quebec sent TransCanada a list of seven conditions the company must respect for both provinces to grant approval to Energy East. Ontario premier Kathleen Wynne then granted support to Quebec's request following a joint meeting with Philippe Couillard later that month. The list included the following elements:

1. Compliance with the highest available technical standards for public safety and environmental protection;
2. Have world-leading contingency planning and emergency response programs;
3. Proponents and governments consult local communities and fulfill their duty to consult with Aboriginal communities;
4. Take into account the contribution to greenhouse gas emissions;
5. Provide demonstrable economic benefits and opportunities to the people of Ontario and Quebec, in particular in the areas of job creation over both the short and long term;
6. Ensure that economic and environmental risks and responsibilities, including remediation, should be borne exclusively by the pipeline companies in the event of a leak or spill on ground or water, and provide financial assurance demonstrating their capability to respond to leaks and spills;
7. Interests of natural gas consumers must be taken into account. (Ontario Office of the Premier 2014)

The provinces' intention to evaluate the pipeline's upstream greenhouse gas emissions was dropped from the conditions a month later, after a visit from Jim Prentice, then Alberta's premier, to Ontario and Quebec (Morrow 2014). In spite of this political shift, both provincial environmental review agencies decided to consider the potential impacts of the pipeline on climate change in their assessment review of Energy East (OEB 2015; BAPE 2016a).

Ontario mandated that the Ontario Energy Board (OEB) review the project in 2013 and hold public consultations in 2014. OEB's final report was published a year later and expressed several concerns about the Energy East project, especially in regard to natural gas supply, water protection, impacts on Aboriginal as well as local communities, and its economic benefits (or lack thereof) (OEB 2015). Ontario premier Kathleen Wynne indicated openness to Energy East after a meeting with Alberta premier Rachel Notley in January 2016, during which Wynne praised the climate plan Alberta adopted under Notley's government. Wynne suggested that "the people of Ontario care a great deal about the national economy and the potential jobs that this proposed pipeline project could create in our province and across the country" (Financial Post 2016). According to one environmentalist, the pathway to blocking the pipeline in Ontario ended when Wynne bought into the Trudeau government's process of "turning down the temperature on the pipeline stuff in exchange for commitments to a number of policy initiatives around federal climate policy" (Scott 2016).

At the municipal level, opposition to Energy East had also emerged. Municipal governments along the pipeline routes debated whether to support the Energy East project, mainly because of concerns about regional safety. For example, the city council of Thunder Bay, Ontario, after two years of divisive debate, decided to support Energy East in May 2017 over concerns about railcar safety (Vis 2017). In contrast, the city of Edmundston, New Brunswick, was concerned about similar issues and eventually took a position against the current pipeline route. The city council was not explicitly opposed to TransCanada's pipeline as long as an alternative route that avoided the regional watershed could be agreed on (Poitras 2016).

Municipal opposition to the project was mostly concentrated in Quebec. Many municipalities in La Belle Province, including four of the five largest cities, had expressed formal opposition to the project. Three of those cities are among the 82 municipalities of the Montreal Metropolitan Community (MMC), which had unanimously decided to stand against the Energy East Pipeline. The MMC decided, based on the results of a public consultation process undertaken in 2015, that the risks of the project outweighed the benefits for the region and that it lacked support from the population. Although the Trans Mountain case discussed in chapter 5 demonstrates that municipal governments don't have the legal capacity to interfere with a cross-provincial pipeline, this concerted opposition from local elected officials sparked passionate reactions in western Canada and raised doubts about the project's political acceptance in Quebec and especially in the Montreal area.

Ideas

Economy, Environment, and Security

Ideas enter the political process by a combination of causal and principled beliefs (Goldstein and Keohane 1993). The oil sands coalition attempted to frame Energy East in terms of economic benefits (mostly job creation, tax revenues, and GDP growth), safety (comparing pipeline safety to the safety of transport by rail was well advised at the time of the Lac-Mégantic accident), and energy security. Former natural resources minister Joe Oliver's public address in support of Energy East summarizes this kind of discourse very well: "Initiatives like this could allow Canadian refineries to process more potentially lower priced Canadian oil, enhancing Canada's energy security and making our country less reliant on foreign oil" (Oliver quoted in Government of Canada 2013). TransCanada's early communication materials about the project emphasized how it would strengthen energy security by displacing significant quantities of Middle Eastern oil: "Eastern Canadian refineries currently import more than 700,000 bbl/d, or 86 per cent of their daily needs, from more expensive overseas sources including Saudi Arabia, Nigeria, Venezuela and Algeria" (quoted in Environmental Defence and Greenpeace 2014).

Jason Kenney, then leader of the Conservative Party in Alberta, made the case for economic benefits and energy security when he commented on the TransCanada termination decision, saying, "It's an attack on our economy and on the energy sector. It's a devastating blow to Canada's economic future. Energy East represented to us an opportunity to make Canada energy independent. Instead, because of uncertainty created by the Trudeau government and the National Energy Board, this decision to cancel Energy East means that Eastern and much of Central Canada will be importing conflict oil through tankers from dictatorships, some of which fund terrorism, rather than buying Canadian oil produced in Alberta and Saskatchewan" (Franklin 2017). Ethical oil champion Ezra Lavant went so far as to describe Energy East as "freedom oil" (Lavant 2013).

In response, the anti-pipeline coalition sought to reframe the pipeline debate in terms of environmental risks (spills and climate change) and challenged the economic and security benefits expected from the project. For example, a petition circulated by Montreal-based environmental NGO Équiterre framed the project as "all risk, no reward" and claimed that the project would not reduce gasoline prices; would not increase energy security, because 93% of the volume would be exported; presents a risk of spills to Quebecers that could threaten drinking water; and would "significantly exacerbate climate change" (Équiterre, n.d.).

Different worldviews and scales of values seem to separate the pro- and anti-pipeline coalitions. While the pipeline advocates perceive the economy and the security of supply as the key issues in favor of the Energy East Pipeline, environmentalists instead emphasize the local and global environmental threats associated with a pipeline project and oil-industry expansion in general. From their perspective, environmental security is more important and should prevail over industrial development and economic considerations.

"Nation Building"

Another key frame for Energy East, also present in the case of Northern Gateway and, to a lesser extent, Trans Mountain, was the pro-pipeline Canadian nation-building discourse. TransCanada has framed Energy East as a symbol of national unity and a project all Canadians should be proud of, as it would carry oil from western provinces to eastern consumers and benefit all Canadians economically.

According to former Alberta premier Alison Redford, "This is truly a nation-building project that will diversify our economy and create new jobs here in Alberta and across the country" (Government of Alberta 2013). Even prior to the announcement of Energy East, former New Brunswick premier Frank McKenna was dreaming of a "pipeline network extending from coast to coast." As he wrote, "This essential infrastructure project would be good for all regions of Canada. It would be an extraordinary catalyst for economic growth. It would be a powerful symbol of Canadian unity" (McKenna 2012). He even characterized the idea as a "bold project, national in scope," comparable to the nineteenth-century Canadian Pacific Railway. David Alward, who was New Brunswick premier from 2010 to 2014, used similar rhetoric to praise a west–east oil pipeline. He argued that such an oil transportation project was "as important to our nation's economic future as the railway was to our past" (Alward quoted in Tedesco 2013). TransCanada's president and CEO, Russ Girling, also compared the Energy East Pipeline to the Canadian Pacific Railway and the Trans-Canada Highway at the public announcement of the project in 2013, saying, "Each of these enterprises required innovative thinking and a strong belief that building critical infrastructure ties our country together, making it stronger and more in control of our own destiny, and this is true of Energy East" (Girling quoted in McCarthy and Jones 2013).

This nation-building frame became particularly salient following the controversial decision made by the MMC to declare its opposition to the Energy East Pipeline, which had inflamed tensions with western members of the oil sands coalition. Montreal mayor Denis Coderre's statement that the MMC stands against Energy East "because it still represents significant environmental threats and too few economic benefits for greater Montreal" (Coderre quoted in CBC 2016b) sparked reactions from many politicians in western Canada and other public figures. Saskatchewan premier Brad Wall reacted this way: "This is a sad day for our country when leaders from a province that benefits from being part of Canada can be this parochial about a project that would benefit all of Canada, including these Quebec municipalities" (Wall quoted in De Souza 2016a). Wall also suggested that there should be a quid pro quo for the interprovincial transfers Quebec receives. In return for the equalization payments (C$10 billion for Quebec in 2015 from "have-provinces"), western Canada should be able to expect "Quebec municipal leaders [to] respond with generous support for a pipeline that supports the very sector that has supported them" (Wall quoted in De Souza 2016a).

CBC television host Rick Mercer also used similar rhetoric to back the Energy East Pipeline, arguing that "we all need this thing [Energy East Pipeline]. It is time for provinces to start asking 'what's in it for Canada?' not just 'what's in it for me?'" (Mercer 2016). Interim Conservative leader Rona Ambrose argued that the debate over Energy East "is affecting national unity," and she also labeled the pipeline a "nation-building project" (Ambrose quoted in Fekete 2016).

Framing Energy East in terms of nation building and Canadian unity had inherent limits, however. First, it was based on an inaccurate image of Energy East as an "all-Canadian pipeline." In fact, only a fraction of Alberta's crude oil flowing through the Energy East Pipeline would have been delivered to eastern refineries. Most of the production would have gone to Saint John's marine terminal for export. As Louis Bergeron, vice president of Energy East, stated during the Bureau d'audiences publiques sur l'environnement Québec (Bureau of Environmental Public Hearings, BAPE) hearings in Quebec, out of the 1.1 million barrels of oil expected to be carried every day by Energy East, 50,000 would have been refined in New Brunswick and between 100,000 and 150,000 in Quebec (Shields 2016a). That leaves about 900,000 barrels per day of unrefined oil, or 80% of the pipeline's volume, flowing through Energy East that would have been destined for export markets, not Canada. In addition to the fact that the project was mostly about exports, the pipeline also was not planned to carry only Canadian oil. Energy East would have helped North Dakota get its oil extracted from the Bakken shale formation out to foreign markets. TransCanada planned to build a connected pipeline to carry up to 300,000 barrels per day from North Dakota through the Upland Pipeline, making this "all-Canadian pipeline" not so Canadian after all (Shields 2016b). Moreover, it is doubtful that this project would have reduced Canada's imports of foreign oil. As Irving Oil president Ian Whitcomb mentioned, Energy East would have helped diversify the company's assets, but it would have continued to import as much oil as it did at the time: "We will add Western Canadian crude to our portfolio as the economics dictate, but probably not at the expense of our Saudi barrels" (Whitcomb quoted in Cattaneo 2016). Finally, the surge in American oil production would mean imports from the United States, not OPEC nations, would dominate Canadian oil imports. Since 2014, imports of US oil have made up more than half of all Canadian imports (National Energy Board 2017b).

Those facts therefore challenge the Canadian oil for Canadians and ethical oil frames industry advocates tried to create. However, these were not the only challenges to the Canadian nation-building narrative promoted by the oil sands coalition. Two competing narratives have also emerged to challenge this core frame: Quebec's nationalist discourse and First Nations' anticolonial discourse.

The Canadian nation-building frame did not have the same resonance within Quebec, where Quebecers' nationalist sentiments are enormously influential in the province's political culture. As Gordon Laxer claims about Quebec, "Its provincial governments, whether federalist or sovereigntist, are put off by the kind of Canadian nationalist rhetoric invoked to support the Energy East line" (Laxer 2015, 177). In contrast to the Canadian nation-building rhetoric, a discourse about self-determination emerged among Quebec sovereigntist political parties. The Parti Quebecois framed the Energy East Pipeline as a nationalist issue. For Parti Quebecois Member of National Assembly Alain Therrien, "The rest of Canada treats us as a colony. They are playing a dangerous game. Québécois don't like being told what to do. It increases nationalist sentiment" (Therrien quoted in Lefebvre 2016). Similarly, the pro-independence Bloc Québécois, the third-largest opposition party in the House of Commons in 2016, framed the Energy East Pipeline debate as an issue of self-determination and territorial sovereignty. The party launched an anti-pipeline campaign with internet ads such as "Énergie Est: C'est à nous de dire non!" (Energy East, it is up to us to say no!) or "Oléoduc de TransCanada: Notre environnement, nos décisions" (TransCanada's pipeline: Our environment, our decisions) (Bloc Québécois 2016, our translations). As Laxer argues, Quebec's "sovereignty narrative runs counter to the appeal to Canadian sovereignty on energy" (Laxer 2015, 177).

But this Quebec nationalism narrative also stretched beyond the traditional sovereigntist base within civil society. The campaign "Coule pas chez nous" against Energy East appealed more broadly to the nationalist sentiment from Quebec's Quiet Revolution. As Erick Lachapelle argued, "This 'chez nous' discourse is a highly effective, powerful slogan that appeals to Quebec values of solidarity and identity, regardless of where one stands on the sovereignty question" (Lachapelle quoted in Valiante 2016). Environmentalists in British Columbia chanting "our coast" appealed to a similar urge for place-based cherished values. But in Quebec similar framing of self-determination has even deeper resonance within the political culture, and the anti-pipeline coalition made skillful use of it.

While the Canadian nation-building frame was at odds with Quebec's nationalism, it was also anathema to Indigenous groups struggling to overcome the legacy of colonialism. For Indigenous groups, it is their people's land and governance that constitute their nation. Explicit appeals to Indigenous nations or the desire for self-determination were explicit in many of the statements of Indigenous groups when expressing their opposition to Energy East. Assembly of First Nations of Quebec and Labrador Regional Chief Ghislain Picard stated that their members "will continue to oppose and fight projects which pose a real danger to their Nations, whether it be

Tar Sands pipelines like Energy East, Tar Sands rail projects like at the Belledune port, uranium mining or offshore oil and gas drilling in the Gulf of Saint Lawrence" (Assembly of First Nations of Quebec and Labrador 2016). Byron Williams and Jared Wheeler of the Assembly of Manitoba Chiefs (AMC) stated their position this way:

> Throughout its active participation in the recent federal reviews of environmental assessment processes, including the modernization of the NEB, AMC has recommended that Settlers reconsider their understanding that Mother Earth can be managed and owned. . . . Mother Earth must be respected. She is a living being, not a "thing" that can be owned or sold. . . . There is an opportunity for the NEB to amend the Draft List of Issues in the Energy East hearing to include the issues of central importance to the nation to nation relationship. Correcting this list now would be an indication that the NEB is committed to its role in the honouring of the nation to nation between Canada and Indigenous peoples and nations. (Williams and Wheeler 2017)

Nowhere was this message clearer than in the text of the Treaty Alliance against Tar Sands Expansion. The preamble reads:

> We have inhabited, protected and governed our territories according to our respective laws and traditions since time immemorial. Sovereign Indigenous Nations entered into solemn treaties with European powers and their successors but Indigenous Nations have an even longer history of treaty making among themselves. Many such treaties between Indigenous Nations concern peace and friendship and the protection of Mother Earth. . . .
>
> As sovereign Indigenous Nations, we enter this treaty pursuant to our inherent legal authority and responsibility to protect our respective territories and coast in connection with the expansion of the production of the Alberta Tar Sands, including for the transport of such expanded production, whether by pipeline, rail or tankers. (Treaty Alliance 2016a)

While the Canadian nation-building frame may have been a resonant concept within the oil sands coalition, it bred alienation among core constituencies from whom the project needed to garner social and political acceptance.

Media Issue-Mention Analysis

Once the formal application for Energy East started in 2014, the project received far more attention in the Canadian media than the other pipelines addressed in the previous chapters, as shown in figure 7.2.[2] By 2015, Canadian media attention to Energy East was greater than the second most

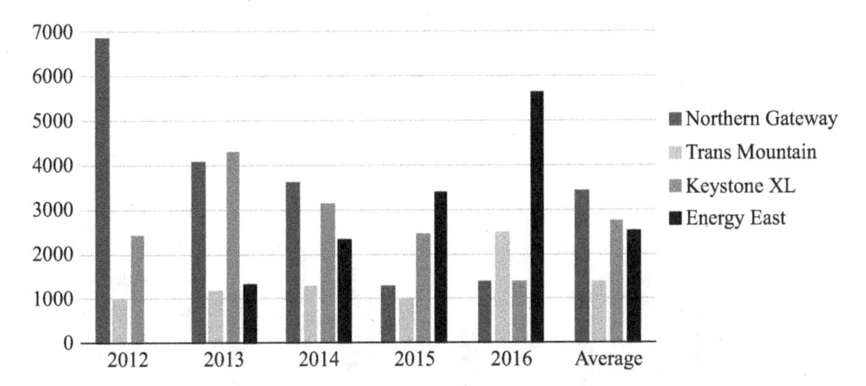

Figure 7.2
Total media mentions of Energy East each year in English Canada, 2012–2016.

mentioned project (Keystone XL), and by 2016, attention to Energy East was more than twice that of its closest competitor (Trans Mountain). No doubt, the pipeline's route through five provinces helps explain the high degree of media attention. The controversy in Ontario and Quebec, Canada's two largest provinces and the seats of the most media, certainly contributed to that attention.

Figure 7.3 analyzes which issues got the most attention from the media from 2013 to 2016.[3] When the Energy East proposal was first announced in August 2013, the attention was squarely on jobs and the economy. In 2013, economic themes were mentioned twice as often as the next closest category—accidents and spills. Climate concerns was the category mentioned least. By the following year, however, climate concerns had nearly doubled as a percentage of mentions and had become increasingly competitive with jobs and the economy. By 2016, climate concerns exceeded jobs and the economy as a percentage of mentions. This trend indicates the enormous success of the counterframing conducted by environmental activists.

Consistent with coverage of the other pipeline cases, media reporting on Energy East raises questions about the relative importance of local versus climate risks in discourse about the pipeline, at least as represented in media reports. While accidents and spills got more mentions than climate in 2013, when the controversy really heated up in 2014, climate mentions exceeded those of local environmental concerns. This shift can be credited in large part to a change in emphasis in environmental discourse resulting from the increased role played by 350.org, an international group focused on climate.

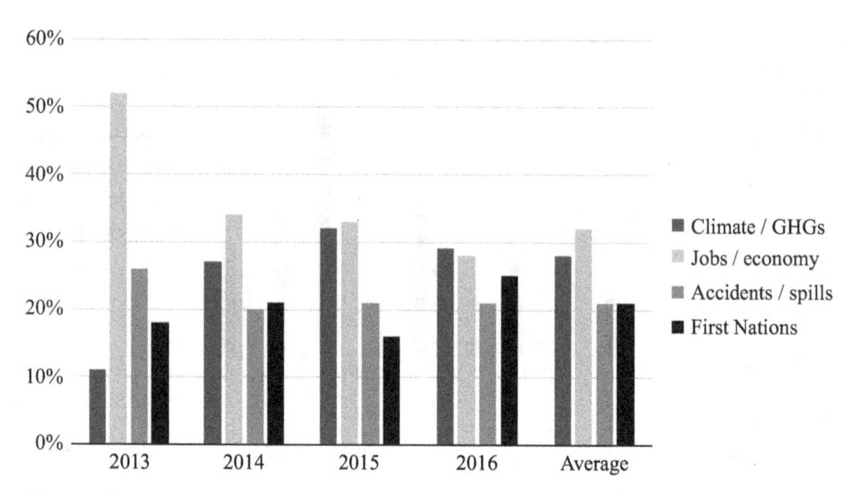

Figure 7.3
Aggregated issue mentions of the Energy East Pipeline in French and English in Canadian media, 2013–2016.

Prior to May 2014, 350.org did not have paid staff in Canada. That changed when Cam Fenton was hired by 350.org to focus on Energy East. From the start, Fenton's strategic goal was to "make the pipeline a climate pipeline" by advocating the adoption of a "climate test" in the pipeline review and approval process (Fenton 2017a). That campaign is described in more detail below.

Figure 7.4 compares the category of issue mentions of Energy East in English- and French-language media in Canada. The differences in the relative balance of economic and environmental concerns are striking. While in the English media jobs was the dominant category over the four years, in the French-language media climate concerns dominated the themes discussed by the media. This result provides another indication of the distinctively strong environmental values in Quebec and the merits and success of the anti-pipeline coalition focusing on resistance to the project in French Quebec.

Public Opinion

Nationally, public opinion consistently showed support for the Energy East Pipeline, but the project also strongly divided the country at the regional level. According to a March 2016 poll by Forum Research (see figure 7.5), 55% of Canadians approved of the project, compared to 32% who disapproved of it. If we look by province, however, the picture is quite different. On one side, approval exceeded opposition in all provinces except Quebec.

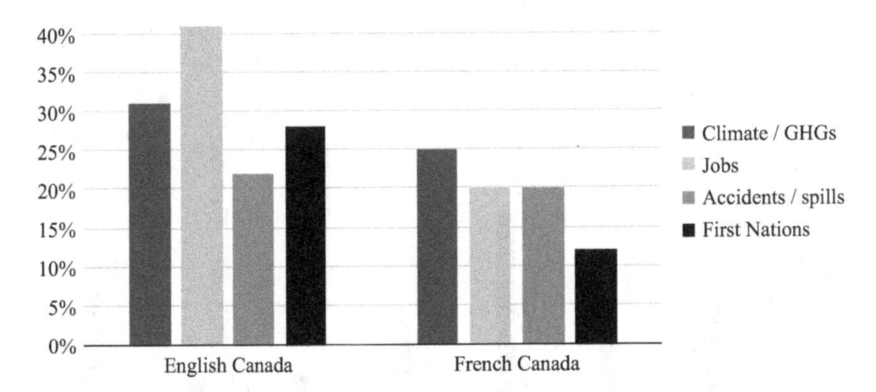

Figure 7.4
Comparison of issue mentions between French- and English-language Canadian media.

Support reached 82% in Alberta and 71% in Saskatchewan, Manitoba, and the Atlantic provinces. There was somewhat less support in Ontario and British Columbia, but supporters still outnumbered opponents (in Ontario, 57% supporting and 28% opposing; in British Columbia, 45% supporting and 39% opposing). On the other side, Quebec was the only province where opposition was higher than support. Only 38% of respondents there supported the project, while 50% stood against it.

Some regional polls seem to confirm this provincial divide. An online survey conducted by CROP-L'Actualité in December 2014 showed similar results: 35% of Quebecers supported the project, whereas 49% opposed it (Castonguay 2015). A poll conducted by Erik Lachapelle in 2015 showed that only 31% of the respondents from Quebec supported Energy East, whereas 51% were against it. In contrast, 71% of Alberta's population supported the pipeline, while opposition reached a mere 15% (Baril 2015).

Institutions and the Politics of Structure

As in the other pipeline cases, much of the conflict over Energy East was a battle over the rules of the game. There was disagreement over the role of provincial governments, especially the Quebec government, as always highly protective of its jurisdiction. Municipal governments also came to play an important role in the Montreal region. Even more so than in the other Canadian cases, the credibility and legitimacy of the regulatory review process led by the National Energy Board came under relentless attack. Some of the

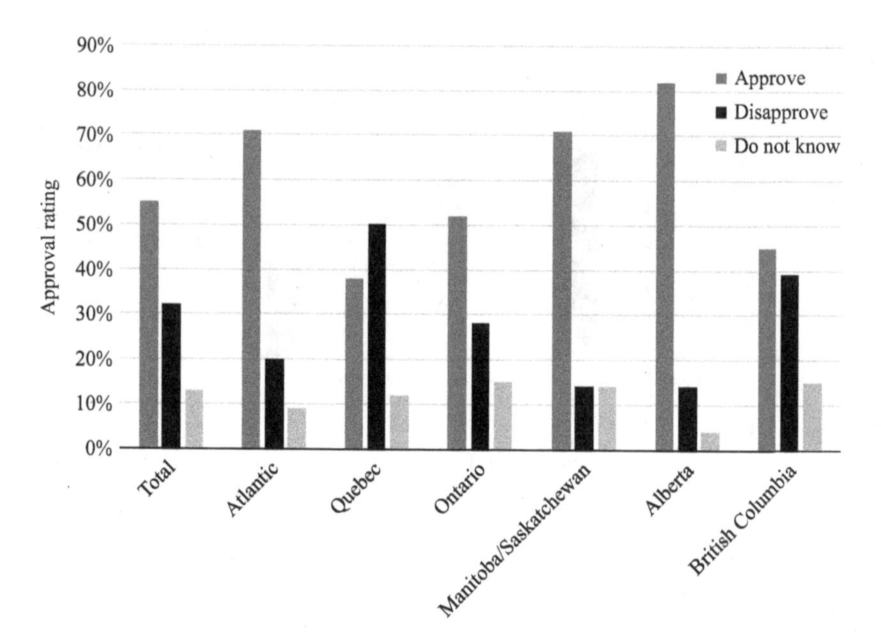

Figure 7.5
Forum Research poll of opinions on the Energy East Pipeline (March 2016).
Source: Forum Research (2016).

wounds were self-inflicted, but environmental opponents mobilized a concerted and remarkably successful campaign to discredit the regulator. Opponents also took to the courts early in this case, blocking one of the two export terminals and forcing Energy East to substantially amend its proposed project.

Assertions of Provincial Authority in the Absence of an Intergovernmental Equivalency Agreement

Unlike the West Coast pipelines, there was no intergovernmental agreement dividing labor or establishing equivalency between the federal government and the provinces. Both Quebec and Ontario chose to conduct their own environmental reviews. In Quebec, the authorization process for interprovincial pipelines is subject to the Environmental Quality Act and includes an assessment and review procedure as well as public hearings under the auspice of BAPE. The provincial government sent TransCanada a letter in late 2014 informing the company that it was expected to comply with provincial laws and undergo a provincial assessment (Government of Quebec 2016a).

TransCanada was conflicted about how to respond, and their position vacillated as events unfolded. On the one hand, they wanted to defend federal supremacy on pipeline regulation and avoid what they considered unnecessary duplication and delay. On the other hand, they were justifiably concerned that formal political opposition from the province of Quebec could doom the project in the federal cabinet. Initially, TransCanada refused to comply, insisting that the supremacy of federal jurisdiction over pipelines means that the company was not required to comply with provincial review requirements.

In response, Quebec initiated a generic review of the project by the BAPE in January 2016 that could proceed even without the company's participation. Before the hearings could begin, however, the province altered course somewhat and, in March 2016, Quebec environment minister David Heurtel filed for an injunction to force TransCanada to comply with the provincial environmental assessment process. Quebec's request was filed two weeks after a coalition of environmental groups filed for an injunction calling for the same thing. A few weeks later, TransCanada, sensing escalating opposition within the province, decided to comply with Quebec's environmental assessment procedures and provide a distinct environmental impact study for Quebec's portion of the pipeline (Government of Quebec 2016b). In return, Quebec agreed to withdraw its injunction (Gralnick 2016). Once the project proposal and environmental impact study were submitted by TransCanada, the provincial environmental review process was taken over by Quebec's Ministry for the Environment (officially the Ministère du Développement durable, Environnement et de la Lutte contre les changements climatiques—MDDELCC). On September 14, 2017, the MDDELCC also suspended its evaluation process for Energy East at the request of TransCanada (Croteau 2017).

Legal scholars have disagreed about the role of provincial government in interprovincial pipelines. In reaction to the seven conditions imposed by Quebec and Ontario, Dwight Newman, professor of law at the University of Saskatchewan and senior fellow at the Macdonald-Laurier Institute, wrote an op-ed in the *Globe and Mail* arguing that "premiers have no constitutional basis to be making conditions, demands or anything else on this pipeline, and they play a dangerous game in attempting to do so. . . . Provinces that want their own provincial jurisdiction respected play a very dangerous game if they do not respect well-established principles of the constitutional division of powers. If Quebec is not willing to respect the clear jurisdiction of the federal government, why should the federal government respect tomorrow the clear provincial jurisdiction of Quebec?" (Newman 2014).

Quebec constitutional scholars disagreed, and reacted to Newman's article by arguing that provinces can impose conditions and sanctions in order to protect the environment, health, and security of their populations even for projects falling under federal jurisdiction (Robitaille et al. 2014). As described in chapter 5, the April 2016 *Coastal First Nations v. British Columbia* decision by the Federal Court of Appeal involving the equivalency agreement about Northern Gateway came to the somewhat confusing conclusion that provinces have the responsibility to review interprovincial pipelines and could attach additional conditions for their approval but could not take any action that was tantamount to rejecting the pipeline (Olszynski 2018). The British Columbia reference case described in chapter 6 clarified the supremacy of federal power over interprovincial pipeline decisions. Provincial authority does not constitute a formal, authoritative veto point. Instead, provincial influence is political. Quebec's position was particularly pivotal to the Trudeau government. Trudeau's electoral support in Quebec was considered essential to his governing majority and could have collapsed if he had approved the pipeline against the will of the Quebec government and a majority of its population.

Municipal Opposition

Municipal governments have less legal leverage than provinces to block a transprovincial pipeline, as the court cases involving Burnaby and the Trans Mountain pipeline reinforced (chapter 6; Olszynski 2018). But the concerted opposition from the Montreal region's mayors threatened the political viability of Energy East. For instance, former Conservative interim leader Rona Ambrose expressed skepticism over TransCanada's ability to build Energy East given the opposition from the MMC. She stated, "I don't see Energy East getting through Montreal, and that's just the political reality of it," adding, "If you think that Denis Coderre in Montreal and the groups that have gotten organized against Energy East are going to give the government social license to go through Montreal, I don't believe it. I'd like to be more optimistic" (Ambrose quoted in Scotti 2017). Reactions to the MMC's anti-pipeline position from western Canadian politicians such as Brad Wall, Brian Jean, and Rachel Notley also illustrate how seriously they considered this opposition from local governments to be. Despite the Montreal region's lack of legal power to veto the Energy East Pipeline, its political power carries significant weight.

Legal Challenge over the Deepwater Terminal in Cacouna, Quebec

Environmentalists were very active in using the courts to attack TransCanada's Energy East proposal. The first critical target was TransCanada's plan to build a deepwater port in Cacouna, along the shore of the St. Lawrence River in Quebec. TransCanada wanted a deepwater port in Quebec for two reasons: it would diversify the points of exportation on Canada's east coast and provide additional economic benefits to Quebec, which TransCanada hoped would foster greater project acceptance in the French-speaking province. However, the choice of Cacouna quickly became a flashpoint for the opposition. The area is known as a nursery for beluga whales living in the St. Lawrence estuary and gulf and therefore represents a highly sensitive habitat for the reproduction of this endangered marine mammal. This created an inviting target for anti-pipeline activists, who then set out to "beat them up with belugas" (Stewart 2016).

Only a few months after TransCanada launched the Energy East project in 2013, environmental groups mobilized against the terminal. Local and regional groups such as the Coalition Bas-Saint-Laurent pour une prospérité sans pétrole (Prosperity without Oil Bas Saint-Laurent Coalition) and the Mouvement Stop Oléoduc (Stop Pipeline Movement) organized protests, marches, and information campaigns in order to prevent TransCanada from conducting seismic surveys in the region that were required in order to advance the proposal for the terminal. They claimed that the drilling would disturb belugas' natural reproductive cycle and harm the already declining population (estimated at 900 in 2012 according to Fisheries and Oceans Canada) (Government of Canada 2017b).

On March 5, 2014, TransCanada announced it would "voluntarily" participate in Quebec's environmental evaluation process for the deepwater port in Cacouna (but not for the entire portion of the pipeline crossing Quebec) while at the same time asserting that the decision to authorize the project fell in the hands of the federal government. On August 21, 2014, Quebec authorized TransCanada to start underwater drilling operations and geotechnical surveys in Cacouna. Five days later, a request for an injunction was filed at the Superior Court of Quebec to stop drilling operations in Cacouna. The request was cowritten by the Centre québécois du droit de l'environnement, David Suzuki Foundation, Nature Quebec, the Canadian Parks and Wilderness Society Quebec, France Dionne, and Pierre Béland (Greenpeace and WWF-Canada also backed the injunction). On September 1, 2014, the superior court rejected the request for a "safeguard order," arguing that it is "alarmist" to affirm that belugas will suffer irreparable harm from drilling activities. In turn, superior court judge Claudine Roy argued that delays in the drilling works would cause "economic harm" to TransCanada (Cour Supérieure du Québec 2014a).

The environmental organizations came back to the court with additional scientific arguments and two weeks later, on September 23, were successful at convincing the court to impose an interlocutory injunction on drilling activities in Cacouna until October 15, when the beluga spawning period would be completed. Judge Claudine Roy found in her judgment that Quebec's MDDELCC, despite spending two months analyzing potential harms on belugas from drilling, could not confirm that mitigation measures proposed by TransCanada would have been sufficient to prevent harm to belugas. The judge also found that Quebec's minister of the environment sought advice from Canada's Department of Fisheries and Oceans (DFO) but that it decided to authorize testing despite not getting a response from DFO. The Superior Court therefore ruled that the applicants were successful in "raising a serious doubt as to the reasonableness of the Minister's decision" and imposed the temporary injunction (Cour Supérieure du Québec 2014b). The ban on drilling was extended beyond October 15 because Quebec's Environment Ministry then refused to issue a permit.

In a coincidental development that elevated the salience of beluga vulnerability, the Committee on the Status of Endangered Wildlife in Canada (COSEWIC) recommended changing the status of St. Lawrence belugas from a "threatened" to an "endangered" species in November 2014, putting additional pressure on TransCanada and Quebec's government to back down on Cacouna's deepwater port proposal (COSEWIC 2016).

As legal, scientific, and political pressure to abandon the Cacouna terminal grew, on December 2, 2014, Premiers Jim Prentice of Alberta and Philippe Couillard of Quebec met to discuss the pipeline. Couillard suggested that TransCanada should seek "alternative sites" for its deepwater port. Prentice emphasized the nation-building benefits of a deepwater port in Quebec, saying, "I hope there will be discussions between the proponents and Premier Couillard about the other alternative locations" to Cacouna. He added, "It would seem to me that the project will be stronger as a piece of nation-building infrastructure if there is a port facility in the province of Quebec" (Prentice quoted in Van Praet and Perreaux 2014).

On April 2, 2015, TransCanada finally decided not to proceed with a marine terminal in Cacouna, citing concerns over belugas' safety in the St. Lawrence River. According to CEO Russ Girling, "This decision is the result of the recommended change in status of the Beluga whales to endangered and ongoing discussions we have had with communities and key stakeholders" (Girling quoted in Penty 2015). This decision was a major setback for the project for two reasons. First, it created significant delays in the project review and approval process because TransCanada would have to amend its application filed with the NEB (Morgan 2015). Second, it increased the political risks to the project by reducing the economic benefit flowing to Quebec. According to Greenpeace's Keith Stewart, "it shifted the risk-benefit calculation for Quebec even further into the 'all risk, no reward' frame" (Stewart 2016).

Delegitimizing the National Energy Board

Mobilizing political and legal pressure against Energy East was critical to the environmental strategy, but there was also a conscious political choice to attack the legitimacy of the regulatory process itself. As described in previous chapters, the NEB has generally been friendly to pipeline projects, but the Harper government chose to increase political control over the regulator by changing the law to transfer the final decision authority on interprovincial pipelines to the federal cabinet. One effect of this change was to politicize the regulatory process further, and in the confrontational energy politics of the 2010s, environmental groups responded by attacking the board and its procedures. The core of the campaign, organized by 350.org, LeadNow, the Council of Canadians, and Greenpeace, was a petition for a "People's Intervention" demanding that the NEB consider climate impacts in its NEB review. The campaign also flooded the hearing process with requests for participation to reiterate that claim. The online petition stated:

> The Energy East pipeline could move over 1 million barrels of tar sands from Alberta to the Atlantic Ocean each day—exporting bitumen and dangerous carbon emissions to the world.
>
> Without considering climate change and listening to people's voices, any review of Energy East will be incomplete and illegitimate. (350.org, n.d.)

The strategy was described by Cam Fenton from 350.org this way:

> The general public didn't see the regulator as illegitimate, but the movement did. This was around the time that Obama had clearly stated that climate will be the measure [on Keystone XL, see chapter 4]. So the US is judging our pipelines on climate, but we are not. It seemed like this big missing piece. It seemed like the NEB was theoretically an arm's length regulator that could make its own determinations. We basically could target this institution and put it in a decision dilemma whereby it either has to include climate change or declare it won't, and further delegitimize itself in public. (Fenton 2017a)

The credibility of the pipeline regulation process was also a major issue in the federal election campaign of 2015. Environmentalists began "bird-dogging" candidates, especially Liberal leader Justin Trudeau, demanding a commitment to a climate test. But the Trudeau camp also saw electoral advantages in criticizing the Harper government's regulatory record. While the Harper government defended the process, Justin Trudeau's Liberal Party platform called for a review of the NEB to "modernize and rebuild trust in the National Energy Board" and changes to the environmental assessment process to "ensure it includes an analysis of upstream impacts and the greenhouse gas emissions" (Liberal Party of Canada 2015b). The New Democratic Party pledged similar changes.

Already tarnished heading into the election, both Harper and Energy East managed to aggravate the NEB's credibility crisis. Harper's appointment of Steven Kelly to the NEB, two days before the start of the 2015 federal electoral campaign that would lead to the Conservative Party's defeat, certainly did not help improve the board's reputation for independence from the industry it regulated. As vice president of IHS Global Canada Ltd., Kelly was hired by Kinder Morgan to defend the economic benefits of the Trans Mountain pipeline expansion in front of the NEB (Cattaneo 2015).

Like the three other pipeline cases, the Energy East case strongly supports the third behavioral hypothesis. Competing interests chose venues that gave them the best chance of success. The anti-pipeline coalition took advantage of the absence of an equivalency agreement to push resistance by provinces, especially Quebec, and politically powerful municipalities. TransCanada fought as hard as it could to keep the review under exclusive federal control but in the end was overwhelmed by the reality of Quebec's power in the federation.

The NEB Review Process

TransCanada announced its intention to proceed with the Energy East proposal in August 2013. Its formal application was submitted to the NEB in October 2014. As a result of the setback with the Cacouna marine terminal, however, the proposal had to be revised. It was resubmitted in December 2015, but the NEB sent the proposal back to the company for revisions. Three months after the Liberals' victory in the federal election campaign, Trudeau's government announced interim principles that would guide both environmental assessment processes and the NEB's ongoing pipeline reviews (described in chapter 6). The government also made it clear that no project would have to start from scratch.

Specifically, three interim measures were identified in regard to the Energy East Pipeline review process:

1. Undertake deeper consultations with Indigenous peoples potentially affected by the project and provide funding to support these consultations;
2. Help facilitate expanded public input into the National Energy Board review process, including public and community engagement activities. The Minister of Natural Resources intends to recommend the appointment of three temporary members to the National Energy Board;
3. Assess the upstream greenhouse gas emissions associated with this project and make this information public. (Government of Canada 2016a)

The NEB officially launched the review process for the Energy East Pipeline on June 16, 2016. It was then supposed to complete its review and submit its report within a 21-month period (National Energy Board 2016e). TransCanada ran into conflicts over its reluctance to translate the application document into French, prompting a legal challenge from Quebec activists. As the company struggled to manage that issue, the NEB process came off the rails as the board descended into crisis in summer 2016. Ethical concerns over the board members, legal challenges, and disruptive protests of the hearings in Montreal led to the cancellation and subsequent reconstitution of the review process and eventually to TransCanada's decision to terminate the project.

Perhaps the biggest blow to the legitimacy of the NEB was the tenacious investigative journalism of Mike De Souza of the *National Observer*. In July 2016, De Souza revealed that in January 2015 former Quebec premier Jean Charest had a private meeting with NEB members reviewing the Energy

East Pipeline (Poitras 2018, 185–189). The NEB confirmed that the meeting occurred, but at first defended itself by arguing that the board members explicitly warned Charest, then a lobbyist for TransCanada, that they could not discuss the Energy East project. The NEB also argued that the board members were not aware of "any contracted work that Charest may have had." "The meeting was set up . . . with the purpose of asking Mr. Charest for his thoughts on how the NEB could effectively engage in Quebec and which stakeholders the NEB might consider meeting," said NEB spokesman Craig Loewen in a statement (De Souza 2016b). Present at the meeting were NEB president Peter Watson and two of the commissioners on the panel reviewing the Energy East proposal, Jacques Gauthier and Lyne Mercier.

In August, an Access to Information Act request pried loose from the NEB the records on the meeting with Charest. Those records revealed that Energy East had in fact been discussed during the meeting, and Charest had given strategic advice for the pipeline's acceptance in Quebec. The NEB was forced to apologize, acknowledging that, "Our response did not accurately reflect the meeting" (De Souza 2016c). These media revelations aggravated the NEB's legitimacy problem just as Energy East hearings were about to begin. In August, two separate requests for motions were filed by environmental groups requesting that the NEB members who participated in the meeting with Charest recuse themselves from the Energy East review (De Souza 2016d).[4] On August 25, former Montreal mayor Denis Coderre added his voice to the growing protest movement and asked the federal government to suspend the NEB's public hearings on Energy East. The Montreal mayor had expressed doubts about the impartiality of the process since the revelations about the private meeting held with Jean Charest (Champagne 2016).

The crisis was heightened when, on August 29, NEB's public hearings held in Montreal were disrupted by a group of protesters. Holding a bilingual banner quoting Justin Trudeau as saying "only communities grant permission," the protesters chanted, "TransCanada on n'en veut pas, ONÉ congédié" (TransCanada we don't want it, NEB fired). The NEB chose to cancel that day's hearing and the one scheduled for the following day as well. "This decision was made in light of a violent disruption in the hearing room this morning which threatened the security of everyone involved in the panel session," according to the NEB (National Energy Board quoted in Shingler and Smith 2016). Natural Resources Minister Jim Carr expressed his concerns about the events as he mentioned that, "Not everyone's going to agree. But everyone should have a right to express themselves, and that's a fundamental Canadian value" (Carr quoted in Shingler and Smith 2016).

A day later, on August 30, the NEB decided to suspend the public hearings on Energy East in the midst of a credibility crisis. On September 9, the NEB panel members finally stepped down and the Energy East hearings were halted until a new panel could be appointed. The NEB said that "all three panel members have decided to recuse themselves in order to preserve the integrity of the National Energy Board and of the Energy East and Eastern Mainline Review. The members acted in good faith" (Marandola 2016). Chair Peter Watson and vice chair Lyne Mercier also recused themselves from administrative duties.

In an effort to salvage the Energy East review process, the federal government appointed four temporary members to the NEB on October 20 (Government of Canada 2016b), but by that point the controversy had strengthened public opposition in Quebec. On November 7, 2016, an SOM poll commissioned by Équiterre, the Coule pas chez nous foundation, the David Suzuki Foundation, Greenpeace Canada, Nature Québec, and the Regroupement vigilance hydrocarbures Québec showed that "a strong majority of Quebeckers (73%) have completely lost trust in the National Energy Board (NEB) following the Charest affair. A strong majority (89%) believe that before the evaluation of the Energy East project starts again the federal government needs to completely reform the environmental assessment process" (Greenpeace Canada 2016). It didn't help the credibility of the NEB review of Energy East when Trudeau's natural resources minister, Jim Carr, announced the creation of the NEB modernization panel on November 8. The very title of the process implied that the NEB was flawed, yet the government was hoping that the Energy East review could still proceed. On January 9, 2017, the NEB announced three new NEB members to review the Energy East Pipeline (National Energy Board 2017a). The newly appointed panel then announced later in the month that it would start the NEB's public hearings from scratch, a decision that added further delays to the project (Snyder 2017).

On May 10, 2017, in a major breakthrough for the anti-pipeline coalition, the NEB announced that it might review upstream and downstream emissions from the Energy East Pipeline since the "Hearing Panel has specifically asked for feedback on this issue" (Government of Canada 2017a; National Energy Board [NEB] 2017c). TransCanada harshly criticized this decision. In a letter published by Blake, Cassels & Graydon LLP on behalf of

TransCanada, the authors asked for more time to reply to public comments and argued that "if the lists of issues for the Projects include the analyses of upstream GHG emissions such analyses will be completely redundant and unnecessary" because the government's interim measures already established that Environment and Climate Change Canada would be doing that analysis (Blake, Cassels & Graydon LLP 2017, 4).

The NEB's credibility crisis deepened when the report of the expert panel on NEB modernization was released in May 2017 and declared a "crisis of confidence" in the NEB and recommended deep reforms, including the creation of new regulatory and information organizations to replace the NEB. A month later, Ontario and Quebec abandoned their formal wait and see positions and asked Jim Carr to stop the Energy East review until the NEB reform process was completed. In a June 27, 2017, joint letter, ministers from Quebec and Ontario—Pierre Arcand and Glenn Thibault—wrote: "Quebec and Ontario recommend that an important outcome of NEB Modernization be a framework for the NEB to review major energy projects that is predictable and increases public confidence" and hope that "the process to modernize the NEB and any new rules that are announced by the federal government provide predictable treatment for major energy projects currently under NEB review" (Arcand and Thibault quoted in Goujard 2017). According to Quebec premier Philippe Couillard, "The process should be completely credible and stabilize the way reviews are done before starting discussions [on the Energy East project]" (Couillard quoted in Goujard 2017).

On August 23, the NEB finally confirmed that it would evaluate upstream and downstream GHG emissions for the Energy East Pipeline (National Energy Board 2017c). This represented a major shift from the NEB's previous approach, in October 2014, not to assess upstream greenhouse gas emissions or the effect of oil production and transport on climate change as part of the Energy East Pipeline review (Shields 2014). It was even an expansion over the Trudeau government's interim measures' commitment to review upstream emissions. This change represented a huge victory for environmental advocates in changing the scope of the regulatory review.

Two weeks later, on September 7, TransCanada requested a "30-day suspension of the Energy East Pipeline and Eastern Mainline Project applications . . . to conduct careful review of recent changes announced by the NEB regarding the list of issues and environmental assessment factors of the project while understanding how these changes impact the projects' costs, schedules and viability" (TransCanada 2017a). In a public statement following TransCanada's request, the NEB announced that "the Board will not issue further decisions or take further process steps relating to the review of the projects" (National Energy Board 2017d). TransCanada also raised the possibility of canceling the project. The company stated: "Should TransCanada decide not to proceed with the projects after a thorough review of the impact of the NEB's amendments, the carrying value of its investment in the projects as well as its ability to recover development costs incurred to date would be negatively impacted" (TransCanada 2017a).

Who Killed Energy East? The Rhetorical Battle to Frame the Termination

On October 5, 2017, four years after it was first proposed and less than a year before it was originally scheduled to come into operation, TransCanada announced that it was terminating the Energy East project. In its media release, the company was notably terse in explaining the decision: "After careful review of changed circumstances, we will be informing the National Energy Board that we will no longer be proceeding with our Energy East and Eastern Mainline applications" (TransCanada 2017b). While the company has remained circumspect in its rationale for the decision, many other strategic actors have jumped into the fray to offer explanations for why the project was terminated.[5]

The Conservative opposition at the federal level blamed the decision on regulatory overreach by the Trudeau government. Conservative MP Lisa Raitt argued that "today is a result of the disastrous energy policies promoted by Justin Trudeau, and his failure to champion the Canadian energy sector. [Trudeau] forced Canadian oil companies to comply with standards that are not required for foreign companies. . . . Everything that Justin Trudeau touches becomes a nightmare" (Ballingall 2017).

Jason Kenney, leader of the United Conservatives in Alberta, took a similar stance, claiming that the decision was "because of uncertainty created by the Trudeau government and the National Energy Board" (Franklin 2017). Trudeau deflected these attacks, claiming that, "It's obvious that market conditions have changed." Natural Resources Minister Jim Carr went even further, characterizing the cancellation as "a business decision" and maintaining that "nothing has changed in the government's decision-making process. Canada is open for business" (Ballingall 2017).

Academics and activists joined the fray. Andrew Leach, an associate professor at the University of Alberta School of Business, also laid the blame squarely on market forces, especially the decline in oil prices and the associated reduction in forecast growth in the oil sands. But Leach also points to Donald Trump's revival of the Keystone XL pipeline, also a TransCanada project, which forced the company to choose which pipeline to lock in shipper's commitments to. Cam Fenton, a climate organizer with 350.org, agrees that changing rules for regulatory review explain TransCanada's decision, but unlike Conservative politicians, Fenton praises the change, saying, "Climate concerns killed Energy East. And that's a good thing" (Fenton 2017b). Climate organizer and Oxford graduate student Bronwen Tucker argued that while market forces were undeniably important, the anti-pipeline social movements played a huge role in Energy East's demise: "Delays won by Indigenous communities, grassroots groups, labour unions and NGOs prevented Energy East from being built when it was still economically and politically feasible" (Tucker 2017).

Conclusion

The regime framework developed here shows how these competing arguments about different drivers can all work together to explain the outcome. The political viability of a major infrastructure project is determined by the balance of power between project proponents and opponents. Power is a product of political resources and strategic actions, both of which can be strongly influenced by market forces, elections, and other background conditions. When it was first proposed, it seemed like TransCanada had the upper hand. An enthusiastic Conservative government held power in Ottawa. Oil prices were over US$100 per barrel, projections were for rapid growth in the oil sands, and alarm was growing within the oil sands coalition about pipeline shortages. Competing projects were struggling: the Keystone XL project had been rejected (the first time) by President Obama, and the Northern Gateway deathwatch was well under way.

Once the project was formally proposed, the resistance movement mobilized and proved to be tenacious and formidable. It used the courts to force delays and even blocked the Quebec terminal. The anti-pipeline coalition skillfully exploited TransCanada's missteps: the revelation of Edelman's heavy-handed strategy, the ill-advised choice of a beluga nursery for the Quebec terminal, the delays in producing French-language documents, and the decision to hire Charest as a lobbyist. The Charest controversy was particularly pivotal in the remarkable success of the anti-pipeline coalition at discrediting the entire NEB process and then expanding the scope of reviews to include upstream and eventually even downstream GHG emissions. Mayors throughout the Montreal region united in opposition, and the Quebec government asserted its jurisdiction to perform its own assessment and make its own decision.

As these points of resistance were emerging, the market changed profoundly. Oil prices crashed in the second half of 2014 and by January 2016 had fallen to less than one-third of what they were in 2013. Oil companies slashed their production forecasts—the Canadian Association of Petroleum Producers' 2017 forecast for pipeline capacity needs in 2030 fell 3 million barrels per day below what had been forecast four years earlier, three times the capacity of Energy East (Leach 2017). The Trudeau government approved competing pipelines, and in 2017 Trump revived Keystone XL.

In deciding whether to continue to pursue the project, TransCanada had to weigh the expected benefits of building and operating the pipeline against not only the costs of construction but also the increasingly costly and time-consuming review process, not to mention the probability of ultimately succeeding in getting government approval to build. In the rhetorical battle to frame who or what to blame for TransCanada's termination decision, it was predictable that Conservatives would blame unduly cumbersome regulatory processes and that Trudeau would deflect all the blame to apolitical market forces. The strategic actions of the anti-pipeline coalition unquestionably raised the process costs of the project, which helped tip the cost-benefit balance to the negative for TransCanada once market conditions had changed so dramatically. Pipeline opponents were proud to celebrate their success, because that is precisely what their relentless legal and procedural strategies were designed to do.

8 The Impact of Pipeline Resistance

Fall 2016 was a very busy time for the Trudeau government on the energy and climate front. On October 3, as environment ministers were meeting to hammer out final details of the Pan-Canadian Framework on Clean Growth and Climate Change, Prime Minister Trudeau rose in the House of Commons to announce the basic architecture of his climate plan. Saying, "We will not walk away from science, and we will not deny the unavoidable," Trudeau announced that the federal government would set a floor price for carbon pollution:

> The price will be set at a level that will help Canada reach its targets for greenhouse gas emissions, while providing businesses with greater stability and improved predictability.
>
> Provinces and territories will have a choice in how they implement this pricing. They can put a direct price on carbon pollution, or they can adopt a cap-and-trade system, with the expectation that it be stringent enough to meet or exceed the federal benchmark.
>
> The government proposes that the price on carbon pollution should start at a minimum of $10 per tonne in 2018, rising by $10 each year to $50 per tonne in 2022.
>
> Provinces and territories that choose cap-and-trade systems will need to decrease emissions in line to both Canada's target and to the reductions expected in jurisdictions that choose a price-based system.
>
> If neither price nor cap and trade is in place by 2018, the Government of Canada will implement a price in that jurisdiction. (Trudeau 2016a)

On November 29, 2016, Trudeau announced that he was rejecting the Northern Gateway Pipeline but approving the Trans Mountain Expansion Project and another project, Line 3 to the American Midwest (Trudeau 2016b). Just over a week later, on December 9, the first ministers formally endorsed the Pan-Canadian Framework on Clean Growth and Climate Change, although

Saskatchewan and Manitoba dissented (Canadian Intergovernmental Conference Secretariat 2016b).

These developments reflected the fruits of a well-organized and resourceful environmental campaign that had been more than a decade in the making. This chapter builds on the previous four pipeline case study chapters to address the first of this book's four core questions: has the strategy of place-based resistance to fossil fuel development been effective at promoting climate action and the reduction of global warming emissions? To address this question, the first section reviews the analytical framework introduced in the first chapter and then reviews and compares the four pipeline case studies to describe and explain the impact of the anti-pipeline campaigns on their proximate goal of stopping pipelines. The second section addresses the anti-pipeline campaigns' broader goal of promoting improved climate policy from governments. The final section reconsiders the state of the oil sands regime after a concerted political backlash to resistance in the latter part of the decade.

Impact of Anti-pipeline Campaigns on Pipeline Expansion

The theory about the influence of infrastructure resistance presented in chapter 1 contained four hypotheses on the relative power of project opponents:

1. The greater the placed-based risks in relation to local economic benefits, the more vulnerable the project is to resistance.
2. The more access opponents have to veto points, the more vulnerable the project is to resistance.
3. The more the project can take advantage of existing infrastructure, the less vulnerable it is to resistance.
4. The greater the geographical separation of risks and benefits, the more vulnerable the project is to resistance.

This framework helps explain the outcomes in our four cases.

With respect to the proximate goal of stopping pipelines, the campaign has been very successful. Of the four major oil sands pipelines targeted by environmentalists, the only one that is currently under construction is Trans Mountain. Keystone XL was canceled by Obama. Trump sought to reverse that decision, but it was then canceled again by Biden. Energy East was canceled by the proponent, frustrated by the relentless opposition, especially in Quebec. Northern Gateway was rejected by the Federal Court of Appeal after deep and sustained resistance and then terminated by the Trudeau government. Trans Mountain was approved by the Trudeau government in 2016 but then abandoned by the proponent, Kinder Morgan, because of political uncertainty created in large part by the place-based resistance against it. It was purchased by the government of Canada but was then blocked by the Federal Court of Appeal. It has now been reapproved by the Trudeau government, a decision that was upheld by the Federal Court of Appeal, and leave to appeal to the Supreme Court were denied. Limited pipeline construction in British Columbia began in spring 2020.

These pipeline controversies are not independent, discrete cases. They are interrelated in two ways. First, they all arise from the same imperative—the growth of the oil sands. That growth imperative gives the oil sands coalition the same strategic objective in each case— increasing market access—but it also gives the anti-pipeline coalition the same strategic objective in each case: constrain oil sands growth in order to limit its environmental consequences.

The second source of interdependence among the pipeline controversies is their relationship to each other in time. Scholars have long noted the importance of timing and sequence in policy agendas and actions (Pierson 2000; Pralle 2006b). Figure 8.1 shows the relationships in time between the four proposals. Northern Gateway was the first one formally proposed, but the first version of its proposal was withdrawn before it became a major controversy. Keystone XL was the first high-profile case. The oil sands coalition was banking on it, with Prime Minister Harper publicly declaring it a "no brainer." When Obama responded to the resistance strategy by delaying the project significantly in 2011, the oil sands coalition quickly ramped up pressure on the reproposed Northern Gateway. The sudden desperation for Northern Gateway is best indicated by Joe Oliver's infamous "foreign radicals" open letter.

As Northern Gateway then stalled in 2012, the Trans Mountain Expansion Project became the sector's new hope. When protests ramped up on Burnaby Mountain late in 2014, the oil sands coalition rapidly shifted to Energy East. Then, when it ran aground in Quebec in 2015, the coalition doubled down on Trans Mountain, at that point the only alternative that still seemed politically viable.

The analytical framework helps explain why Trans Mountain thus far has been more successful than three other pipelines. All four pipelines have salient, concentrated risks—the risk of tanker spills in the Northern

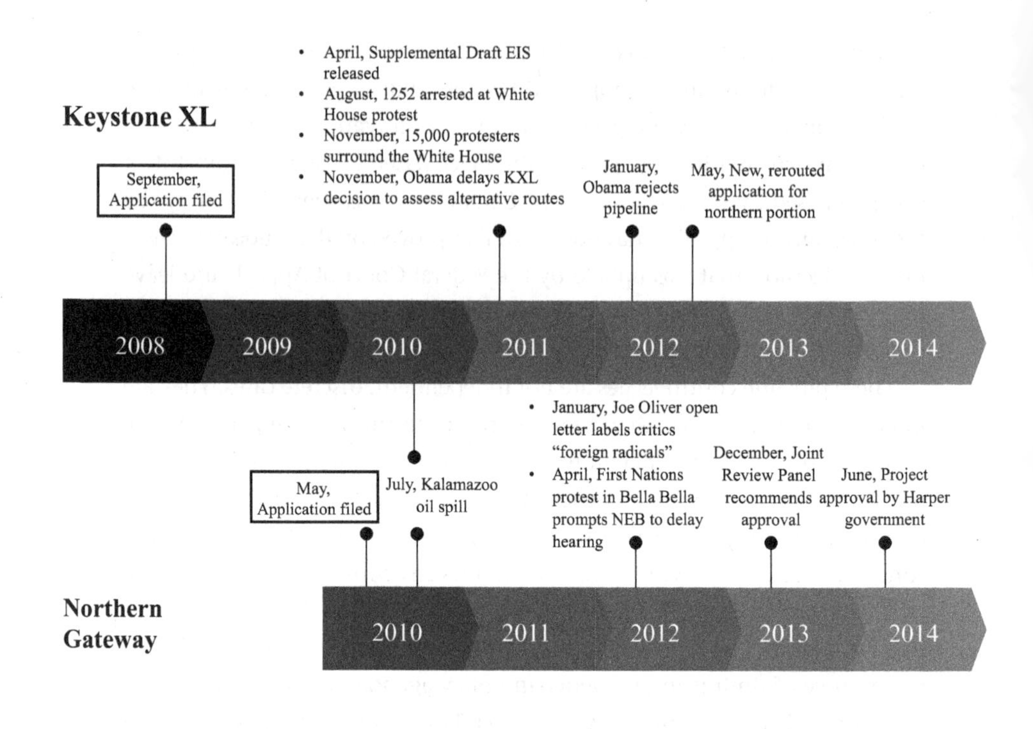
Keystone XL
September, Application filed
April, Supplemental Draft EIS released
August, 1252 arrested at White House protest
November, 15,000 protesters surround the White House
November, Obama delays KXL decision to assess alternative routes
January, Obama rejects pipeline
May, New, rerouted application for northern portion
2008
2009
2010
2011
2012
2013
2014
January, Joe Oliver open letter labels critics "foreign radicals"
April, First Nations protest in Bella Bella prompts NEB to delay hearing
May, Application filed
July, Kalamazoo oil spill
December, Joint Review Panel recommends approval
June, Project approval by Harper government
Northern Gateway
2010
2011
2012
2013
2014

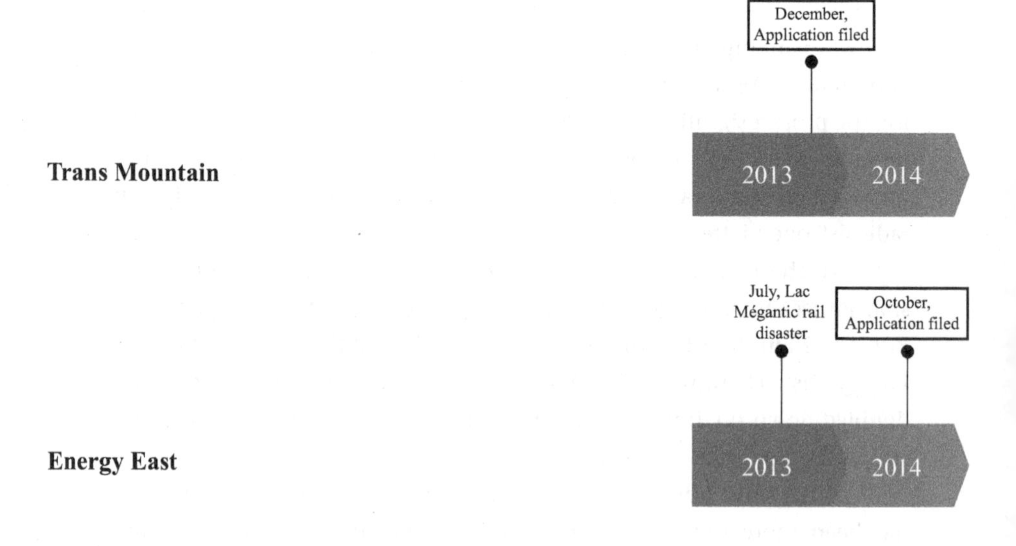
December, Application filed
Trans Mountain
2013
2014
July, Lac Mégantic rail disaster
October, Application filed
Energy East
2013
2014

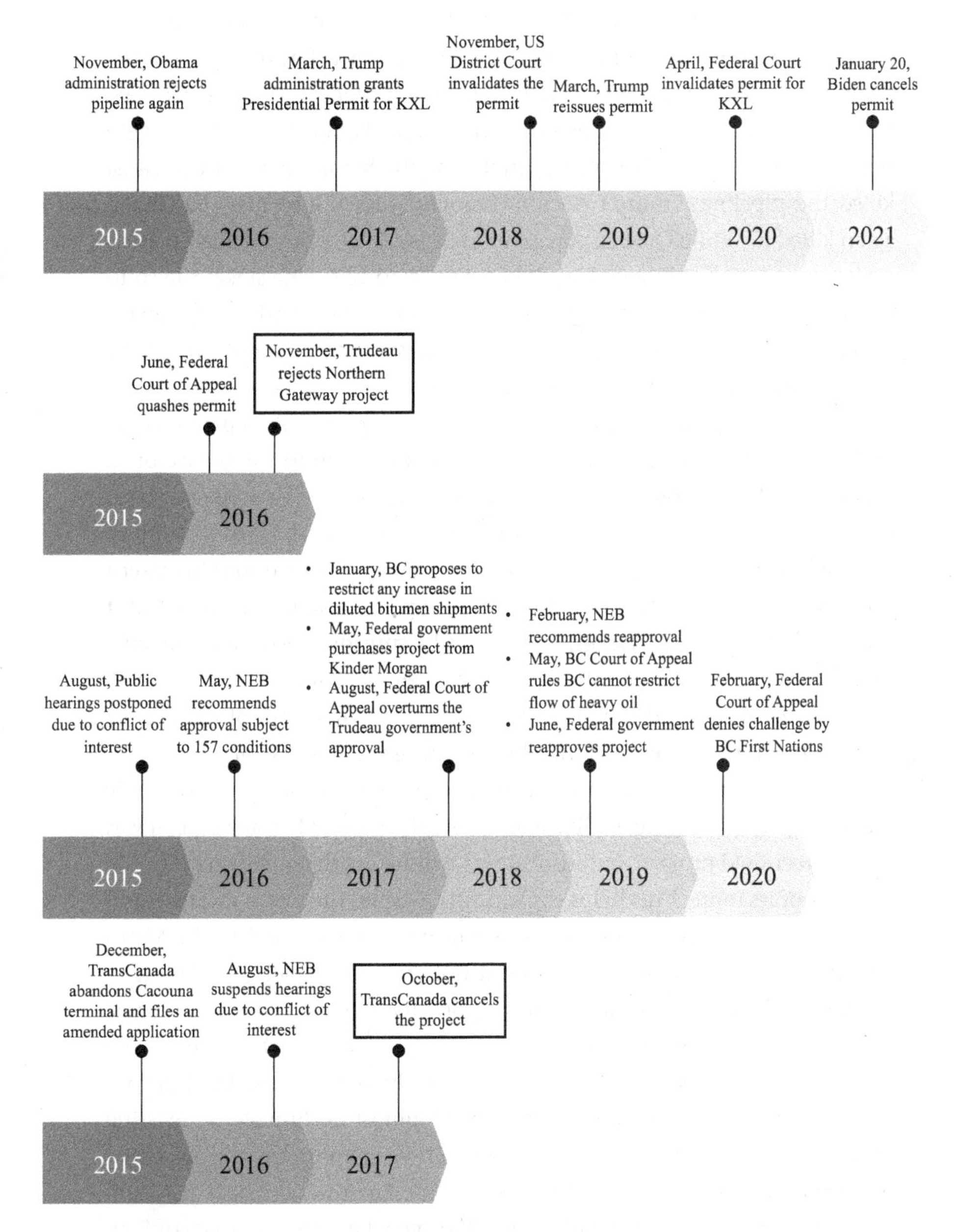

Figure 8.1

Temporal relationship between the four pipeline projects.

Gateway and Trans Mountain cases and the risk of pipeline spill to precious sources of clean water in the Keystone XL and Energy East cases.

In all four cases, the anti-pipeline coalition had access to critical veto points. Environmental and Indigenous access to courts has been a vital part of the resistance. Opponents certainly have not won all the judicial decisions, but in each case courts played a critical role. Political control over veto points is pivotal, a fact best demonstrated by the Keystone XL case: Obama killed the pipeline, Trump's election brought it back into play, but Biden again killed it. While Quebec may not have a legal veto, in Canadian politics on issues as grand as Energy East, it had a de facto veto. The stakes the competing coalitions have regarding what values are represented by those with formal decision authority explain why election strategies and outcomes have been so important to these competing coalitions.

It is no surprise that all four pipelines have geographically separated risks and benefits. Most of the economic benefits would go to the oil sector in Alberta, while the environmental risks of pipelines are elsewhere—across the Continental Divide in the Northern Gateway and Trans Mountain cases, across the forty-ninth parallel in the Keystone XL case, and in eastern Canada in the Energy East case. The great hope for Energy East was that it might avoid the fate of the other pipelines by spreading economic benefits more widely across the country, but the cancellation of the Cacouna deepwater terminal stripped the major Quebec-based benefit out of the project and rendered it even more vulnerable to Quebecers' opposition.

Trans Mountain got its greatest advantage from being able to rely mainly on existing infrastructure. Unlike Northern Gateway and Keystone XL, it was not a "greenfield project" but instead was twinned with an existing pipeline for most of its route. This helps explain why, despite the much greater number of people potentially affected by a pipeline or tanker spill in the Metro Vancouver region, Trans Mountain did not attract as much opposition in opinion polls. Energy East also had the potential to benefit from being able to repurpose an existing pipeline, but not where it probably mattered the most: east of Montreal in Quebec. It is noteworthy that two smaller, less contested oil sands pipeline projects that relied almost exclusively on existing infrastructure—Line 9 and Line 3—have both been approved and are either in operation or almost ready (Janzwood 2020). Environmentalists did campaign actively against Line 9 but not until the very late stages of that conflict, and they lost.

This explanatory framework is not intended to suggest that these four factors, singly or in combination, somehow determine the fate of a pipeline. What they do is shift the odds either for or against pipeline advocates or opponents. Keystone XL has confronted enormous barriers but could have been approved if it withstood judicial challenges and Biden hadn't won the 2020 election. Trans Mountain has restarted construction but could still be thwarted by on-the-ground opposition.

Climate Impacts of Place-Based Resistance

How effective has the strategy of place-based resistance to fossil fuel development been at promoting climate action and the reduction of global warming emissions? Prior to the anti-pipeline campaign, place-based resistance against fossil fuel infrastructure in North America was focused on coal. The anti-coal campaign was built on the same core strategy as the anti-pipeline movement: environmental groups committed to climate action allied themselves with place-based groups focused on local risks. In some cases, this was air pollution and related health impacts caused by coal-fired power plants. In others, it was the impacts on land and water of mountaintop removal mining (Cheon and Urpelainen 2018, chapter 6).

Between 2005 and 2017, coal use for electricity generation in the United States declined by 40%, leading to a reduction of 600 million tonnes of carbon dioxide emissions (Gruenspecht 2019). Place-based resistance is not the only reason why coal has been in such steep decline in the United States. The most influential factor has been the shale gas revolution that undermined the relative competitiveness of coal as a power source. But according to Cheon and Urpelainen, "it is indisputable that the vibrant opposition to coal power plants makes investments in coal riskier and less profitable" (Cheon and Urpelainen 2018, 148). The most direct impact of the anti-coal campaign was the prevention of construction of new coal plants, contributing to the cancellation of 132 coal plant proposals by 2010 (132). The social mobilization against coal also contributed to the Obama administration's mercury regulations and Clean Power Plans, which further undermined the viability of coal.

The choice by North American environmentalists to focus on oil sands pipelines was more contested. When Keystone XL was targeted, some climate policy analysts were highly skeptical of the strategy (e.g., Leach 2011; Revkin 2011; Levi 2013, chapter 4). When Energy East and Trans Mountain became nationally divisive issues in Canadian politics, the University of Calgary's Trevor Tombe was harshly critical of the cost-effectiveness of choosing to block pipelines rather than working through carbon pricing, saying, "Climate change is a problem, but blocking pipelines is not the solution. Such efforts may distract from good policy at best, and jeopardize it at worst" (Tombe 2016).

But there's a fundamental difference between the logic of the climate policy analyst and that of the climate policy advocate. The analyst focuses on the cost-effectiveness of different policy tools but overlooks the prime justification that climate advocates see in blocking infrastructure. By allying with place-based interests, advocates were able to mobilize a resistance movement that abstract support for economy-wide policies simply could not. As one campaigner in British Columbia who wished to remain anonymous explained, "Try going door to door campaigning on carbon pricing." And while blocking individual projects was a vital part of the strategy, the higher-level objective was to motivate governments to adopt more ambitious climate policies. By that criterion, the anti-pipeline movement has unquestionably been successful.

The pathway to success was from the environmental campaigns in the media and on the ground, to climate policy reform in Alberta, to climate policy reform by the government of Canada. The unbridled expansionism of the oil sands met with growing environmental resistance starting in 2005. Environmentalists launched a coordinated, multipronged campaign to undermine the economic and political rationale for oil sands growth. First, they adopted an ambitious framing campaign to shift the discourse about the oil sands, labeling them "dirty oil" (Nikiforek 2010). Second, building on their successful marketing campaigns in forestry, they targeted foreign buyers in the United States and Europe. Third, and ultimately most importantly, they sought to contain expansion by blocking the approval of new pipelines to get product to market.

Greenhouse gas emissions from the oil sands have been affected by these cancellations and delays. Oil sands production has increased faster than pipeline takeaway capacity and created a gap in prices between oil sands crude and North American and global oil prices. Chapter 2 described these market dynamics. Pipeline constraints lead producers to rely on more expansive rail and create higher price differentials—both of which mean lower profits and revenues for producers and therefore less money to invest in new production. This has the effect of reducing industry growth and the emissions that accompany it.

The National Energy Board estimated in 2016 that by 2025 oil sands production would be 450,000 tonnes per day lower than if production were unconstrained by pipelines (National Energy Board 2016c, chapter 10). This amount of forgone production is equivalent to about 10 million tonnes of greenhouse gas emissions per year (Tombe 2016), compared to 67 million tonnes from the oil sands in 2014. Even without new pipelines, emissions would grow, but they would be substantially lower a decade from now than they would have been without pipeline cancellations and delays.

In addition to slowing or stopping pipelines, constraining oil production and consequently GHG emissions, the anti-pipeline movement also contributed to notable advancements for climate policy in Canada. The year 2015 was a watershed year in Canadian energy and climate policy, first with the stunning election of the NDP in Alberta in May (Bratt et al. 2019; Sharpe and Braid 2016) and then the Trudeau Liberals' defeat of Harper's Conservative government in October (Pammett and Dornan 2016). Perhaps the most direct indicator of success of the place-based resistance campaign was the express motivations for Alberta's Climate Leadership Plan. For the previous decade, the response by the Conservative Party in Alberta to the environmental campaign was to make modest policy reforms and substantial investments in public relations and lobbying in foreign capitals (Hoberg and Phillips 2011; Urquhart 2018). The Alberta NDP adopted a strategic approach that differed from that of their Conservative predecessors and embarked on developing a climate plan they believed would give them greater market and political credibility in international markets and the rest of Canada. According to Andrew Leach, the University of Alberta professor Notley tapped to chair the Climate Leadership Panel, the motivation of his panel was not exclusively pipeline approval but more about addressing the image problems of the oil sands: "This was more than just about pipelines. It was about having a policy in Alberta where you could credibly say, 'We have a policy that stands up well to everybody else's so there shouldn't be discriminatory policies over and above those that are aimed at industries. . . . It was really about carving out an equitable treatment of Alberta and primarily the oil sands, but its industry and growing population as a whole. . . . But it wasn't 'what's the policy that gets us a pipeline?' Not by any stretch" (Leach 2017).

But Leach believes that the direct link between pipelines and the climate plan "was certainly much more in the premier's take." As he explains, "To me it is pretty obvious that by the time we got to the end of our process you weren't going to have federal government support for a pipeline if Rachel Notley stood up and said 'it's not our time to act on climate change.' You're going to make it really easy for other provincial governments to oppose pipelines, if that is your approach. . . . So there were a lot of things that by taking on good policy you could probably change the probability of a pipeline approval and construction" (Leach 2017).

Well before Notley's stunning election, a group of environmentalists began having facilitated discussions with a group of oil sands executives in an effort to broker a compromise on oil sands growth and pipelines. This process was separate from the work of the Climate Leadership Team led by Leach. The link between pipeline opposition and climate policy was central to this process. In fact, oil sands companies agreed to a 100 million tonne cap on oil emissions in exchange for the environmental groups' apparent agreement to stand down their opposition to new pipelines (Urquhart 2018, 281). The companies were represented not by the Canadian Association of Petroleum Producers but by four companies in the sector—Suncor, Cenovus, CNRL, and Shell—who believed that a more proactive approach to climate policy was necessary to improve the sector's political legitimacy at home and in foreign markets. These companies and the group of environmental leaders came to an agreement and reported the results to the Notley government, which decided to add the emission cap proposal to the recommendations coming from the Climate Leadership Team (Leach 2017).

Alberta's Climate Leadership Plan was released on November 22, 2015, on the eve of the first conference of first ministers in a decade and just before the Paris summit. In addition to capping oil sands emissions at 100 million tonnes, the plan committed to phasing out coal by 2030, increasing renewable electricity production to 30% of the total by 2030, implementing carbon pricing, and regulating methane (Government of Alberta 2015). When Premier Notley made the announcement, she made explicit reference to how the environmental campaign against the oil sands had damaged the province's international reputation and, as a result, market access. Fresh in her mind was President Obama's November 6 announcement rejecting the Keystone XL pipeline:

> In our role as Canada's principal energy producer we need to step up to the climate change issues. Thoughtful people in the energy industry, including the industry leaders standing with me here today, have been saying for a long time that we can and must do a better job. We got a major wakeup call on this a few weeks ago in the form of a kick in the teeth. Unfairly, in my view, the President of the United States claimed that our production is some of the dirtiest oil in the world. That is the reputation that mistaken government policies in the past have earned for us. We are a landlocked energy producer with a single market. A single market that just took a very hard run at us. So we need to do better. And we are going to do better. (Notley 2015)

After describing the main components of the plan, including the 100 million tonne emission cap on the oil sands, she continued, "I'm hopeful that these policies, taken overall, will lead to a new collaborative conversation about Canada's infrastructure on its merits. And to a significant de-escalation of conflict worldwide about the Alberta oil sands."

Justin Trudeau won the Liberal leadership in April 2013. His first major speech on energy and climate issues was in October of that year, at the Calgary Petroleum Club. Castigating Harper for needlessly antagonizing domestic opponents and the Obama administration, he clearly articulated his belief that stronger climate policy was the means to gain approval of pipelines: "Let me be clear on this. If we had stronger environmental policy in this country: stronger oversight, tougher penalties, and yes, some sort of means to price carbon pollution, then I believe the Keystone XL pipeline would have been approved already" (Liberal Party of Canada 2013).

On March 3, 2016, in a speech to the delegates of the Globe conference of clean energy firms and advocates, Trudeau strengthened the pipeline-climate linkage by introducing the argument that new oil pipelines would help finance the clean energy transition, saying, "The choice between pipelines and wind turbines is a false one. We need both to reach our goal, and as we continue to ensure there is a market for our natural resources, our deepening commitment to a cleaner future will be a valuable advantage" (Smith 2016).

That same day, Trudeau met with premiers to discuss the Pan-Canadian Framework on Clean Growth and Climate Change. The resulting Vancouver Declaration committed the federal government and provinces to "Implement GHG mitigation policies in support of meeting or exceeding Canada's 2030 target of a 30% reduction below 2005 levels of emissions, including specific provincial and territorial targets and objectives."

The Trudeau approach to climate policy was clearly reflected in its wording: "Transition to a low carbon economy by adopting a broad range of domestic measures, including carbon pricing mechanisms, adapted to each province's and territory's specific circumstances, in particular the realities of Canada's Indigenous peoples and Arctic and sub-Arctic regions. The transition also requires that Canada engage internationally" (Canadian Intergovernmental Conference Secretariat 2016a).

The first ministers agreed to establish a series of working groups and meet to finalize the plan in fall 2016. The declaration was less explicit about the pipeline-climate linkage. The only reference to that argument is in the clause stating "Recognizing the economic importance of Canada's energy and resource sectors, and their sustainable development as Canada transitions to a low carbon economy" (Canadian Intergovernmental Conference Secretariat 2016a).

This link between pipeline and climate policy couldn't have been made more clearly than when Prime Minister Trudeau himself announced his government's approval of the Trans Mountain Expansion Project in November 2016: "And let me say this definitively: We could not have approved this project without the leadership of Premier Notley, and Alberta's *Climate Leadership Plan*—a plan that commits to pricing carbon and capping oilsands emissions at 100 megatonnes per year" (Trudeau 2016b). Two years later, he was still making that link explicit: "In order to get the national climate change plan—to get Alberta to be part of it, and we need Alberta to be part of it—we agreed to twin an existing pipeline. Yes, they were linked to each other" (Harper 2018).

When British Columbia's opposition became a significant threat to the project in 2018, Trudeau lashed out at its premier, John Horgan, for threatening to unravel the grand bargain that made the advance in climate policy possible, saying, "John Horgan is actually trying to scuttle our national plan for fighting climate [change]. By blocking the Kinder Morgan pipeline, he's putting at risk the entire national climate change plan, because Alberta will not be able to stay on if the Kinder Morgan pipeline doesn't go through" (McSheffrey 2018).

Given the power of the oil industry and its political allies, the victories of the anti-pipeline coalition are impressive. Place-based resistance was effective at delaying pipeline projects and in some cases leading to their outright cancellation. In concert with the media strategy that damaged the brand of Alberta oil, those actions inflicted economic pain on the oil sands industry. When the Alberta and Canadian governments were run by conservative parties, they retrenched and defied environmental critics. In 2015, they were occupied by more progressive parties, who believed a more strategic approach was to advance climate policy to increase their legitimacy in politics and markets. They forced the government of Alberta to price carbon, paving the way for the government of Canada to enact nationwide carbon pricing. The resistance strategy reduced oil sands production and emissions below what they would have been and eventually forced Canadian governments to adopt stronger climate policies than they would have absent the effects of that resistance strategy.

These victories represent a significant policy shift resulting from a change in the balance of power between the oil sands coalition and the anti-pipeline coalition. Significant policy changes such as this are considered unlikely without "significant perturbations external to the subsystem" (Sabatier and Jenkins-Smith 1993), but, in this case, it was the climate movement's shift to embracing "keep it in the ground" supply-side strategies that seemed to be the most important force driving the policy change. The climate movement, remarkably, transformed oil sands pipelines to the south, east, and west into lines in the sand on climate and it worked, producing a shift in power and a corresponding shift in policy.

The Backlash against the Pipeline Resistance: Revisiting the Oil Sands Regime

The anti-pipeline coalition unquestionably created a formidable shift in political power in North America and especially north of the forty-ninth parallel. Environmentalists knew that to make progress in reducing the stranglehold the oil sector had on politics and policy, they were going to have to cause them economic pain. They set out to do so by blocking infrastructure, and it worked in getting the attention of the industry and its allies. It shifted policy at the provincial and federal levels.

The Rise of a Canadian Petro-Populism

But the oil sands coalition did not sit idly by. Coalitions are made up of strategic actors that are constantly reassessing their strategies in response to their environment (Meyer and Staggenborg 1996; Hochstetler 2011). The resistance movement's rise in power elicited a formidable political counterattack by a political giant caught off guard. When the environmental campaign against the oil sands emerged around 2005, the initial response of the oil sands coalition was, at first, co-optation of adversaries and dramatically increased investment in public relations and government lobbying (Hoberg and Phillips 2011; Urquhart 2018). When pipeline resistance first started to bite, the oil sands coalition doubled down on these tactics but failed to gain much traction.

By 2015, however, the oil sands coalition realized it needed a new strategy, and it chose to emulate the practices of its adversaries. According to Shane Gunster,

> The remarkable success of Indigenous, environmental and local community resistance to pipeline projects (especially Northern Gateway) created a lot of anxiety in the C-suite of the oil and gas industry. Industry groups such as the Canadian Association of Petroleum Producers (CAPP) worried that their traditional tools of corporate power—back-door lobbying, influence over corporate media, big-budget ad campaigns—were no longer as effective in shaping political discourse and opinion around oil and gas. Instead (and inspired by similar initiatives by the American Petroleum Institute), they needed to aggressively mobilize those constituencies most likely to support their agenda—oil industry workers, resource dependent communities, conservatives. (Gunster 2019)

The Canadian Association of Petroleum Producers created "Canada's Energy Citizens" to recruit ideologically sympathetic Canadians to engage in supporting the industry by attending rallies and speaking out on both mainstream and social media (Wood 2018). Energy Citizen was joined by other social media campaigns supportive of the oil sector, including Oil Sands Action, Oil Respect, and Oil Sands Strong (Gunster 2019). While much of this activity has the characteristics of an "astroturf" campaign of corporate manipulation of putatively populist campaign activities, it has tapped into a deep cultural vein in resource-dependent communities (Gunster et al., 2021). The "petro-nationalism" reached a crescendo during February 2019, when a truck convoy set out from Red Deer, Alberta, to Ottawa to protest the Trudeau government's carbon tax, proposed environmental assessment reforms, and tanker ban on the north coast of British Columbia. Under the banner of United We Roll, the convoy garnered massive attention from the Canadian media (Blewett 2019).

When Jason Kenney's United Conservatives trounced Rachel Notley's New Democrats in the 2019 Alberta election, this movement came into the hall of government power. In his election victory speech, Kenney proclaimed the oil sands coalition had had enough:

> But we have been targeted by a foreign funded campaign of special interests seeking to landlock Canadian energy.
>
> This means that we Canadians have become captive to the United States as the only market for Canada's largest export product: our energy. . . .
>
> And now I have a message to those foreign funded special interests who have been leading a campaign of economic sabotage against this great province.
>
> To the Rockefeller Brothers Fund, the Tides Foundation, Lead Now, the David Suzuki Foundation and all of the others:
>
> Your days of pushing around Albertans with impunity just ended.
>
> We Albertans are patient and fair minded, but we have had enough of your campaign of defamation and double standards. Today, we begin to stand up for ourselves, for our jobs, for our future. Today we begin to fight back. (Kenney 2019)

Implementing a campaign promise, Kenney established a C$30 million "energy war room" to "respond in real time to the lies and myths told about Alberta's energy industry through paid, earned, and social media" (United Conservative Party 2019). It also went so far as to launch a public inquiry into the "well-funded foreign campaign [that] has defamed Alberta's energy industry and sought to land-lock our oil" (Government of Alberta 2019a).

Kenney's victory strengthened an increasingly conservative political landscape across Canada. Trudeau had lost his most important ally, the Ontario government, when Doug Ford's Conservatives defeated the province's long-ruling Liberal Party. Ford immediately abandoned the province's cap and trade regime, withdrew from the Pan-Canadian Framework, and challenged the constitutionality of the federal carbon pricing backstop. Kenney's victory meant Alberta joined Ontario and the two other prairie provinces, Saskatchewan and Manitoba, as a Conservative bulwark from the Ottawa River to the Continental Divide. Like Ford, he dismantled his predecessor's climate policies, withdrew from the Pan-Canadian Framework, and joined the legal fight against the federal climate framework. The inevitable result of this changing political landscape is that Trudeau's much-vaunted climate policy, the core achievement of the pipeline resistance campaign, would be front and center in the 2019 federal election battle.

Changing Political Economy

In addition to the climate policy advances described in the previous section, some extraordinary changes in the governance, or political economy, of the Canadian oil sector also resulted. Relentless resistance to Trans Mountain essentially forced the federal government to nationalize the project. When the government of Pierre Trudeau created Petro-Canada as a state-owned enterprise in 1975 and enlarged the National Energy Program in 1980, the oil coalition erupted in protest. When the government of Justin Trudeau nationalized a major midstream asset, the oil sands coalition simply cheered. When skyrocketing price differentials, caused in part by pipeline resistance, forced Rachel Notley to curtail oil sands production, the oil sands coalition simply went along with the dramatic extension of state power. These changes did not alter the composition of the oil sands coalition, but they did change the relationship of the private and public sector actors within the coalition. The Trudeau government, at first a reluctant participant in the coalition, had become an owner of a major oil sands asset.

Shifting Power Resources for the Anti-pipeline Coalition

The anti-pipeline coalition has won some vital victories beyond the delays and cancellations of pipelines. Helping to defeat the Harper government created the conditions for Canadian progress on climate policy. Helping John Horgan's NDP win the 2017 election in British Columbia brought them a reliable ally outside Quebec. But at the end of the decade, the anti-pipeline coalition faced the growing petro-populist backlash with shifting political resources at their disposal. For social movements, even a modicum of success can undercut momentum if it demobilizes actors or resources. For the anti-pipeline coalition, their success with Alberta's Climate Leadership Plan and Trudeau's Pan-Canadian Framework meant that support for the resistance movement from the US foundations, which had been so maligned by conservatives, was largely withdrawn.[1]

While the movement lost some financial support, it had a tremendous new resource heading into the October 2019 federal election: a renewed salience in public opinion for climate and environmental issues. As noted in chapter 2, environmental issues spiked in salience in 2007 but then got hammered by the Great Recession and didn't recover through 2015. There were multiple indications that the issue reemerged as top of mind among Canadian voters. An Environics poll in April 2019 put environmental and climate issues as the second most important problem for Canadian voters after the economy (Environics Institute 2019). Trudeau's carbon policies were at the center of the 2019 election campaign, as Andrew Scheer, leader of the Conservative Party of Canada, made criticism of Trudeau's carbon tax a centerpiece of its campaign.

The election resulted in a minority Liberal government. Although the Liberal Party of Canada lost its majority in Parliament, the election was a strong repudiation of the Conservatives, who received only 32% of the vote compared to the 39% share received by Trudeau's party. When considering the vote shares of the New Democratic Party and the Green Party, both of whom had stronger climate policy platforms than the Liberals, 63% of Canadian voters chose parties pushing for stronger climate policies (Meyer 2019; *Washington Post* Editorial Board 2019). In December 2020, Trudeau seized on this mandate and announced an ambitious new climate plan that, for the first time, contained credible policy commitments that could meet the nation's 2030 target and establish a path to its newly legislated requirement to attain net-zero greenhouse gas emissions by 2050 (Jaccard 2020; Environment and Climate Change Canada 2020).

Conclusion

This chapter has built on the previous four pipeline case study chapters to address the first of this book's four core questions: has the strategy of place-based resistance to fossil fuel development been effective at promoting climate action and the reduction of global warming emissions? This chapter has shown that the anti-pipeline resistance has contributed to the adoption of stronger climate policies. The political backlash that began with Doug Ford's election has threatened this progress, but as of June 2020, the major policy advancements remain in place. Trudeau's Liberal government was reduced to minority status in the fall 2019 election, but overall the salience of the climate issue during the campaign, and the resulting distribution of seats across pro-climate political parties, helped protect the policy progress against the backlash from oil sands supporters.

Chapters 9 and 10 turn to the book's second question: does the place-based resistance strategy against fossil fuels risk the unintended consequence of feeding place-based resistance to the clean energy transformation? This is done in two stages. Chapter 9 examines a "clean energy" megaproject, the Site C Dam in northeastern British Columbia. Then chapter 10 takes a broader look at the resistance to renewable energy projects by examining a variety of cases across North America.

III The Resistance Dilemma

9 The Site C Dam and the Political Barriers to Renewable Energy

Overview

Before entering politics, Andrew Weaver was a University of Victoria climate scientist with a considerable global reputation. When the government of British Columbia was making plans to announce its intention to proceed with a large hydroelectric dam in northeastern British Columbia in 2010, staff invited Weaver to attend the elaborate 2010 media event in Hudson's Hope, in the Peace River region. As a climate expert who was an enthusiastic supporter of the project, the staging put Andrew Weaver just behind the premier. But he changed his position upon becoming British Columbia's first elected legislator from the Green Party.

Weaver gave two reasons for the change. First, he said that the economic case has changed as the project has become more expensive, demand growth has not materialized as predicted, and the costs of alternatives have fallen. Second, he developed an appreciation of the impacts on the region's Indigenous people. He told *Globe and Mail* reporter Justine Hunter: "To be blunt, one of the things I didn't consider back when I was a climate scientist, thinking about nothing but climate science, was the issue of First Nations' rights and title" (Hunter 2017b).

Addressing the climate crisis involves a rapid phaseout of carbon-emitting fossil fuels and an accelerated adoption of clean energy technologies. Environmentalists and First Nations have focused much of their energy on resisting new fossil fuel infrastructure, under the banner of "keep it in the ground." In the cases presented thus far in this book, these conflicts have pitted the government of Alberta and the oil sector, which seek expanded market access, against environmentalists and First Nations, who are concerned about climate change, local and regional environmental impacts, and

Indigenous rights and title. This coalition has proven to be surprisingly formidable at halting or delaying new pipelines.

But does this "keep it in the ground" movement, paradoxically, risk the unintended consequence of feeding place-based resistance to renewable energy infrastructure and thus to the necessary clean energy transition? Hydropower dams, utility-scale solar and wind facilities, and high-voltage transmission lines necessary to integrate renewables into the grid have all faced determined resistance from place-based groups (Cleland et al. 2016; Batel et al. 2013; Devine-Wright 2009; Fast 2013). Does the decision to ally so closely with place-based opponents risk creating a "resistance dilemma" by legitimizing place-based resistance that can then be mobilized to thwart needed clean energy infrastructure? This question is addressed in two parts. First, this chapter does a deep dive into one high-profile conflict that links the western Canadian pipeline conflicts to the clean energy transition debate. Second, chapter 11 will explore this potential resistance dilemma in more detail by examining a broader range of cases.

This chapter begins the exploration of this question by examining the case of a large "clean energy project," the Site C Dam, which is a C$10 billion, 1,100-megawatt project in northeastern British Columbia. Despite the potentially enormous benefit of producing low-carbon electricity for generations, this dam has been contested by virtually the same coalition that opposes new oil sands pipelines. Despite this opposition, the project was approved by the governments of British Columbia and Canada in 2014, and construction started in mid-2015. That approval was confirmed after the 2017 British Columbia election brought to power a new government highly skeptical of the Site C Dam that later went on to endorse the project. A number of legal challenges to the project have been rejected by Canadian courts. As of December 2020, the project was threatened by revelations of new geo-technical problems.

The first section of this chapter reconsiders the book's analytical framework as focus shifts from oil sands pipelines to renewable energy projects. The second section provides a brief overview of the project and the underlying controversies. The third section provides an overview of the major actors in the controversy. In the fourth section, the content of ideas at work in this case is explored through a discussion of public opinion and media framing of the dispute, followed by an overview of the institutional rules and conflicts. The fifth section provides an analysis of how the environmental assessment processes dealt with the most important economic and environmental issues. After that, the implications of the 2017 British Columbia election are addressed, along with the evolution and status of court cases against the project. The chapter concludes by considering what this controversy reveals about the political challenges of the clean energy transition.

Analytical Framework

One of the main ideas guiding this book is that, in the case of oil pipelines at least, it is most useful to think about the relative power of project opponents as a function of four variables:

1. The salience of place-based, concentrated risks and benefits;
2. Whether opposition groups have access to institutional veto points;
3. Whether the project can take advantage of existing infrastructure;
4. The geographical separation of risks and benefits.

Are these same four factors at work in the case of low-carbon energy infrastructure? For the first three factors, the answer is a definite yes. Whether renewable projects pose risks to salient, place-based values is also very relevant to the magnitude and intensity of opposition they could activate. As chapter 10 will show, much of the controversy regarding wind and transmission lines is precisely about how they would alter places valued by communities. Solar and wind plants have also generated opposition because of sensitive ecological habitats.

With respect to the accessibility of institutional veto points, the rules and decision-making structures surrounding clean energy projects are critically important. Indeed, as we'll see in chapter 10, one of the most important institutional conflicts over renewable energy projects has been whether local governments, which are understandably likely to be concerned about localized project impacts, have the ability to block projects through zoning or other policies. Third, like pipelines, if renewable energy projects can take advantage of existing infrastructure, their visual and landscape impacts are likely to be less.

The application of the fourth factor, the geographical separation of risks and benefits, is more complex but an important feature of the project's politics. Many renewable energy projects have the potential to be sited in proximity to where the power will be used, which would concentrate the risks and benefits in the same place. But the desire to take advantage of the most favorable locations for renewable power generation means that they are often distant from the source of demand, which creates a clear and potentially divisive separation of risks and benefits.

Transmission lines are essentially pipelines for electrons. Their risks to water and land are much lower, but their aesthetic impacts, if they are built above ground, are typically much greater. Like oil pipelines, transmission lines are long, thin, linear projects that therefore affect a number of communities and potentially different subnational or even national jurisdictions. In many cases, transmissions lines have attracted more resistance than new renewable power facilities themselves, so indeed all four of these factors are also very important in determining the strength of resistance to renewable energy infrastructure.

While many aspects of a large dam like Site C are different from the long, linear infrastructure of pipelines or transmission lines, this project is also characterized by opposition access to veto points; the presence of existing development and infrastructure; salient, place-based risks; and a significant geographic separation of risks and benefits. Project opponents worked hard to influence elections so that sympathetic parties could control the British Columbia government, lobbied relentlessly to have the project referred to an independent regulator, and made liberal use of the courts. In comparison to other renewable energy projects, big dams have a distinctively large impact in one region, potentially intensifying the degree of opposition from affected interests. The adverse environmental impacts of this dam are focused in the Peace River region of northeastern British Columbia, while the benefits will be felt in distant load centers. Unlike many of the pipeline projects, however, both the risks and benefits are largely felt within one subnational jurisdiction.

Background on the Project and Controversy

The Site C project would be the third dam in a system of hydro reservoirs on the Peace River in northeastern British Columbia (Cox 2018, 80–88). The WAC Bennett Dam was completed in 1968 and provides 2,730 megawatts of power. Downstream, a second dam, the Peace Canyon Dam, was completed in 1980 and provides 700 megawatts of power. The Site C Dam has been on and off the province's energy agenda several times since 1980, when BC Hydro first formally submitted an application to proceed with the project.

But that effort was rebuffed when the project was rejected by the British Columbia Utilities Commission in 1982 because BC Hydro had not provided "(1) an acceptable forecast that demonstrates that construction must begin immediately in order to avoid supply deficiencies and (2) a comparison of feasible alternative system plans demonstrates, from a social benefit-cost point of view, that Site C is the best project to meet the anticipated supply deficiency" (British Columbia Utilities Commission 1983, 9–10).

Formal interest in the project was rejuvenated by Premier Gordon Campbell in the 2007 Energy Plan, which said the province "will enter into initial discussions with First Nations, the Province of Alberta and communities to discuss Site C" (Government of British Columbia 2007). In April 2010, Campbell announced that he was instructing BC Hydro to move forward with the project (Government of British Columbia 2010; Cox 2018, 7–17). BC Hydro submitted its project description to the British Columbia Environmental Assessment Office (BC EAO) in May 2011, and in August 2011 both the BC EAO and the Canadian Environmental Assessment Agency accepted the project for review.

Extensive hearings conducted by the Joint Review Panel revealed three significant sources of opposition. First, Aboriginal groups in the region have long been strongly opposed to the project and have raised concerns in multiple venues about whether the decision process that led to the project's approval is a violation of their treaty rights. Second, local and provincial environmental groups have opposed the project because it would flood regionally valuable agricultural land and have significant impacts on fish and wildlife habitat in the region. Third, a variety of groups, including clean energy firms, major industrial consumers, environmental groups, opposition Members of the Legislative Assembly (MLAs), and a variety of independent experts, have raised concerns about whether the project is justified, given future projections of electricity demand, and whether it is excessively costly in comparison to alternatives (Behn and Bakker 2019).

The Joint Review Panel issued its report on May 1, 2014. Both the federal and the provincial governments issued approvals for the project in October 2014, and the British Columbia government approved the project for construction in December 2014 (Government of British Columbia 2014). Despite a number of legal challenges, BC Hydro initiated construction in late July 2015, with an expected completion date of 2024. Premier Clark, who was in power at the time, famously vowed to push the project "past the point of no return" (Palmer 2017). A significant number of legal challenges to the approval decision have been dismissed. Despite the defeat of the Clark government, the New Democratic Party (NDP) agreed to continue construction after an agonizing period of review. However, one First Nations' claim in court that their treaty rights have been infringed has yet to be resolved.

Actors

The political economy of this project is distinct from many others because the proponent, BC Hydro, is a wholly state-owned enterprise of the government of British Columbia. This makes the government the primary proponent. In addition to the usual critics of megaprojects within the environmental and Aboriginal communities, the project has also been opposed by private-sector industrial interests.

Government

The most important government actor in this case is the government of British Columbia, as both the authoritative decision-maker for the project and owner of BC Hydro. The core interest of the political branch of government is reelection. An affordable, reliable electricity supply is critical to maintaining public confidence in the government. As the project proponent, the provincial government has a vital stake in the economic benefits flowing from the project and the avoidance of negative financial, environmental, or social impacts that could provoke voter backlash. In its modern incarnation, the project was proposed by Premier Gordon Campbell, who had an ambitious agenda to be a recognized leader on climate change and clean energy (Cox 2018, 80). When Clark took over in 2013, she didn't advance the climate agenda but enthusiastically embraced the Site C Dam.

When Clark's BC Liberals fell from power after the May 2017 election, the position of the government of British Columbia softened. The NDP, skeptical of the project's economic merits, campaigned to have the project reviewed by the British Columbia Utilities Commission, an independent regulatory body. Their minority government was supported by the British Columbia Green Party, which wanted the project canceled outright (Hunter 2017a). The Confidence and Supply Agreement between the two parties sided with the NDP position (BC Green Caucus and BC New Democrat Caucus 2017), but in the end, as will be discussed, the NDP reluctantly chose to continue with the project.

The government of Canada has jurisdiction over fisheries and navigable waters, so it shared regulatory responsibility for the project's review and approval. Given its stance favoring resource development, the Harper government was generally supportive of the project, and it never became a source of tension between the federal and provincial governments. The government of Alberta has stakes in the project because of the potential for downstream impacts to the spectacular Peace-Athabasca Delta. While it did register as an intervenor, significant interprovincial conflict never emerged. The province's submission raised concerns about downstream changes in hydrological flows and impacts on fish, but these concerns were addressed through required minimum flow requirements and commitments to consultation (Joint Review Panel 2014). Antidam opponents sought to elevate the importance of the downstream impacts of the project on Wood Buffalo National Park, a UNESCO World Heritage Site (Cox 2018, 108; UNESCO 2017). But neither the government of Alberta nor the federal government were sufficiently concerned about those impacts to challenge the project's approval. Local governments in the Peace region have stakes in the project but have not emerged as antagonists in this case.

Environmentalists

The core interest of environmentalists is to minimize the environmental effects of providing energy services. The Site C project had the potential to be a divisive wedge within the British Columbia environmental community. On the one hand, it is a renewable, non–fossil fuel energy source that has the potential to generate large quantities of low-carbon electricity, with the added benefit of being able to store electricity to help manage intermittent renewables. For that reason, it could attract support from climate activists and clean energy advocates, which play a substantial role in the British Columbia environmental movement (Shaw 2011). But it is also a large dam that inevitably alters river flow and habitat for both aquatic and terrestrial species, so it is no surprise that many environmentalists would have significant concerns about it (Cox 2018, chapter 4).[1]

However, Site C has not created these divisions. Despite being a "clean energy" project, Site C has provoked widespread opposition from many of the same actors that oppose new "dirty energy" oil sands pipelines. Table 9.1 lists the main British Columbia environmental groups opposing the

Table 9.1
Environmental groups opposing the Site C Dam, March 2017

Group name	Facebook likes	Twitter followers
Peace Valley Environmental Association	2,538	3,088
Sierra Club of British Columbia	9,813	7,705
David Suzuki Foundation	480,977	146,000
Wilderness Committee	7,924	9,519
British Columbia Sustainable Energy Association	3,254	5,669
West Coast Environmental Law	6,493	13,200
Pembina Institute	5,737	20,576
Dogwood British Columbia	30,963	12,354

project. The lead local group in opposition is the Peace Valley Environmental Association, which receives strong support from the broader British Columbia environmental community.[2] The project is also opposed by the Wilderness Committee, Sierra Club of British Columbia, LeadNow, the David Suzuki Foundation, and the British Columbia Sustainable Energy Association. In an email to supporters, LeadNow denounced the project this way: "Government and industry are calling this dirty energy export strategy a 'clean' energy plan. In reality, it's anything but clean. Their plan would use taxpayer dollars to flood critical wildlife habitat, destroy prime agricultural land, poison groundwater across BC and drive up our province's dangerous carbon pollution." The Wilderness Committee's statement of opposition reads as follows: "We can't let this happen. The Site C Dam would destroy critical ungulate habitat that has sustained wildlife that has supplied generations of First Nations people with food and cultural sustenance for thousands of years. It will destroy one of the largest and most important wildlife corridors on the continent, and submerge valuable carbon sinks instead of promoting food security and the need to adapt to climate change" (Wilderness Committee, n.d.).

The British Columbia environmental community is not opposed to new energy projects, clean or dirty, across the board. Indeed, in 2009, a conflict erupted among British Columbia environmentalists over small "run of river" hydro projects proposed by the private sector. Many groups opposed the new "independent power projects," both because they threatened public control over the electricity system and because of their risks to fish habitats and recreational resources. But when the 2009 provincial election campaign started, a coalition of three influential environmental groups—ForestEthics, the David Suzuki Foundation, and the Pembina Institute—made statements supporting the climate policies of Gordon Campbell. Tzeporah Berman, of then ForestEthics, explicitly endorsed the need for some new energy projects to foster the clean energy transition (Berman 2011, 253–259). This came as a huge surprise to many other British Columbia environmentalists who were launching campaigns to support the British Columbia NDP because of Campbell's support for independent power projects and created a significant rift within the environmental community (Shaw 2011).

But Site C is different for three reasons. It is a large dam that concentrates environmental effects in one area. As a result, it magnifies place-based impacts more than smaller independent power projects did. Second, at various times British Columbia government officials justified the project as a source of power for expanded fossil fuel production, either liquefied natural gas (LNG) within British Columbia (Cox 2020a) or the oil sands across the border in Alberta (Linnitt 2016). It is strongly opposed by the local First Nations. In their pipeline resistance campaigns, environmentalists have regularly used the threat to First Nations rights and title as a critical argument against the projects. It would undermine their credibility, and their alliance with First Nations, if they turned a blind eye to the same concerns expressed about a clean energy project (as several people told me in personal interviews).

Industry

Industry interests in energy projects are divided between those who supply energy services and those who are large consumers. Suppliers in the energy industry have an interest in expanding revenues and profits. For major electricity consumers, their core interest is in minimizing the cost and maximizing the reliability of energy services. The Site C Dam comes with a different political economy than industry-led energy projects. In this case, the project's proponent is the government of British Columbia, through BC Hydro. Major industry groups, in contrast, have either been generally skeptical or explicitly opposed.

The province's construction industry has been understandably enthusiastic about the project (Penner 2014). The project was a major blow to the clean energy industry, which had benefited so strongly from Premier Campbell's promotion of independent power projects throughout the first decade of the new millennium (Hoberg 2010). Premier Clark's choice to privilege a government-led project over soliciting proposals from the private sector virtually eliminated the prospects for new independent power projects for the foreseeable future. This sector's trade association, Clean Energy BC, has been strongly opposed to the dam project from the outset. It commissioned and publicized several studies emphasizing the economic and social benefits of relying on dispersed renewables instead of a hydro megaproject (Palmer 2014; London Economics International 2014). The Association of Major Power Customers of British Columbia, which represents "about 20 of the largest employers and industrial customers in the province," has strongly opposed the project out of concern that its high costs will put further pressure on electricity rates (Lavoie 2014).

First Nations

The Site C Dam has been strongly opposed by area First Nations for some time. The affected region is covered by Treaty 8, established in 1899. Both the Treaty 8 Tribal Association and individual First Nations under the treaty have undertaken legal action against the project. The two most active groups have been the Prophet River First Nation and the West Moberly First Nation. The treaty provides for the Aboriginal right to hunt, trap, and fish throughout the territory but also provides for land to be "taken up from time to time for settlement, mining, lumbering, trading or other purposes" by the Crown (Treaty 8 Tribal Association 1899).

The Treaty 8 Tribal Association issued a declaration opposing Site C in 2010 (Treaty 8 Tribal Association 2011). Prophet River and West Moberly have challenged the approval decision in both provincial and federal courts, but thus far with no success (see the discussion of court actions). West Moberly has also been very concerned about the pace of natural gas development in the Fort St John region of the province as a result of the Clark government's aggressive push to establish an LNG industry. West Moberly's chief, Roland Willson, has sought to leverage the Supreme Court's *Tsilhqot'in* decision to issue an ultimatum to the British Columbia government. He explained that his people are not opposed to resource development but that the province was pushing too many projects in his territory: "I've said you can't have both. If you want to push Site C, we're not going to be in favour of any LNG projects, any of the pipeline projects up there. We don't want to be there but if that's the case, we don't have any other choice" (Moore 2014).

The provincial and federal governments did carry out an extensive consultation process with First Nations in the area, beginning before the environmental assessment process was triggered and continuing through the approval decisions by the two governments. A summary of the governments' efforts is contained in the joint Consultation and Accommodation report, released in September 2014 (Government of Canada and Government of British Columbia 2014). The report describes the Crown's obligation in these terms: "When intending to take up lands, the Crown must exercise its powers in accordance with the Crown obligations owed to the Treaty 8 First Nations, which includes being informed of the impact of the project on the exercise of the rights to hunt, trap and fish, communicate such findings to the First Nations, deal with the First Nations in good faith, and with the intention of substantially addressing their concerns" (Government of Canada and Government of British Columbia 2014, 29).

The report concluded that "consultation has been carried out in good faith and that the process was appropriate and reasonable in the circumstances." It argues that accommodation has been provided through project modification and conditions, and offers of impact-benefit agreement that include land transfers and multimillion-dollar compensation payments. BC Hydro claims that "these measures would offset the residual effects of the proposed Project if it is authorized by Governments to proceed" (Government of Canada and Government of BC British Columbia 2014, 67).

Despite the extensive engagement with First Nations, there is no indication in the record that the government ever offered area First Nations leadership or equity in the project. It is not clear whether such efforts would have changed the position of the most affected First Nations, but a comparative study of contested energy infrastructure projects across Canada concluded that, of the projects it considered, one of the few projects to gain social acceptance gained this acceptance because the government proponent made local First Nations partners in the project. In developing the Wuskwatim hydroelectric project, the government of Manitoba was able to shift the position of the Nisichawayasihk Cree Nation from strongly opposed to accepting of the project by making them partners and co-owners in the project (Cleland et al. 2016).

In the case of the Site C Dam, the West Moberly was not moved by the government's offers of compensation. Chief Roland Willson responded that, "We maintain our view that it is simply not possible to adequately compensate our community for the permanent destruction of the Peace River Valley" (Government of Canada and Government of BC British Columbia 2014, 67). Some First Nations, however, have accepted compensation packages. In March 2017, BC Hydro announced it had entered agreements with the Doig River

First Nation and Halfway River First Nation, both of which had originally opposed the project. The agreements include "a lump sum cash payment, a payment stream over 70 years, procurement opportunities, the selection and transfer of provincial Crown lands and commitments respecting certain land management initiatives" (BC Hydro 2017). Similar agreements have been signed with the McLeod Lake Indian Band, Saulteau First Nations, and Dene Tha' First Nations (BC Hydro 2016a; BC Hydro 2016b; BC Hydro 2015).

Academics

As is becoming increasingly common in the contested sphere of Canadian energy and environmental politics, groups of academics have also chosen to weigh in on the Site C controversy, with strong opposition to the project. An initiative was led by University of British Columbia (UBC) professor of geography Karen Bakker, who chairs the Program on Water Governance. Working with several consultants and UBC law professor Gordon Christie, Bakker was able to assemble over 200 scientists to sign a letter opposing the project. The statement focused mostly on issues of Aboriginal rights and the economic justification for the project. It called for the federal government to revisit its decision and to make a determination on whether Aboriginal treaty rights were infringed and, if so, whether that infringement was justified. It called on "both governments to explain why the unprecedented imposition of numerous significant adverse environmental effects is justified by a Project whose electricity output is presently unnecessary and for which less expensive and less environmentally damaging alternatives exist." It called on the British Columbia government to refer the project for review to the British Columbia Utilities Commission, and it urged both governments to suspend issuing any additional permits until the First Nations' challenges have been resolved by courts (Site C: Statement by Concerned Scholars 2016).

In what seems to be an unprecedented move, the Royal Society of Canada added its weight to the letter. The supporting letter to Prime Minister Trudeau is authored by Maryse Lassonde, the president of the Royal Society, and states: "A group of Canadian scholars, including several members of the Royal Society of Canada, have raised serious concerns regarding the process used for approval. . . . As President of the Royal Society, I am in agreement with the key issues they raise" (Lassonde 2016). That letter suggests her support was as an individual in her capacity as society president, but another page on the society's website suggests that it is an action of the society as a whole: "A group of leading Canadian scholars has raised serious concerns regarding the process used to approve the mega dam called Site C. The Royal Society of Canada has taken the unusual step of issuing a separate supporting letter addressed to Prime Minister Justin Trudeau" (Royal Society of Canada, n.d.).

While it is always a challenge to assess the influence of such letters, the statement was widely reported in the media and added weight to the message of environmentalists, First Nations, and local activists questioning the project's justification.

Public Opinion

While opposition to the Site C project among First Nations and environmentalists has been very strong, that opposition does not seem to have resonated with the British Columbia public in the same way that the oil sands pipelines have. Unfortunately, there does not seem to be any publicly available polling on Site C that is independent of the proponents or known critics prior to the 2017 election. However, BC Hydro commissioned surveys from Abacus from 2013 to 2016. As figure 9.1 shows, awareness of the project has been relatively high, especially since it was authorized in late 2014. Figure 9.2

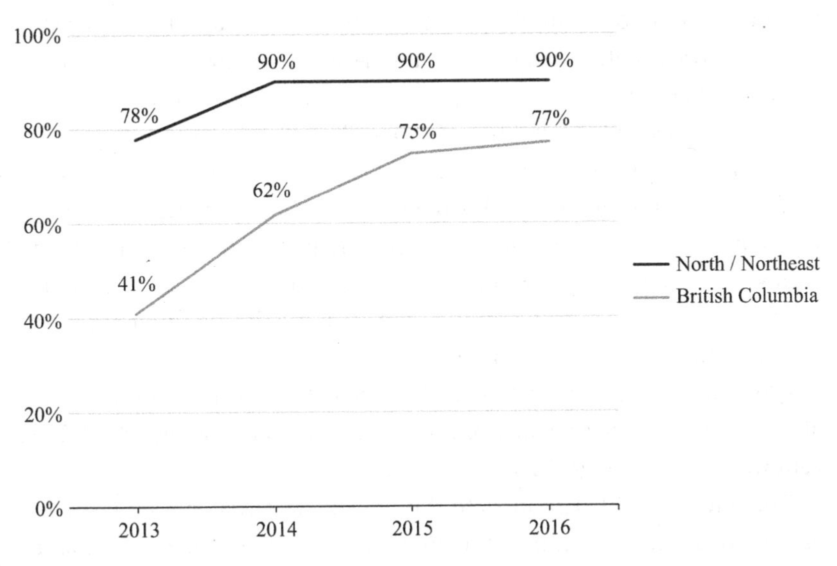

Figure 9.1
Awareness of Site C Dam project measured by responses to the following question: "Have you seen, read or heard anything about BC Hydro's proposed new dam, known as Site C, near Fort St. John?" *Yes Only.
Source: Abacus Data (2016b).

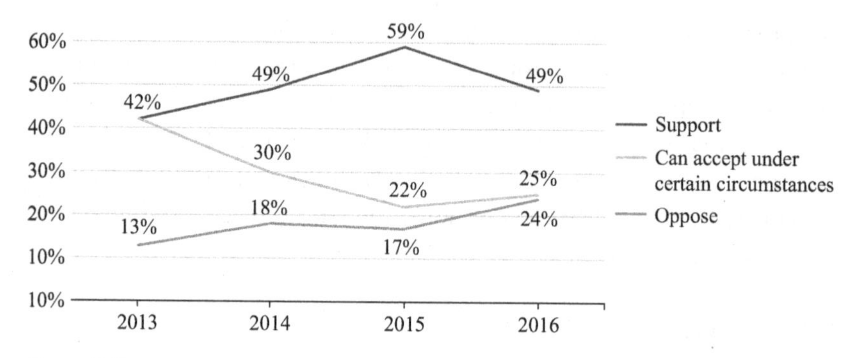

Figure 9.2
Support for Site C Dam project measured by responses to the following question: "Is the idea of building Site C, a new hydroelectric dam, to help meet the rising demand for electricity in BC, an idea you support, can accept under certain circumstances, or oppose?"
Source: Abacus Data (2016b).

shows levels of support. The project received support from 59% of the public in 2015, although it dropped back down to 49% in 2016. Opposition rose slightly from 18% in 2014 to 24% in 2016 (Abacus Data 2016b).

The other pre-2017 publicly available poll was commissioned by DeSmog Canada, an environmentally oriented online news magazine. Figure 9.3 shows its core result. When respondents are informed about changes in BC Hydro's demand projections, support for referring the project to independent review was quite high. Province-wide, 73% supported an independent review, while only 18% opposed it (Insights West 2016b).

The poll "found more British Columbians outright oppose the dam (44 per cent, 21 per cent strongly) than support it (39 per cent, 11 per cent strongly)" (Insights West 2016b). BC Hydro issued a media release denouncing the poll for errors. In particular, it challenged the core framing of the question that new power would not be needed until 2028, saying, "In fact, our load forecast indicates that without Site C, British Columbia would already have a capacity deficit of 8-per cent and an energy deficit of 2-per cent within 10 years" (BC Hydro 2016c).

The dueling polls demonstrate the obvious fact that survey responses are highly dependent on how the questions are posed. When the support/oppose question is framed in terms of meeting future electricity needs, support can be quite high. When the question is framed in a way that suggests the electricity will not be needed, support can be exceeded by opposition. In comparison to the pipeline cases, only the most critical framing can get

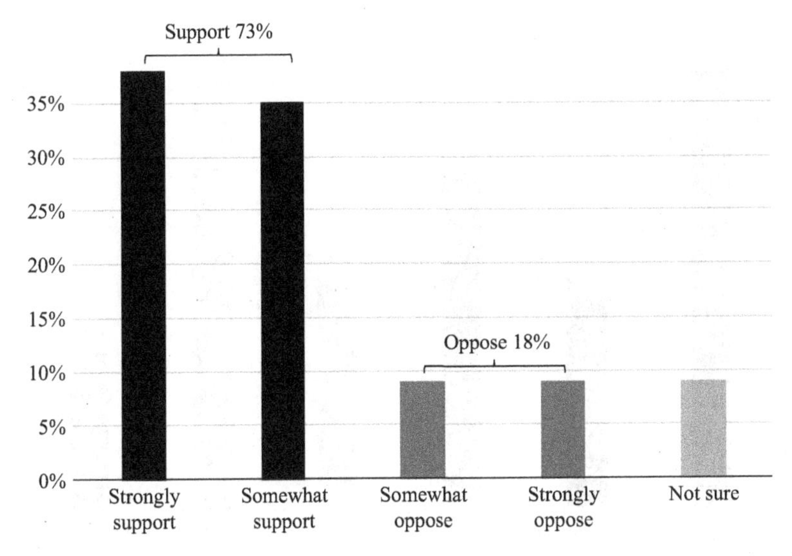

Figure 9.3
Results of DeSmog Canada poll of support for independent review of Site C Dam in response to the following question: "BC Hydro's projections indicate that the province will not need new power until 2028 at the earliest. The panel that reviewed Site C called for the demand case and cost to be examined in detail by the province's independent regulator. Thinking about this, do you support or oppose sending the Site C dam for an independent review of both costs and demand?"
Source: Insights West (2016b).

opposition up to levels found in more "neutral" surveys about the pipelines. Opposition to the Trans Mountain Expansion Project has vacillated between 36% and 57% (Insights West 2018).

After the 2017 provincial election, a plurality remained in favor of continuing construction. A September 2017 poll showed those in favor of completing the project exceeded those opposed by a 45% to 27% margin (Angus Reid Institute 2017). A poll shortly after the NDP's decision to continue with the project showed that 52% thought the government made the right decision and 26% thought it was the wrong decision (Angus Reid Institute 2018a).

Issue Framing

Like the two West Coast oil sands pipelines, Site C has been a major issue in provincial politics and has received an enormous amount of media

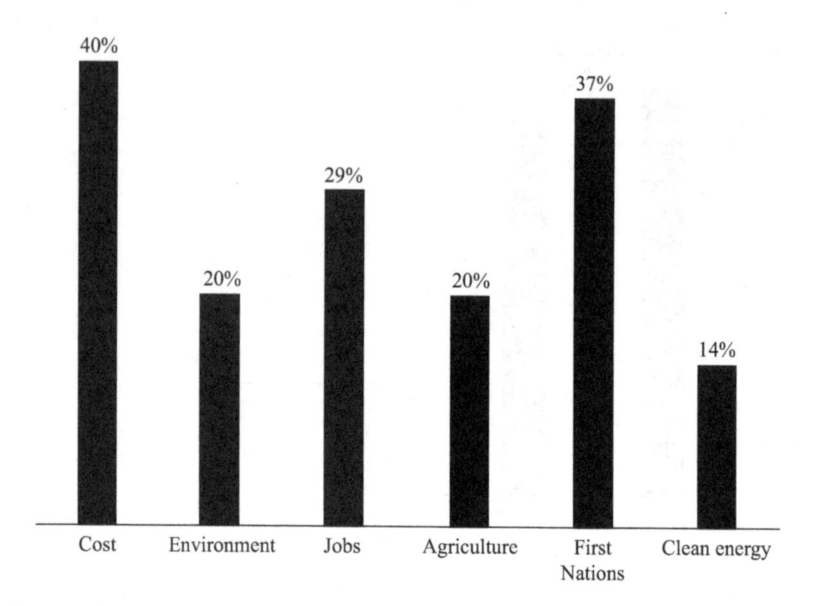

Figure 9.4
Percentage of total media mentions of the Site C Dam that name specific issues.

attention. The highest number of annual Site C mentions in the Canadian Newsstream database was 1,166 in 2016. That's very close to how prominent the Trans Mountain Expansion Project was in its peak year (1,860 mentions in 2016) but not nearly as prominent as Northern Gateway was at its peak in 2012, at 6,842 mentions.

Figure 9.4 analyzes what issues have been emphasized by the media in its reporting on Site C. The figure shows the percentage of media stories in the Canadian Newsstream database mentioning Site C that also mention six categories of issues: "cost"; "environment" or "environmental"; "agriculture," "agricultural," or "farm"; "First Nations"; and "clean energy." The results show that the two most prominent issues for the media have been project costs and issues with First Nations, with job benefits coming in third. Issues about environmental impact and loss of farmland were about half as prominent compared to costs and First Nations.

Institutions and the Politics of Structure

As discussed in previous chapters, institutional arrangements are critical to the balance of power over public policy because they specify who has decision-making authority and the rules of decision-making and participation. Because of their importance, strategic actors frequently struggle to reshape institutions to advance their interests (Baumgartner and Jones 2010). In the case of Site C, this "politics of structure" (Moe and Wilson 1994) centered around three major institutional issues: (1) the allocation of authority between the province and the federal government, (2) whether the project would be subject to review by an independent regulator, and (3) First Nations rights and title.

Interjurisdictional Agreement

The Site C Dam project was reviewed and approved in the Harper era. That government's emphasis on "responsible resource development" encouraged interjurisdictional cooperation in the regulatory review of major energy projects (Hoberg 2016). In the case of Site C, there was an agreement that the British Columbia Environmental Assessment Office and the Canadian Environmental Assessment Agency would cooperate on the project's review. In February 2012, the two agencies signed an intergovernmental agreement to conduct a joint review of the project.

The Joint Review Panel agreement provided that the provincial and the federal governments would each appoint one member, and the two governments would jointly agree on the chairperson. The terms of reference charged the panel with, among other things, assessing

- the purpose of and need for the project;
- alternatives to the project; and
- the environmental, economic, social, health, and heritage effects of the project, including the cumulative effects.

The agreement required the panel to report on issues of Aboriginal concern but expressly forbade the panel from making any findings about Aboriginal rights and title and the government's duty to consult. Section 2.5 of the agreement states:

> The Joint Review Panel will not make any conclusions or recommendations as to:
>
> (a) the nature and scope of asserted Aboriginal rights or the strength of those asserted rights;
>
> (b) the scope of the Crown's duty to consult Aboriginal Groups;
>
> (c) whether the Crown has met its duty to consult Aboriginal Groups and, where appropriate, accommodate their interests in respect of the potential adverse effects of the Project on asserted or established Aboriginal rights or treaty rights;

(d) whether the Project is an infringement of Treaty No. 8; and
(e) any matter of treaty interpretation. (Minister of the Environment, Canada and Minister of the Environment, British Columbia 2012)

Because the risks and benefits of the project are largely contained within the province of British Columbia, intergovernmental conflicts have not emerged as a significant issue on the project.

Insulation from Independent Regulator

In electricity regulation, the respective roles of elected politicians and independent regulators in energy planning and regulation are frequently fought over. British Columbia has an independent regulatory agency, the British Columbia Utilities Commission, charged with reviewing rates, long-term plans, and project justification and financing. As mentioned, the British Columbia Utilities Commission reviewed and rejected the Site C Dam in 1982. The agency became a thorn in the side of the government of Premier Gordon Campbell when it rejected the government's Long Term Acquisition Plan (LTAP) in 2008. The British Columbia Utilities Commission ruled that several critical parts of the plan were not adequately supported by rigorous economic analysis (Hoberg 2010).

Campbell responded by stripping the British Columbia Utilities Commission of much of its authority in the 2010 Clean Energy Act. The commission's authority to review electricity rates and long-term plans was removed, as was its authority to review several major projects—among them the Site C Dam. Because of this, the government has more discretion to move ahead with desired projects even if they can't be justified on economic grounds. Understandably, since costs have been such a dominant issue for the project, project opponents have been highly critical of this decision and have demanded that the project be submitted to the British Columbia Utilities Commission for review. But Premier Clark maintained her predecessor's position of refusing to have the project reviewed by the commission. When the NDP came to power in 2017, it made good on its campaign commitment to refer the project for a nonbinding review by the British Columbia Utilities Commission.

Shifting the Venue to the Courts

As in most major energy infrastructure issues in Canada, the role of First Nations in project decision-making has been enormously controversial in the case of the Site C Dam. Several of the region's First Nations remain adamantly opposed to the project and are challenging it in court. The evolution and status of these cases is discussed later in the chapter.

Project Decision-Making and Status

The Joint Review Panel's review process was highly formal and involved extensive documentation and detailed hearings. Recall that three major issues have dominated the Site C controversy: environmental impacts, including the loss of farmland, the economic justification for the project, and First Nations issues. This section reviews how environmental and economic issues were addressed throughout the project review process. First Nations issues were dealt with earlier, in the section on actors, and will be revisited in the section on litigation.

Environmental Impacts

Large dams flood valley bottoms and inevitably cause significant environmental impacts. There's no denying that fact, and BC Hydro did not deny this. Their environmental impact statement states that, "The conclusion of the substantial work undertaken to date indicates that the effects of the Project can largely be mitigated through careful project planning, comprehensive mitigation programs and ongoing monitoring during construction and operations. As a result, the Project is unlikely to result in a significant adverse effect on most of the [valued components—VCs]. However, a determination of significance has been made for the following VCs." The environmental impact assessment then lists the following significant adverse environmental effects:

- Fish and fish habitat, particularly to three distinct sub-groups of species, "the migratory Arctic grayling in the Moberly River, the migratory bull trout that spawn in the Halfway River and mountain whitefish that rely on Peace River habitat"
- Wildlife Resources Habitat for certain endangered migratory birds: Canada, Cape May and Bay-breasted Warblers, Yellow Rail and Nelson's Sparrow
- Vegetation and Ecological Communities including a marl fen, tufa seeps, old and mature riparian and floodplain forests, and species at risk plants

- Current Use of Lands and Resources for Traditional Purposes including "the loss of some important multi-use, cultural areas and valued landscapes, including sites at Attachie, Bear Flats and Farrell Creek." (BC Hydro 2013, 34)

The Joint Review Panel received a number of submissions that emphasized the severity of the environmental impacts of the dam. For the most part, the panel supported BC Hydro's conclusions about whether effects were significant. One exception was on wetlands. In that case, the panel disagreed with the proponent's claims that effects would not be significant (Joint Review Panel 2014, 64). The panel also concluded that the project was likely to pose significant risks to more sensitive and migratory species than BC Hydro claimed. In no case did the panel downgrade BC Hydro's assessment of significance, and it upgraded it several times.

Economic Justification

BC Hydro centered its argument for the project on a forecast that electricity demand is "expected to increase by about 40 per cent over the next 20 years" and that Site C was required to meet this demand, even taking into account the province's ambitious requirement of meeting two-thirds of new demand with conservation and efficiency. BC Hydro's environmental impact statement compared the Site C option with two other portfolios: a "clean generation" portfolio consisting of wind, biomass, and run-of-the-river power and a "clean+thermal generation" portfolio that would build new natural gas plants. Its cost analysis concluded that the "Site C portfolio provided material ratepayer savings" compared to the two other options. Adjusted unit energy costs for Site C were C$110 per megawatt-hour (2013 dollars) compared to the "clean generation" portfolio costs of C$181 per megawatt-hour and "clean+thermal generation" costs of $156 per megawatt-hour.

BC Hydro summarized the economic benefits this way:

> As a clean, renewable resource, the Project would deliver electricity with very low GHG emissions per unit of energy produced. Emissions would be comparable to other renewable sources such as wind and run-of-river hydro. As such, the Project will support both provincial and federal GHG reduction strategies. In addition, the dependable capacity provided by the Project will facilitate the integration of additional renewables into BC Hydro's system, supporting the Province's clean energy strategy. (BC Hydro 2013, 37)

The utility made the case for the net benefits of the project as follows:

> BC Hydro concludes that while the Project has the potential to result in some significant residual effects, they are justified by (1) the public interest served by delivering long term, reliable electricity to meet growing demand, (2) the employment, economic development, ratepayer, taxpayer, and community benefits that would result, (3) the ability of the Project to meet this need for electricity with lower GHG impact than other resource options, and (4) because the Project would take advantage of water already stored in the upstream reservoirs to generate over 35 per cent of the energy generated by BC Hydro's largest facility with only 5 per cent of the reservoir area. (BC Hydro 2013, 37)

It is important to note that neither of these concluding statements emphasizes the potential for Site C to add to British Columbia's capacity to support the integration of renewables in neighboring jurisdictions. The Joint Review Panel, however, did note these benefits at one point in describing BC Hydro's case for the dam's benefits (Joint Review Panel 2014, 273).[3]

The Joint Review Panel summarized the participants' views of the justification for the project, many of them critical. In a stunning assertion, it stated that "the Panel cannot conclude on the likely accuracy of Project cost estimates because it does not have the information, time, or resources. This affects all further calculations of unit costs, revenue requirements, and rates." It seems that it was the panel's job to perform this analysis, but they clearly felt the capacity to do so was limited. This conclusion was used as a justification for its recommendation 46: "If it is decided that the Project should proceed, a first step should be the referral of Project costs and hence unit energy costs and revenue requirements to the BC Utilities Commission for detailed examination" (Joint Review Panel 2014, 280).

The panel viewed BC Hydro's load forecasting as "sound" but was critical of the fact that it was not accompanied by a "long-term pricing scenario for electricity and its substitutes" (Joint Review Panel 2014, 287). Reminiscent of the British Columbia Utilities Commission's critique of the 2008 LTAP, the panel also concluded that "demand management does not appear to command the same degree of analytic effort as does new supply" (Joint Review Panel 2014, 291). On balance, the panel did raise questions about the methodology used to compare alternatives but seemed satisfied in the end with the strength of the economic argument for the project, saying that "in the long term, Site C would produce less expensive power than any alternative" (Joint Review Panel 2014, 298).

The panel's final conclusions were, depending on your perspective, either contradictory or nuanced. On the one hand, the panel emphasized the economic need for and benefit from the project: "The Panel concludes that B.C. will need new energy and new capacity at some point. Site C would be the least expensive of the alternatives, and its cost advantages would increase with the passing decades as inflation makes alternatives more costly." On the other hand, it questioned the timing of the decision: "The Panel concludes that the Proponent has not fully demonstrated the need for the Project on the timetable set forth" (Joint Review Panel 2014, 305–306).

Government Response

Despite the mixed blessing of the Joint Review Panel, both the federal and the provincial governments issued approvals for the project. The British Columbia government issued an environmental assessment certificate in October 2014, and approved the project for construction in December 2014 (Government of British Columbia 2014). It did so despite the Joint Review Panel's acknowledgment that it did not have the capacity to scrutinize the economic justification for the project and its recommendation to refer it to the British Columbia Utilities Commission. The government of Canada approved the project (with 80 conditions) in October 2014 (Canadian Environmental Assessment Agency 2014). As is now customary in federal government decisions on major projects where the assessment concludes that there are significant environmental impacts, the decision statement simply asserts that the adverse effects are "justified in the circumstances." Despite a number of legal challenges, BC Hydro initiated construction in late July 2015, with an expected completion date of 2024.

2017 Election and the NDP Government

The May 2017 election produced a minority NDP government, supported by the three-member caucus of the Green Party. Acting on its campaign commitment and postelection agreement with the Greens, the government referred the project to the British Columbia Utilities Commission to perform an independent review in August 2017. The British Columbia Utilities Commission was not asked to redo a full assessment of the project's merits but rather to assess the economic consequences of three options: (1) completing the project as planned, (2) suspending construction with the option to resume it later, and (3) terminating construction and remediating the site.

Despite the compressed timeframe, the commission heard from 304 speakers and received 620 written submissions. The process revealed the same pattern of interests and positions as the Joint Review Panel had found. Environmentalists (and the academics supporting their arguments) emphasized that BC Hydro's demand forecast was outdated and exaggerated, and that the cost of alternatives had declined significantly. The clean energy industry pushed the same arguments, and major power consumers continued to stress their opposition to the project because of concerns about increased industrial rates. The construction industry continued its enthusiastic support for the project, and BC Hydro did its best to defend the project's merits.

As anticipated, the commission was not impressed with the project's economics. It did not think the project would be completed on time or within its C$8.3 billion budget, projecting that completion would cost over $10 billion. It warned against taking the suspend-and-restart option seriously, concluding that it was by far the least desirable of the options economically and practically. The costs to terminate and remediate were calculated to be C$1.8 billion, on top of which the costs of replacement energy would need to be added. The commission agreed with many of those submitting evidence that BC Hydro's demand forecast was "excessively optimistic." The key conclusion, and the closest it came to a recommendation, was the statement that "the Panel believes increasingly viable alternative energy sources such as wind, geothermal, and industrial curtailment could provide similar benefits to ratepayers as the Site C project at equal or lower Unit Energy Costs" (British Columbia Utilities Commission 2017, 3).

Making the Best of a Bad Deal: The NDP Government Commits to Site C Completion

In the wake of the British Columbia Utilities Commission's report, a number of commentators were confident it gave the NDP government the ammunition it needed to terminate the project (Cox 2018, 250). But on December 11, 2017, NDP premier John Horgan, appearing uncharacteristically dour, announced that his government had decided to complete the Site C Dam project.[4] In making the announcement, Horgan made clear his opinion of the previous government and the "terrible situation" in which his government found itself:

> It's clear that Site C should never have been started. But to cancel it would add billions to the Province's debt—putting at risk our ability to deliver housing, child care, schools and hospitals for families across B.C. And that's a price we're not willing to pay.

> We will not ask British Columbians to take on $4 billion in debt with nothing in return for the people of this province and, even worse, with massive cuts to the services they count on.
>
> The old government recklessly pushed Site C past the point of no return, committing billions of dollars to this project without appropriate planning and oversight. Our job now is to make the best of a bad deal and do everything possible to turn Site C into a positive contributor to our energy future. (Government of British Columbia 2018)

Critics objected to Horgan's assertion that the project's sunk costs (C$2.1 billion spent and $1.8 billion in remediation costs) would need to be imposed on ratepayers or taxpayers in the short term, but Horgan obviously felt boxed in by how far the project had progressed before he formed a government. Politically and fiscally, Christy Clark had succeeded in pushing the project "past the point of no return." Andrew Weaver of the Green Party opposed the decision but decided not to remove the Greens' support for the NDP minority government (Bailey 2017).

The Enduring Battle in the Courts

Site C opponents have filed multiple lawsuits against the approval decisions by the governments of British Columbia and Canada, but thus far Canadian courts have been decidedly unfriendly to project opponents. Table 9.2 provides an overview of the litigation and its status. Thus far, all lawsuits, whether by First Nations or environmental groups, that challenged the project's approval have been dismissed. The only case decided for the plaintiff was when BC Hydro applied to have an injunction enforced to remove protesters from disrupting construction.

The Peace Valley Landowner Association challenged the British Columbia government`s decision not to, among other things, refer the project to the British Columbia Utilities Commission, but the British Columbia Supreme Court and Court of Appeal ruled it was within the cabinet's prerogative to decide. The group also challenged the federal government's decision for not sufficiently justifying its finding that the "significant adverse environmental effects were justified in the circumstances" (*Peace Valley Landowner Association v. Canada (Attorney General)* 2015 FC 1027). But as in other cases that have sought to challenge terse federal cabinet decisions under the Canadian Environmental Assessment Act, the Federal Court ruled that the cabinet's justification was sufficient.

Table 9.2

Status of Site C litigation

Case	Court and citation	Subject	Most recent action
Prophet River First Nation and West Moberly First Nation v. British Columbia et al.	British Columbia Supreme Court	Civil claim that Site C would infringe on Treaty 8 rights; request for permanent injunction	Request for interim injunction denied October 25, 2019; negotiations under way while parties prepare for trial
Prophet River First Nation and West Moberly First Nation v. BC Hydro	British Columbia Court of Appeal (on appeal from British Columbia Supreme Court), 2017 BCCA 58	Lack of an unjustified infringement determination and adequacy of consultation	Dismissed February 2, 2017
West Moberly First Nation and Prophet River First Nation v. BC FLNRO	British Columbia Supreme Court, 2016 BCSC 2007	Adequacy of consultation	Dismissed October 13, 2016
Prophet River First Nation v. Canada	Federal Court of Appeal (on appeal of Federal Court decision), 2017 FCA 15	Whether the federal cabinet was required to consider whether environmental effects constitute infringement	Dismissed January 23, 2017
BC Hydro v. Ken Boon et al.	British Columbia Supreme Court, 2016 BCSC 355	Injunction against protesters blocking construction	Injunction granted February 29, 2016
Peace Valley Landowner Association v. British Columbia Ministry of Environment	British Columbia Court of Appeal, 2016 BCCA 377 (on appeal from the British Columbia Supreme Court)	Whether the government of British Columbia could choose not to consider Joint Review Panel recommendations for future government regulation of BC Hydro	Dismissed September 15, 2016
Peace Valley Landowner Association v. Canada (Attorney General)	Federal Court, 2015 FC 1027	Whether the federal cabinet sufficiently justified the significant adverse effects	Dismissed August 28, 2015

The Prophet River First Nation and West Moberly First Nation challenged the British Columbia and federal decisions on the grounds that the environmental effects of the dam infringed on their treaty rights and that the governments did not consult sufficiently. Four decisions, two in British Columbia courts and two in federal courts, completely rejected the First Nations' arguments in the form in which they had been presented. On the question of infringement, the courts ruled that in making its decision the cabinet (in the federal case) and ministers (in the provincial case) were under no obligation to make a determination of infringement on treaty rights. The courts ruled that the proper vehicle for considering the question of infringement was through a civil action, not a judicial review.

Several of the decisions did carefully scrutinize the record on consultation and ruled that it was sufficient, rejecting what they saw as the First Nations' effort to transform the duty to consult into a veto of the project. The British Columbia Court of Appeal's decision ruling is worth quoting at some length:

> The duty to consult and accommodate does not afford First Nations a "veto" over the proposed activity: Mikisew at para. 66. Here, the appellants have not been open to any accommodation short of selecting an alternative to the project; such a position amounts to seeking a "veto." They rightly contend that a meaningful process of consultation requires working collaboratively to find a compromise that balances the conflicting interests at issue, in a manner that minimally impairs the exercise of treaty rights. But that becomes unworkable when, as here, the only compromise acceptable to them is to abandon the entire project. (*Prophet River First Nation v. British Columbia (Environment)* 2017 BCCA 58)

The two First Nations applied for leave to appeal to the Supreme Court of Canada, but in June 2017 those applications were rejected, bringing that line of litigation to finality. These decisions revealed again how far current Canadian law is from the standard of consent many First Nations expect from the Crown's commitments to fully implement the UN Declaration on the Rights of Indigenous Peoples.

After the NDP government's decision to proceed with the Site C Dam project, Prophet River and West Moberly launched a new line of legal attack. In January 2018, the two groups filed a notice of civil claim, arguing that the dam would be an infringement on their Treaty 8 rights and asking for a permanent injunction. The proceedings would involve a lengthy trial process, so the two groups applied for an interim injunction as well. The injunction request was denied on October 25, 2018. The judge ruled that the First

Nations' request failed the "balance of convenience" test, arguing that "the massive scale of the Project—that is, the very thing that makes its impact on West Moberly's treaty rights so significant—also amplifies commensurately the harm weighing on the other side of the scales if the Project were to be enjoined" (*West Moberly First Nations v. British Columbia*, paragraph 334).

Conclusion

Keeping global warming within manageable limits requires a massive and rapid transformation of the energy system to virtually eliminate greenhouse gas emissions, yet renewable energy projects are frequently stymied by conflict and delay. This chapter has examined the case of a large "clean energy project," the Site C Dam, a C$10 billion, 1,100-megawatt project in northeastern British Columbia. Despite the potential benefit of producing low-carbon power for generations, the project has been vehemently opposed by a broad coalition of environmental groups, First Nations, and academics. Even if some environmentalists might have been inclined to support the project because of its contribution to the clean energy transition, their ideological and strategic commitment to honoring First Nations' demands for consent on major project decision-making prevented them from doing so.

While BC Hydro unquestionably engaged in extensive discussions with First Nations, there is no evidence that their engagement efforts went deep enough to be successful with the most directly affected First Nations. While it may be that no offer could have overcome the opposition of the West Moberly and Prophet River First Nations, it is telling that there is no indication in the record of discussions about providing First Nations partnership or equity in the project, a strategy that has been successful in other highly contentious resource projects. First Nations' lawsuits trying to block the project have received one setback after another.

While Site C would unquestionably result in significant environmental impacts in the Peace River region, it is striking how little recognition there has been of the project's contribution to the broader decarbonization agenda. What's missing in much of the opposition critiques, especially with respect to economic justification, is the broader continental view. The fact that power may not be needed in British Columbia at the time of project completion misses the point about the project's broader potential contribution to decarbonization. British Columbia is part of the Western Electricity Coordinating

Council, which includes Alberta and the western United States. Site C's most valuable contribution might well be to provide storage to firm up intermittent renewables as their penetration increases to replace polluting fossil fuel power plants (Jaccard 2017). While passingly mentioned in the Joint Review Panel report, this benefit was barely mentioned in BC Hydro's submissions and was completely ignored by environmental critics of the project.

The case is a likely indication of the power that committed governments retain to overcome resistance to new infrastructure. Given the analytical framework used in this book, project opponents would be expected to be quite formidable. Risks were highly concentrated in the Peace River region. Opponents had access points, first to the courts and then to the Horgan government after 2017. They put their case to every potential obstacle available to the project—the review itself, the courts, the British Columbia Utilities Commission, and a new, more environmentally oriented government. While it was downstream from two large dams, significant new infrastructure was required: a valley needed to be flooded, and a large dam had to be constructed with associated turbines, substations, and high-voltage transmission lines. The risks to the Peace River region are clearly separated from the benefits of faraway power consumers.

The Site C Dam seems to have overcome the political and legal resistance to it. Christy Clark gave the Site C project such enormous momentum that the May 2017 election had limited impact on its trajectory. One advantage that the BC Liberals had is that Site C resistance—virtually the same coalition of actors fighting the pipelines in British Columbia—has not been able to capture mass public opinion in the same way that the campaigns against Northern Gateway and Trans Mountain did. That made it easier for Christy Clark's government to forge ahead despite the resistance campaign and, once construction outlays had gotten big enough, for the Horgan government to reluctantly continue. Delays have already cost the project at least four years, and new revelations of geo-technical risks in the summer of 2020 created new concerns about the viability of the project. After another investigation, Horgan announced again that he would continue the project, now with a staggering C$16 billion cost and a delay of completion by one year, till 2025.

10 How Resistance to Renewable Energy Infrastructure Might Frustrate Climate Solutions

Large hydroelectric dams have significant environmental impacts by their very nature, and it is not surprising that a dam like Site C would attract considerable opposition from environmental groups. The resistance challenge is not confined to large dams. Many other types of renewable energy generation and transmission facilities have confronted determined opposition from local groups. Solar and wind power projects, vital to replacing fossil fuels for electricity generation, have generated controversy from local groups concerned about property values, changes to species habitats, landscapes, aesthetics, and human health. New high-voltage electric transmission lines have also attracted significant resistance. Renewable energy projects are frequently in quite different locations than fossil fuel infrastructure, so new transmission lines are usually required to supplement the buildout of new renewable energy sources. In addition, the integration of intermittent renewables into the electricity grid is projected to require significant new transmission capacity and deeper integration across larger geographical areas.

Chapter 9 reviewed the analytical framework in light of the differences between fossil fuel and renewable energy projects. This chapter continues the investigation into this book's second question: does the decision to ally so closely with place-based opponents risk creating a "resistance dilemma" by legitimizing place-based resistance that can then be mobilized to thwart needed clean energy infrastructure? It does so by examining a broader range of cases of resistance to renewable energy. After a brief overview of the literature on the social acceptance of renewables, the chapter will address, in turn, conflicts over wind power projects in central Canada and New England, solar power projects in California, and transmission line projects within California and between Quebec and New England states.

Most of this analysis will be based on a review of published literature and government documents. This chapter examines what motivated resistance campaigns and how much impact they have had in thwarting or altering proposed renewable power infrastructure projects.

The purpose of highlighting this resistance dilemma is not to question the legitimacy, sincerity, or efficacy of place-based resistance. Nor is it to cast blame on the climate movement or challenge the wisdom or merits of its place-based resistance to fossil fuel infrastructure. Place-based resistance has a long tradition predating its adoption by climate activists, and place-based resistance to renewable energy infrastructure—be it against large dams or rural area wind farms—was a widely used strategy before Bill McKibben discovered the Keystone XL pipeline. Rather, the reason to highlight the resistance dilemma is because there is an urgent need to build massive amounts of clean and renewable energy infrastructure, and our institutions have historically not been effective at resolving the tensions between local desires to minimize impacts and the broader public interest in establishing needed infrastructure. In short, we have a process legitimacy crisis, and we need to address that in order to effectively address the climate crisis.

Literature on Social Acceptance of Renewables

Widespread resistance has spawned a substantial body of literature on the social acceptance of renewable energy (e.g., Wustenhagen, Wolsink, and Bürer 2007; Cleland et al. 2016; Batel et al. 2013; Devine-Wright 2009; Fast 2013). One pervasive theme is the importance of local values. Most scholars writing in the field reject the "not in my backyard" (NIMBY) framing of place-based opposition, insisting instead on the imperative of respecting the attachment of people to place (Devine-Wright, Devine-Wright, and Cowell 2016; Fast et al. 2016; Sovacool and Ratan 2012; Hyland and Bertsch 2018). One systematic review of the literature concludes that "local communities may be more willing to accept projects if developers site and design them in ways that work with, rather than against, local identities and people's attachment to specific places" (Devine-Wright, Devine-Wright, and Cowell 2016, 5). In addition, there is a significant amount of evidence that project resistance can be effective, either by contributing to the demise of undesirable projects or by leading to concrete project improvements that address the concerns of local critics (Hager and Haddad 2015).

Another virtually universal theme throughout the social acceptance literature is an emphasis on engaging host communities meaningfully and early in the process and demonstrating how community input influenced project design (Devine-Wright, Devine-Wright, and Cowell 2016; Fast et al. 2016). Failure to do so frequently leads to "public enquiry, prolonged planning delays, additional expense, and local community distrust in network organisations" (Cotton and Devine-Wright 2012). Cotton and Devine-Wright (2013) argue that "stronger collaborative or partnership planning approaches, devolved power arrangements and stronger local community scrutiny of developer applications are justified, both on ethical grounds to support procedural justice, and on strategic grounds to ameliorate public opposition and the risk of planning failure."

A third theme in the social acceptance of renewables literature is the importance of providing economic benefits to those affected by the project. Some studies emphasize the importance of community ownership or shared ownership in fostering public acceptance (Cleland et al. 2016; Devine-Wright, Devine-Wright, and Cowell 2016). Others find that local economic benefits are more important than actual ownership per se (Hyland and Bertsch 2018). Regardless, there is a general consensus that some form of substantial community benefit is essential. All three of these themes are apparent in the following case studies of resistance to renewable energy infrastructure.

The fact that there is significant literature about local resistance to new renewable energy infrastructure does not mean that all, or even most, new projects face resistance. Recent work by Giordano et al. (2018) uses an effective research design that moves beyond the opinion surveys or small-number case studies that have dominated the literature. Their study examined 53 proposed wind projects in the United States and "found evidence of at least one type of opposition mobilization activity in 43 of the 53 proposals (81%), suggesting that some oppositional activity was fairly common among proposed wind projects in the selected states. However, only 19 of the 53 proposals (36%) experienced three or more types of opposition mobilization activities (thus scoring in the set of opposed cases by our measure), suggesting that more involved opposition efforts were rarer" (Giordano et al. 2018, 126).

They categorize resistance to wind project proposals as "on the whole, relatively moderate." They also emphasize that resistance activities increase when more levels of government are involved, because they increase access points for project opponents (Giordano et al. 2018, 129).

All these themes are consistent with the analytical framework presented in chapter 1 and applied thus far to the five large projects. The projects most likely to attract strong opposition are those with salient place-based risks and few or no economic benefits to those same communities. The greater the local benefits, the easier it is for projects to overcome concerns about local impacts. Opposition is more formidable when it has access to multiple veto points. The remainder of this chapter examines six cases of renewable energy resistance in Canada and the United States.

Conflicts over Wind Power in Ontario

Ontario Decarbonization Policy

In Canada, the most significant resistance to renewable energy infrastructure has been to wind power in Ontario. Beginning in 2004, the government of Ontario, controlled by the Ontario Liberal Party under Premier Dalton McGuinty, undertook a bold decarbonization initiative to phase out coal-fired electricity generation, which in that year made up about one-fifth of the province's electricity supply. To diversify its low-carbon supply mix, Ontario initiated a feed-in tariff program in 2006, offering a guaranteed price for hydro, wind, solar, and biomass facilities for a 20-year contract. The province became more ambitious in 2009, when the McGuinty government enacted the Green Energy and Green Economy Act (hereafter the Green Energy Act).

The Green Energy Act expanded the feed-in tariff program. In addition to increasing the subsidized rates, the Green Energy Act made a number of other changes designed to reduce barriers to rapid development of renewable energy. Local transmission companies were required to connect renewable projects to the grid and grant them priority access (Fast et al. 2016; Loudermilk 2016).[1] In order to expedite approvals and installation, the renewable energy approval process was created with the goal of having decisions within six months of project submission. Approved projects would be entitled to certain exemptions from the permit requirements under the Environmental Protection Act and Ontario Water Resources Act. Modest community consultation requirements were also included (Walker 2010).

Most controversially, the province also amended the Planning Act to remove direct control over land-use decisions from municipal governments (Fast and Mabee 2015; Fast et al. 2016). In a speech to the London Chamber of Commerce, Premier McGuinty justified the change as necessary to avoid the "not-in-my-backyard" syndrome thwarting development of renewable energy, saying:

> We're going to find a way through this new legislation to make it perfectly clear that NIMBYism will no longer prevail when it comes to putting up wind turbines, solar panels and bio-fuel plants. . . . Our new law will uphold rigorous safety and environmental standards, but once those standards have been met, we intend to assert the greater public interest in clean, green electricity and the jobs that come with it. Municipalities will no longer be able to reject wind turbines, solar panels or bio-fuel plants because they don't like them. We can't allow interests to oppose these simply because they don't like them. (CBC 2009)

At the time, these changes were enormously popular with the public. A poll shortly before the Green Energy Act was enacted showed 87% of respondents approved of the proposed act. Support was high even where resistance to wind power had been reported as a result of projects proposed under the 2006 policy (CNW 2009).

The Rise of Resistance to Wind Power

As communities learned about proposed wind projects, however, "a fierce and well-organized backlash" emerged (Mulvihill, Winfield, and Etcheverry 2013, 10). Not all rural residents in areas where projects were proposed were opposed to them. Many residents were either neutral toward the projects or supportive, seeing their "green" development attributes as consistent with their rural lifestyle and promoting their livelihoods (Fast, Mabee, and Blair 2015). But intense, organized, and vocal opposition also emerged. By 2011, local wind resistance groups had emerged in every provincial electoral district with a wind turbine (Stokes 2013). Wind Concerns Ontario was created as a coalition of community groups, and by 2011 it had 50 local chapters (Stokes 2013). The groups were successful at mobilizing municipal politicians. By 2011, 78 municipalities had passed resolutions against wind turbines (Stokes 2016).

Opposition resulted from a combination of concerns. One significant trigger was visual and cultural: to many in rural areas, wind turbines reflected an industrialization of the landscape that was anathema to their sense of place (Fast and Mabee 2015). Second, there were also more pecuniary concerns about property values. A 2013–2014 study found a significant reduction in housing prices within 5 kilometers of turbine sites in two communities with turbines along Lake Ontario, but interestingly not for those within

1 kilometer. Properties closer to the turbines may have lease agreements with developers, so they would benefit financially in a way that more distant properties would not (Fast, Mabee, and Blair 2015; Christidis and Law 2012).

Third, human health concerns have been one of the biggest issues in the Ontario conflict. "Wind turbine syndrome," as it came to be called, emerged as nearby residents reported concerns with sleep interruption, headaches, fatigue, dizziness, ear irritation, concentration problems, and irritability. Wind opponents argued that these impacts were the result of a combination of mild noise (a whirring sound from turbine blade movements), vibrations, and visual light flickering based on sun position and shadow effects. Health criticisms have persisted despite the absence of any credible evidence linking proximity to wind turbines with any physical ailments (Christidis and Law 2012; Knopper and Ollson 2011). In 2010, the Ontario Chief Medical Officer of Health published a comprehensive review of the evidence. The report reinforced that no known links exist between wind turbine noise and sleep issues, dizziness, or headaches but did acknowledge that residents may find it annoying. The report stated that improved community engagement may alleviate concerns about proposed wind turbine projects and that community attitudes and perceptions are related to perceived levels of annoyance (Ontario Chief Medical Officer of Health 2010; Fast et al. 2016).

While concerns over visual impacts, property values, and wind turbine syndrome have dominated Ontario wind resistance discourses, various scholars have emphasized how the institutional arrangement around wind power contributes to the resistance, both directly by creating a backlash against those who feel excluded and indirectly in how alienation and annoyance contribute to perceived health impacts or a more general reduction in well-being. The two most consequential features contributing to resistance are the stripping of planning authority from local governments and the dearth of community-owned projects (Fast et al. 2016; Mulvihill, Winfield, and Etcheverry 2013; Walker and Baxter 2017). Wind Concerns Ontario denounced the accelerated approval process for "tearing apart the fabric of rural Ontario" (Stokes 2013, 495). Chapter 11 will address these and other contributors to social acceptability of renewable energy technologies.

The resistance movement was effective at mobilizing for the 2011 provincial election, in which wind turbines became a highly contested issue. Despite the quality of the wind resource, proposals for offshore turbines in the Great Lakes were met with vehement resistance. In advance of the election, the governing Liberals placed a moratorium on offshore siting of wind turbines (Mulvillhill, Winfield, and Etcheverry 2013). Premier Dalton McGuinty's governing Liberal Party lost nearly all their rural seats and lost their majority, but remained in power with a minority government. Stokes (2016) estimated that the opposition to wind power cost the governing Liberals between 4% and 10% of the vote from residents living within 3 kilometers of a proposed or operational wind turbine. In the 2014 provincial election, the Liberals, by then led by Kathleen Wynne, succeeded in recovering their majority by winning an additional 10 seats.

Policy Revisions

In moving forward with feed-in- tariff contracts after 2011, the government attempted to remedy issues with wind turbine resistance by prioritizing projects with a clear demonstration of community backing via municipal council resolutions. This, however, turned out to work against the initiative, as nearly a quarter of the province's municipalities, 89 in total, passed resolutions stating that they were "unwilling hosts." The provincial government ended the feed-in tariff program for projects over 500 kilowatts in 2013 amid widespread criticism (Fast et al. 2016).

During its existence from 2009 to 2013, the feed-in tariff program resulted in 61 contracts for large (>500 kilowatts) wind facilities, creating 3,100 megawatts of capacity (Fast et al. 2016). Ownership was skewed toward large, foreign-owned wind energy companies. According to Fast et al. (2016), "there is only one feed-in tariff project with cooperative ownership and several with partial aboriginal ownership, despite the existence of incentives for cooperative and aboriginal-owned projects."

In 2015, the provincial government introduced a new program for wind development, this time through a competitive bid process rather than a feed-in tariff. This system lent preference to bids that clearly demonstrated prearranged positive commitment from local governments and at least 75% of local landowners in signed agreements (Fast et al. 2016).

Estimated Impact of Resistance

There is no comprehensive analysis examining the impact of wind resistance on project cancellations, delays, or costs. One indicator of community resistance can be found in both appeals of permitting decisions and political mobilization. According to Fast (2016), up to the time of his study, of the 29 wind projects approved, 26 of them had been appealed to the Environmental Review Tribunal. While only one of the appeals led to the project being canceled, the other appeals resulted in delays to the projects.

One revealing study examined the impact of the process reforms on project timing. Part of the Green Energy Act's express purpose was to facilitate project development by streamlining review processes and by taking authority away from local governments, avoiding local resistance that could lead to project delays and cancellations. A study by Margaret Loudermilk shows that process reforms do not seem to have worked as intended. Despite all the measures to facilitate project approval, the time elapsed between project application and operation was no faster after the reforms than before (Loudermilk 2017). The study does not explicitly examine how much community opposition contributed to the failure of the process reforms to expedite project development.

Despite the explicit intent of preventing place-based resistance from thwarting development of renewable energy projects, the decision to take approval authority away from local government seems to have had the exact opposite effect: increasing local resistance to wind turbines. According to Fast and Mabee (2015, 9), "removing local planning authority over wind projects has had the most negative repercussions" for community support for project development.

Despite this resistance, Ontario has made enormous strides in decarbonizing its electricity sector since it began phasing out coal. In 2005, coal made up 21% of capacity and 19% of energy generation. It was completely phased out in 2014 (IISD 2015). Wind made up less than 0.1% of capacity in 2005 and had grown to 12% of installed capacity and 8% of electrical energy generation by 2016 (National Energy Board 2017e, 20). As of December 2017, Ontario had 94 wind installations with a total of 2,515 turbines, for a total installed capacity of 4,900 megawatts (Canadian Wind Energy Association, n.d.). Electricity-sector greenhouse gas emissions declined from 32 million tonnes in 2005 to 4 million tonnes in 2017, a remarkable 88% reduction (Government of Ontario, n.d.).

Election-Induced Policy Reversal

In June 2018, a Conservative majority government was elected, making Doug Ford Ontario's premier. Climate policy and renewable energy were among the salient issues in the campaign. Ford has been quite hostile to

Table 10.1
Issue ranking and party advantage, Ontario 2018 election

Issue	Percentage of Ontarians ranking issue in top three	Which party is best at dealing with the issue? Party and percentage advantage (second-closest party)
Health care	54%	NDP +11% (Liberals)
Economy and jobs	36%	Conservatives +19% (NDP)
Lower taxes	29%	Conservatives +35% (NDP)
Lower energy costs	28%	Conservatives +19% (NDP)
Debt repayment, balanced budget	19%	Conservatives +36% (NDP)

Source: IPSOS (2018a).

the green energy agenda. His election platform directly linked the Green Energy Act to higher electricity prices (referred to as "hydro" in much of Canada), saying, "For too long, well-connected insiders have been getting rich off your hydro bills. The Green Energy Act alone represents Ontario's largest-ever wealth transfer from the poor and middle class to the rich" (Ontario Progressive Conservatives 2018). As table 10.1 shows, energy costs were a significant issue in the campaign, ranking fourth among issues in a preelection poll, with 28% of respondents saying energy costs were among the top three election issues. Of those who believed energy costs were a significant issue, Ford's party had a 19% advantage over the second-place NDP and a 38% advantage over the governing Liberal Party (IPSOS 2018a). A poll taken later that month found that 61% of respondents said high electricity prices would affect their vote, and among those the Conservatives had a modest advantage over the NDP and a large advantage over the Liberals (IPSOS 2018b). Ford's first act after becoming premier was to dismantle the province's cap and trade program. Several days later, he canceled 759 renewable energy contracts that were in the works.

Offshore Wind in New England

The Cape Wind Project was a proposed offshore wind farm of 130 wind turbines in the Horseshoe Shoal region of Nantucket Sound, off Cape Cod, Massachusetts, in the United States. The project, which would be the first offshore wind facility, was proposed by Cape Wind Associates, LLC, and developed by Jim Gordon in 2001 as part of the US offshore wind power development plan intended to generate 1,500 gigawatt-hours of electricity per year. The project was expected to generate a maximum electricity capacity of 468 megawatts, with an average output of 174 megawatts (BOEM, n.d.).

In November 2001, Cape Wind Associates filed a permit application for the wind farm with the Army Corps of Engineers, the federal agency regulating offshore wind power projects. In 2005, regulatory authority over offshore wind energy projects was delegated to the Bureau of Ocean Energy Management (BOEM) in the Department of the Interior. Because of these changes in the regulatory authority, the Cape Wind Project suffered a setback in completing its environmental impact statement, which was finally published in January 2009. In October 2010, Cape Wind Associates signed its commercial offshore renewable energy lease after the Department of the Interior approved its issuance in April 2010.

However, the Cape Wind Project faced relentless opposition and protracted court challenges for over 10 years. In July 2016, the US Court of Appeals for the District of Columbia rejected the government approvals for the project on the basis that Cape Wind Associates had not been able to obtain "sufficient site-specific data on seafloor and subsurface harzards" (Cassell 2016). Eventually, the shifting regulatory hurdles and legal challenges resulted in the failure of the Cape Wind Project to meet its contract commitments to sell power to local utilities, the National Grid, and NSTAR,[2] and thus the project was terminated in December 2017 (Seelye 2017; BOEM, n.d.).

The following analysis consists of two parts. The first part examines what motivated the resistance campaign against the project, including aesthetic concerns, environmental impacts, and decreased values of shorefront estates. The second part discusses the demise of the Cape Wind Project. Its slow death was caused by the initial absence of a regulatory framework for offshore renewable energy projects and strategic use of the American court system by the small but well-funded and highly effective opposition, the Alliance to Protect Nantucket Sound.

What Had Motivated the Resistance Campaign?

First, the resistance campaign was led by the Alliance to Protect Nantucket Sound, a nonprofit environmental organization established in 2002 in response to the proposed Cape Wind Project and dedicated to preserving

Nantucket Sound as a protected area. The opposition also included Public Employees for Environmental Responsibility, the Cape Cod Chamber of Commerce, which successfully galvanized support from local businesses, the Humane Society, and Barnstable Land Trust, which were powerful local conservation organizations (Watson and Courtney 2004). The proposed location of the project had ushered in aesthetic and cultural concerns.

Project opponents were concerned that the 130 wind turbines would jeopardize the tourism value of the region and turn Nantucket Sound into an "industrialized" site (Watson and Courtney 2004; Ejima et al. 2015). Walter Cronkite, the late legendary broadcaster, denounced the project by proclaiming, "Our national treasures should be off limits to industrialization" (Burkett 2003). The resistance campaign also gained support from the Wampanoag Tribe of Gay Head (Aquinnah). The Aquinnah claimed that Nantucket Sound should be protected as a sacred area because their ancestors once lived on land that is now covered by the waters of Nantucket Sound (Love 2014). On July 6, 2011, they filed a lawsuit against the federal government, given that Secretary of the Interior Kenneth Salazar had issued the federal approval of the project in April 2010 (Toensing 2011; *Town of Barnstable, Massachusetts et al. v. Ann G. Berwick et al.* 2014).

Second, because the Cape Wind Project was the first offshore wind farm in the United States, the opposition raised concerns about navigation and potential environmental impacts to marine life, migratory birds, and especially seafloor and subsurface hazards (Ejima et al. 2015; Cassell 2016). The opposition expressed these environmental concerns through the Bureau of Ocean Energy Management within the Department of the Interior, which was the regulatory authority of the project, at a US district court in March 2014. Despite the dismissal of the case by the district court in November 2014, the Alliance was successful at the court of appeals in further delaying the Cape Wind Project because the BOEM was required to undertake adequate geological surveys before any construction could begin (Cassell 2016).

Third, the Alliance's relentless resistance campaign had a strong economic motive because of concerns over the potential decrease in value of shorefront estates of wealthy families. These properties were owned by the Kennedys, billionaire William Koch, former secretary of state John Kerry, and former governor Mitt Romney, so it was not surprising that they were the most adamant opponents of the project (Seelye 2017; Eckhouse and Ryan 2017). In his *New York Times* op-ed in 2005, Robert F. Kennedy, Jr., stated, "I do believe that some places should be off limits to any sort of industrial development. I wouldn't build a wind farm in Yosemite National Park. Nor would I build one on Nantucket Sound" (Kennedy 2005). The Alliance had raised approximately $40 million, of which William Koch was known to have donated $1.5 million (Seelye 2017). The huge donation allowed the Alliance to constantly challenge the Cape Wind Project in court, thereby making the permitting process costly and exhaustive for Cape Wind Associates and Jim Gordon.

Impacts of the Resistance Campaign in Thwarting the Cape Wind Project

The eventual demise of the Cape Wind Project was caused by the prolonged court battle waged by the highly effective and well-funded Alliance. However, the absence of an established framework for reviewing offshore renewable energy projects created regulatory hurdles, thereby allowing the opposition to take advantage of the hurdles and exacerbating the legal burdens on project proponent Cape Wind Associates.

When Cape Wind Associates proposed the wind power facility in November 2001, the US Army Corps of Engineers was the regulatory body responsible for granting the permit. It took the Corps of Engineers three years to publish a draft of the environmental impact statement (EIS) for the construction (BOEM, n.d.). Then, the Energy Policy Act of 2005 changed the regulatory authority from the Corps of Engineers to the BOEM. Similar to the procedure under the Corps of Engineers, the processing of the EIS could not be achieved any faster under the authority of the BOEM: the draft and final EIS versions were published in January 2008 and January 2009, respectively (BOEM, n.d.).

The BOEM was also criticized by the opposition for not conducting adequate geophysical and geotechnical surveys to gather data about the seafloor, resulting in a violation of its obligations under the National Environmental Policy Act (NEPA). Although the insufficient conduct of geological surveys could be attributed to the BOEM's lack of experience in handling offshore energy projects, it provided ammunition to the opposition to challenge the BOEM in court, further delaying the construction of the wind farm. Hence, the lack of a regulatory framework led to the protracted permitting procedures by both the Corps of Engineers and BOEM and the inadequate handling of obligations under the NEPA by the BOEM.

Lastly, the 130 wind turbines, which were to be located more than three miles from shore and required new infrastructure, including roads and transmission lines, were subject to regulation by federal, state, and local jurisdictions (Zeller 2013). This allowed constant litigation against the Cape Wind Project at every level. From 2001 to 2014, the opposition had challenged the project in court, notably against the Army Corps of Engineers, the BOEM, and the Massachusetts Energy Facilities Siting Board, for the approval of two undersea transmission cables from the proposed facility to the regional power grid and against the Department of Public Utilities (DPU) over the above-market power purchase agreements between Cape Wind Associates and its two partners (*Town of Barnstable, Massachusetts, et al. v. Ann G. Berwick, et al.* 2014).

The Cape Wind Project enjoyed enormous support from major environmental groups, including the Sierra Club, the Natural Resources Defense Council, and Greenpeace (Zeller 2017). More importantly, the project received approvals at all federal, state, and local levels. Nonetheless, it faced fierce opposition from the highly effective and well-funded Alliance and other groups. The protracted, costly, and exhaustive court battles led to the failure of Cape Wind Associates to meet its contract commitments by December 31, 2014. Given the cancellation of contracts by the National Grid and NSTAR, the Cape Wind Project was no longer financially feasible and had to be abandoned in December 2017.

Solar Controversies in California

In 2008, California's Renewables Portfolio Standard was strengthened to require 33% of the state's retail electricity to come from renewable sources by 2020 (Hunold and Leitner 2011; Cain and Nelson 2013). In 2015, it was further strengthened to require 50% renewable power by 2030. This policy has led to the development of large-scale renewable energy projects, including solar projects made possible by opening up public lands in remote areas of the Mojave Desert. This "Solar Renaissance" (Hunold and Leitner 2011) exposed the trade-off between the protection of wildlife and renewable energy development to reduce the threats of climate change. The following analysis focuses on two major solar energy projects of the "Solar Renaissance" era, the Ivanpah Solar Electric Generating System and Soda Mountain Solar Project.

Ivanpah Solar Electric Generating System

Located in the Mojave Desert, the Ivanpah Solar Electric Generating System is a 377-megawatt concentrated solar power facility built on 3,400 acres of public land near the California-Nevada border (Moore and Hackett 2016; BrightSource Energy, n.d.). The $2.2 billion project was developed by BrightSource Energy, NRG Energy, and Google. In April 2011, BrightSource Energy received a $1.6 billion loan guarantee from the Department of Energy (Garthwaite 2013; Wiener-Bronner 2014). The facility consists of three separate heliostat fields with more than 170,000 12-foot heliostats and three 450-foot power towers (Metcalfe 2016; Danelski 2017). The Ivanpah solar project was the top priority of the Obama administration's push to reduce America's carbon footprint and move toward a green energy economy.

When BrightSource Energy proposed the project to the California Energy Commission in October 2007, the initial design included a 400-megawatt plant to be constructed on 3,400 acres of land and having 272,000 heliostats arranged in 10 circular fields, each with a central power tower (Moore and Hackett 2016). After redesigning the facility four times, the California Energy Commission granted the siting permit to the current version. The draft environmental impact statement was published in late 2009, and the California Energy Commission held public hearings in early 2010 (Moore and Hackett 2016).

In October 2010, the California Energy Commission approved the project, and construction was completed in 2013 (Moore and Hackett 2016). The facility officially opened on February 13, 2014, and it was the largest concentrated solar power station in the world. The Ivanpah plant has been in operation since its inauguration in 2014.

Soda Mountain Solar Project

The Soda Mountain Solar Project is a proposed 287-megawatt solar photovoltaic power facility built on 1,767 acres of public land along Interstate 15 and less than a mile from the Mojave National Preserve in San Bernardino County, California (Steinberg 2016; Press-Enterprise 2016). The project, which would provide power to more than 86,000 homes, was part of the Obama administration's Climate Action Plan to develop 20,000 megawatts of renewable energy on public lands by 2020 (Steinberg 2016; Press-Enterprise 2016). In June 2015, the city of Los Angeles decided not to purchase electricity from the Soda Mountain Solar Project, delivering a blow to the project's former developer, Bechtel Corporation (Sahagun 2015). In March 2016, the project received approval from the US Department of the Interior (Press-Enterprise 2016).

On August 23, 2016, however, the Soda Mountain Solar Project was unable to obtain final approval from San Bernardino County to start construction activities (Sahagun 2016a). By a 3–2 vote, the county board of supervisors declined to authorize a county permit, with Vice Chairman Robert Lovingood saying, "We endorse renewable energy, but this was the wrong project in the wrong location" (Sahagun 2016a). By this time, Regenerate Power had bought the project from Bechtel Corporation. After the rejection from San Bernardino County, Regenerate Power was determined to overcome the final hurdle and push the project forward (Steinberg 2016). Nevertheless, construction has not been initiated at the time of writing.

Motivations for Resistance

As both the Ivanpah and Soda Mountain solar projects are in close proximity to the Mojave Wilderness, the primary motivation for resistance to the siting of the two projects concerns impacts on wildlife species such as desert tortoise, birds, and bighorn sheep in the Mojave National Preserve. In the case of the Ivanpah solar project, there were additional concerns about loss of a spiritual place and spots for recreational activities.

Desert tortoises have lived in the Ivanpah Valley region for millions of years and are listed as a threatened species under the Endangered Species Act (Kerlin 2018; Moore and Hackett 2016). The Ivanpah tortoises are considered a genetically distinct population, and the Ivanpah Valley region is an important habitat for the survival of the species (Moore and Hackett 2016). Furthermore, the desert tortoises are vulnerable to human development.

Desert conservationists and biologists opposed the siting because the project would encroach on tortoise habitat. Surveys found more than 150 tortoises near the proposed location for the facility (Garthwaite 2013). The Ivanpah project site was also a refuge for migratory birds traveling along the Pacific flyway (Sahagun 2016b). The intense radiation created by thousands of the heliostat mirrors has actually resulted in birds being burned alive while flying through the facility (Sahagun 2016b; Sweet 2015; San Bernardino Sun 2014). Estimates of the number of deaths per year varied greatly, ranging from a low of 1,000 by BrightSource Energy, to 3,500 in a *Wall Street Journal* report, to 6,000 by federal biologists, to a high of 28,000 by the environmental group Center for Biological Diversity (Sahagun 2016b; Sweet 2015; San Bernardino Sun 2014).

The majority of dead birds consisted of hummingbirds, warblers, doves, sparrows, and swallows. Plumes of smoke appeared as the birds were incinerated in midair, which led the birds to be given the name "streamers" (Sahagun 2016b). Because of the high number of bird deaths, federal wildlife experts referred to the Ivanpah project site as "a mega-trap" for wildlife species (San Bernardino Sun 2014). Major opponents of the siting of the Ivanpah solar project included the Sierra Club, which argued for the resiting of the power facility to a place that was not a habitat for the desert tortoise, and the National Parks Conservation Association, which stated that the proposed siting would "degrade the federally protected resources of Mojave National Preserve" (Moore and Hackett 2016).

In the case of the Soda Mountain Solar Project, opponents expressed similar environmental concerns over habitat for bighorn sheep, foxes, owls, and migratory birds. This underdeveloped Soda Mountain region was an important habitat for the bighorn sheep, but they were separated between North Soda Mountain and South Soda Mountain by Interstate 15 (Sahagun 2016a; Steinberg 2016; Press-Enterprise 2016). As the bighorn sheep population had experienced strong growth in recent years, biologists proposed to restore migration corridors to avoid having the species become genetically isolated (Sahagun 2016a; Steinberg 2016; Press-Enterprise 2016). The proposed power facility would undermine the effort to reestablish the key migration routes and thus have inadvertent impacts on the growth of the bighorn sheep.

In addition to its value as a wildlife habitat, the Ivanpah Valley was a spiritual place for several Native American tribes in the region. A prayer site and an altar were on the hill above the project site (Moore and Hackett 2016). Also, the Native American tribes believed that the spiritual powers originated in the absence of human development in the area (Moore and Hackett 2016). Hence, the siting of the project triggered relentless resistance from the Native American peoples. They organized a 14-mile relay run, the Ivanpah Spirit Run, and turned it into an online documentary, *Solar Gold*, by Robert Lundahl (Moore and Hackett 2016).

Furthermore, opponents claimed that the Ivanpah Valley region was a treasured place for hiking, camping, and bird watching. An activist said, "This is big energy taking public lands that we own" (Moore and Hackett 2016). Indeed, the message that reverberated throughout the resistance campaigns was that the project demonstrated the "privatization of public wildlands . . . by transforming multiuse places into single-use industrial zones" (Moore and Hackett 2016). Activists held two protest hikes, in 2008 and 2010, to uphold the right of the public to hike and camp on the Ivanpah land (Moore and Hackett 2016).

Impacts of the Resistance Campaigns

Despite the resistance campaigns, construction of the Ivanpah Solar Electric Generating System was eventually completed, and it began operating in 2013. The Ivanpah project received multiple awards, such as the Concentrated Solar Power Project of the Year by Solar Power Generation USA in February 2012 and the Plant of the Year by *Power Magazine* in August 2014 (Overton 2014; Wind Energy and Electric Vehicle Review 2012).

While the strong resistance of opponents was not able to stop the project, it did result in several significant changes. First, the developers had to scale back from the original 400-megawatt design to the current 377-megawatt version to reduce the disturbance to desert tortoise habitat. Second, the Bureau of Land Management (BLM) ordered a temporary suspension of construction in April 2011 to gauge the impacts on the desert tortoises (California Desert District, Bureau of Land Management 2011). In June 2011, the BLM lifted the suspension order as the US Fish and Wildlife Service "found the project [was] not likely to jeopardize the endangered desert tortoise" (Bureau of Land Management 2011). Third, BrightSource Energy has spent more than $56 million on mitigation efforts for desert tortoises, including the care program for juvenile tortoises, providing the nurseries, and relocation programs (Wiener-Bronner 2014; BrightSource Energy, n.d.). Without the relentless pressure from the environmentalists, desert conservationists, and biologists, such mitigation efforts might not have been implemented.

Unlike the Ivanpah Solar Electric Generating System, the Soda Mountain Solar Project has not been able to overcome the resistance. The strong opposition campaigns led to the cancelation of power purchase plans by its major customer, the city of Los Angeles, in June 2015. The Sierra Club was strongly in favor of the city's decision, saying, "The Sierra Club is delighted to see the city do the right thing and choose not to sign a power purchase agreement with this harmful project" (Sahagun 2015). In addition, project opponents had successfully lobbied the San Bernardino County Board of Supervisors to rule against the project by not granting the final permit that the developer needed to proceed with construction. The project has been halted until now.

Both projects experienced the relentless resistance campaigns during the siting process because of the negative impacts on wildlife in the Mojave Desert, but the outcomes were different. Two factors may explain the failure of the Soda Mountain Solar Project. First, the city of Los Angeles was expected to be the key customer to purchase electricity from the project. It turned out that the Los Angeles Department of Water and Power found other proposed renewable energy projects that would charge the city less for electricity. Second, although the project was in the federal plan to reduce the country's reliance on fossil fuels, it did not receive as strong support from the federal government as the Ivanpah solar project did. In addition, the Soda Mountain Solar Project experienced a change in the project developer, which may have complicated its ability to surmount opposition. In stark contrast, with assistance from the Obama administration, reinforced by its powerful developers, the Ivanpah solar project successfully overcame all the roadblocks to completing its construction phase.

Transmission Line Conflicts in California

The Tehachapi Renewable Transmission Project is a 173-mile transmission project developed by Southern California Edison to bring up to 4,500 megawatts of renewable energy (enough to supply three million homes) from wind farms in Kern County to substations in Los Angeles and San Bernardino Counties (Southern California Edison, n.d.a). The project, with an estimated cost of $2.1–$2.5 billion, was designed to contribute to California's renewable portfolio standard's requirement to obtain 33% of its energy from renewable sources by 2020 (Cain and Nelson 2013).

As part of Decision 09-12-044, granted in December 2009 by the California Public Utility Commission (CPUC), Southern California Edison received approval for the construction of a 3.5-mile segment of 500-kilovolt overhead transmission facilities, Segment 8A, through a residential area of Chino Hills (CPUC 2013). This segment triggered vehement opposition from residents of the city. In October 2011, the city of Chino Hills formally requested that the segment planned through their community be "undergrounded." In July 2013, the California Public Utility Commission granted the petition of Chino Hills and ordered the undergrounding of the 3.5-mile transmission line.

Motivations for Resistance

The resistance was motivated by concerns about visual disruption, decreased property values, and health and safety concerns. Opposition within Chino Hills resulted in the formation in 2007 of Hope for the Hills, a nonprofit grassroots organization of about 1,500 residents in Chino Hills established to raise awareness about their concerns over the Tehachapi Renewable Transmission Project. The 3.5-mile segment of overhead power lines would consist of transmission towers reaching 195–198 feet in height and occupying a 150-foot right-of-way (CPUC 2013). In comparison to other cities along the project's route, Segment 8A in Chino Hills had the narrowest right-of-way. Thus, the towers would be located very close to residential structures, exacerbating the visual impact of the transmission lines. Chino Hills had 200 residential structures affected by the narrow right-of-way, which was more than in the towns of Duarte (94) and Ontario (36) (CPUC 2013).

Hope for the Hills and the city of Chino Hills relentlessly advocated undergrounding the lines because of concerns that the proximity of the transmission towers could reduce homeowners' property values (Tasci 2013). Most importantly, Chino Hills had become part of the identity of residents, since they had grown attached to the city. Hence, the visual disruption by the tall towers would lead to a disruption of this sense of place and an impingement on the community's identity.

Hope for the Hills and the city of Chino Hills were concerned that the proximity of the lines would expose residents to electromagnetic radiation and therefore increased risk of cancer (Tasci 2013; Nisperos 2016). Although evidence for health risks from high-voltage transmission lines has not been proven definitively, the perceived health risks certainly intensified the community-based stigma toward Segment 8A. Another aspect of perceived risks in this case was the concern over earthquakes. Chino Hills is located in an earthquake-prone zone, so residents were worried about whether the tall structures could collapse in a disaster (Tasci 2013; Nisperos 2016). For instance, Garcia, a registered nurse who has lived in Chino Hills with his family since 1997, said, "We live in an earthquake zone. If a disaster strikes, that thing could fall right through my house" (Willon 2011).

Impacts of the Resistance Campaigns

Hope for the Hills had utilized protests, social media, and the internet to amplify the perceptions of risk in the community (Cain and Nelson 2013).

The city of Chino Hills was also an active opponent of the project, committing $4.7 million in legal fees to force Southern California Edison to put the power line underground. Although the California Supreme Court refused to hear the challenge against Southern California Edison, the two parties were successful in lobbying the California Public Utility Commission (Dombek 2011). On November 11, 2011, the California Public Utility Commission ordered that the utility halt the construction of Segment 8A and required it to submit alternatives for Segment 8A in response to an application for rehearing and motion for partial stay filed by the city of Chino Hills (CPUC 2011). On July 11, 2013, the CPUC ruled against Southern California Edison, voting 3–2 in favor of undergrounding Segment 8A in Chino Hills, though the lines remained above ground in other cities.

Cost estimates of undergrounding Segment 8A in Chino Hills ranged from $300 million to $800 million, compared to the cost estimate of $170 million to build the overhead transmission line (Dombek 2012; Southern California Edison, n.d.b). However, this seemed to be a better option than the alternative suggested by Hope for the Hills and the city of Chino Hills. The city had suggested an alternative route in which the transmission lines would run through the existing rights-of-way of Chino Hills State Park. This alternative would have required an amendment to the Land Use General Plan, which could have delayed construction for 8 to 15 months. In summer 2014, Southern California Edison began construction of the underground line in Chino Hills. The Tehachapi Renewable Transmission Project has been in operation since December 2016, though it was originally scheduled to be operational in 2015 (Tweed 2010).

The Northern Pass between Quebec and New England

The Northern Pass project is a proposed US$1.6 billion system of high-voltage transmission lines to bring 1,090 megawatts of Canadian hydropower produced by Hydro-Quebec to New Hampshire and the rest of New England (Northern Pass Transmission, n.d.; Pentland 2018). The project, developed by Eversource Energy (hereafter Eversource), comprises 192 miles of 80- to 135-foot towers and transmission lines running from the border town of Pittsburg, New Hampshire, where it would connect to the Quebec Hydro grid, and ending in Deerfield, New Hampshire, where it would connect to the grid of New England (Keir and Ali 2014; Tierney and Darling

2017). One-third of the proposed transmission lines would be underground lines, given that 80% of the facilities are on existing transmission rights-of-way or under public roadways (Tierney and Darling 2017).

The Northern Pass was expected to generate up to C$500 million in annual revenues for Hydro-Quebec (CBC 2018a). More importantly, the Northern Pass could help New England substantially reduce carbon emissions, by up to 3.2 million tons per year (Northern Pass Transmission, n.d.). In November 2017, Hydro-Quebec and Eversource received a presidential permit for the project from the US Department of Energy (US Department of Energy 2017). In January 2018, they continued to receive approval from Massachusetts for the Northern Pass by winning the biggest 20-year energy deal in the history of Quebec's public utility (CBC 2018a).

However, the project was rejected in February 2018 by New Hampshire's Site Evaluation Committee, which is the state's key permitting authority over the project (CBC 2018b; Pentland 2018). Eversource challenged the decision in court, but on July 19, 2019, the New Hampshire Supreme Court upheld the rejection. As a result, Eversource announced it was terminating the project (NHPR 2019). With Northern Pass rejected, Massachusetts decided to proceed with a revised plan, the New England Clean Energy Connect project, to import hydropower from Quebec through Maine. The project has been authorized to proceed and, absent legal challenges, construction was expected to begin in late fall of 2020 (New England Clean Energy Connect 2020).

The following analysis consists of two parts. The first part examines what has motivated the opposition to Northern Pass. The second part discusses the impacts of the resistance campaign over the Northern Pass, in which the key players include the Society for the Protection of New Hampshire Forests (the Forest Society), the Appalachian Mountain Club, and SOS Mont Hereford.

Motivations for the Resistance Campaign

The resistance campaign against the project was motivated by visual impacts, decreased property values, environmental impacts, and economic impacts. The project proposed to run through the tourism region of New Hampshire, the Great North Woods region (also known as the North Country), which is home to Franconia Notch State Park, Pawtuckaway State Park, the Appalachian National Scenic Trail, and White Mountain National Forest. The construction of thousands of new towers through the North Country would obstruct the scenic landscapes in these natural tourist attractions, according to the visual impact analysis by the Appalachian Mountain Club (Difley 2011; Burbank 2012).

Opponents are concerned that the tower's visibility may reduce the attractiveness of the scenery and have detrimental impacts on tourism, which is the second-largest industry in New Hampshire (Tierney and Darling 2017; Difley 2011). Indeed, studies have found that the Northern Pass could lead to a 9% reduction in tourism-related spending, which translates to average annual losses of $13 million to the gross state product and approximately 200 jobs between 2020 and 2030 (Tierney and Darling 2017).

Project proponents have emphasized job creation and increased tax payments. Gary A. Long, president and chief operating officer of Public Service of New Hampshire, has stated that the Northern Pass would create an annual average of 1,200 jobs during the three-year construction period and an estimated $24.5 million in state, local, and county tax payments in New Hampshire (Long 2011).

On the other hand, opponents have highlighted the temporary nature of construction jobs and the export of economic profits from New Hampshire (Keir and Ali 2014). The State Energy Strategy of New Hampshire, published in 2014, has called for energy independence, increasing use of in-state renewable energy resources, and circulation of energy revenues within the state's economy (Tierney and Darling 2017). Hence, opponents have raised the concern that the benefits of the Northern Pass would be exported to large companies such as Hydro-Quebec and the project's developer, Eversource, based in Hartford, Connecticut, and Boston, Massachusetts, while New Hampshire would bear the most burden from the project but receive few benefits from it.

Although various studies have produced mixed evidence on whether transmission lines cause a decrease in property values, local residents in towns along the proposed route have strongly opposed the project (Evans-Brown 2014). The visual impact would undoubtedly reduce the attractiveness of properties located near the towers.

Concerns were also raised that the transmission facilities would cause forest fragmentation on the protected conservation lands owned by the Forest Society in New Hampshire. Since 1901, the Forest Society has had a mission of protecting the landscapes of New Hampshire and a goal of "[protecting] sustainably-managed forests to support our forest-based economy" in the face of growing commercial development pressure (Forest Society, n.d.). Unsurprisingly, the Forest Society is the most relentless opponent of the Northern Pass, mobilizing its reputation and finances for its opposition campaign "Trees Not Towers: Bury Northern Pass" (Forest Society, n.d.).

The Northern Pass project has also faced strong opposition in Quebec, Canada. The transmission line of the Quebec portion would run through the conservation area of Hereford Mountain, part of the White Mountains of the Appalachians (Montreal Gazette 2017). The SOS Mont Hereford group, which is comprised of Nature Québec, Estrie Regional Environmental Council, and the Appalachian Corridor and Protected Natural Environments Network, has called on Hydro-Quebec to reconsider the route (Montreal Gazette 2017). Because of the location of the transmission facilities in the heavily forested regions, opponents have expressed concerns over the decline of biodiversity, including environmental degradation of wetlands and forests and disruption of wildlife habitat.

As with the Cape Wind Project, there is also an Indigenous resistance movement against the Northern Pass. Dams, reservoirs, and power stations of Hydro-Quebec that would produce the energy for New England are constructed on the traditional territory of the Pessamit Innu, a tribal nation in Quebec (Casey 2017). The Innus have opposed the project because of concerns that their salmon fishery and traditional hunting grounds could be affected (Casey 2017). Although the Innus voiced their opposition during the public hearing session of the Site Evaluation Committee in July 2017, the impact of the allegation on the Site Evaluation Committee's decision is unclear.

Impacts of the Resistance Campaign

The Forest Society and SOS Mont Hereford have called for all the power lines to be buried underground. Because of the high cost of burying all the lines, Eversource and Hydro-Quebec have only agreed to have 60 miles of underground lines (Northern Pass Transmission, n.d.). Because the two sides had uncompromising stances on the location of the transmission lines, the Forest Society was able to raise $850,000 to secure a 5,800-acre conservation easement on a property that would be on a potential route of the Northern Pass (Keir and Ali 2014; Forest Society, n.d.; State Impact New Hampshire, n.d.).

The opposition also resorted to personal criticism in the media. The Balsams Grand Resort Hotel, located in the northernmost part of New Hampshire, has been under a redevelopment plan spearheaded by Les Otten, who has received a $2 million loan from the $200 million development fund managed by Eversource (Tracy 2016; Difley and Webb 2016). Otten has been denounced for his ties to Eversource and criticized for pressuring the North Country Chamber of Commerce to change its opposition to the project (Tracy 2016; Difley and Webb 2016). Although Otten has denied the allegation, the opposition's condemnation has put a stain on Eversource's reputation, further exacerbating the unpopularity of the Northern Pass in New Hampshire.

Because of the relentless opposition, the project and Eversource's appeal were rejected in February and May 2018, respectively, by the Site Evaluation Committee (Casey 2018; CBC 2018b). On October 12, 2018, the New Hampshire Supreme Court accepted the appeal of Eversource, which was expected to be heard in early 2019 (Concord Monitor 2018). Given the reputation and resources of the project's opponents, particularly the Forest Society, and disagreement over the underground transmission lines, the Northern Pass is expected to endure a protracted litigation process. The project is likely dead as a result of Massachusetts's March 2018 decision to cancel the Northern Pass project and instead pursue the competing Maine transmission line project of Avangrid (Chesto 2018; CBC 2018c).

Conclusion

The cases reviewed in this chapter clearly demonstrate that place-based resistance has the potential to frustrate the implementation of renewable energy infrastructure required for decarbonization. Not all, or even most, renewable energy projects attract opposition (Giordano et al. 2018), and even when they do, in many cases opposition can be surmounted. The record nevertheless contains a significant number of cases where place-based resistance has resulted in costly delays and/or project modifications, or, most dramatically, outright project cancellations. Table 10.2 provides a capsule overview of the cases presented here.

In the case of wind power in Ontario, place-based opposition led to a number of delays, modifications, and even cancellations of projects. The 2018 election resulted in a humiliating loss for the governing Liberal Party and a reversal of many of its climate and renewable energy policies. Place-based resistance did not play a direct role in the 2018 election results, but the extreme politicization of the province's energy and climate policies did

Table 10.2
Summary of renewable energy controversies

Project (jurisdiction)	Outcome
Ontario wind	Substantial opposition produced many costly delays and cancellations; program scrapped after 2018 election
Cape Wind Project (MA)	Canceled after protracted resistance campaign
Ivanpah Solar Electric Generating System (CA)	Operating—approved after modifications to address environmental concerns
Soda Mountain Solar Project (CA)	Canceled after protracted resistance campaign
Tehachapi Renewable Transmission Project (CA)	Operating—approved after opposition forced expensive "undergrounding" of critical segment
Northern Pass (NH)	Canceled after protracted resistance campaign

contribute to the election result. In the Cape Wind case off the Massachusetts coast, place-based resistance contributed directly to the project's cancellation.

The chapter also reviewed efforts to site two concentrated solar power projects in the Mojave Desert region of California. One of the projects has been blocked by environmental concerns about wildlife habitat. The other is under operation after some delay and redesign of the project to reduce habitat disturbance. The case of the California transmission line, proposed explicitly to connect new wind farms to load centers, was able to surmount opposition but only after delays and costly project modifications to place a segment through Chino Hills underground. The Northern Pass Transmission project, which would have helped New England reduce carbon emissions by importing hydropower from Quebec, has been canceled as a result of vehement place-based resistance.

These cases also reveal the importance of the four factors emphasized by the analytical framework. The salience of place-based, concentrated risks and benefits is apparent in all these cases, from treasured rural landscapes in Ontario; to desert tortoises, bighorn sheep, and migrating birds in the Mojave Desert; to cherished forested mountains in New Hampshire; and precious views of unspoiled Nantucket Sound. Impacts on special place-values play a critical role in all these cases. Projects that have been able to surmount place-based resistance have found ways to tailor the project to reduce the risk to treasured values sufficiently, as shown by the Ivanpah solar project and Tehachapi Renewable Transmission Project.

Opposition groups' access to institutional veto points is a very important element of the power of project opponents, but in complex ways. The multiple veto points of the American federal system were especially apparent in the Cape Wind and Northern Pass cases, where opposition groups seemed to try every venue possible to block the project, including courts and federal and state regulatory processes. In the Soda Mountain solar case in California, it was the San Bernardino County Board of Supervisors that rejected the project. In the Ontario wind case, in the early years of resistance, community groups also sought to use the zoning authority of local governments to block projects, but the provincial government stripped them of that authority. While that removed the capacity of local governments to thwart projects, it also decreased the sense of community empowerment, which has aggravated the degree of resistance. We will return to this dilemma shortly.

The more a project can take advantage of existing infrastructure, the less resistance it is likely to encounter. Power lines, for example, have a smaller marginal impact on a landscape if they can be sited in, or adjacent to, existing rights-of-way, but projects that have that advantage are by no means guaranteed to be successful. A very high fraction of the Northern Pass Transmission project would have taken advantage of existing infrastructure, but some portions could not—and those segments generated enough resistance to thwart the project.

The final factor is the geographical separation of risks and benefits. All these cases reveal the importance of this variable as well. While renewable energy creates greater potential to concentrate risks and benefits in the same location, it frequently does not. Rural communities' resistance to wind power in Ontario was so strong because the benefits of the development were typically far away. Transmission lines, pipelines for electrons, inherently impose impacts on the communities they pass through for the benefit of those at one or both ends of the line.

This chapter has demonstrated that like new fossil fuel infrastructure, renewable energy infrastructure has attracted significant place-based resistance, which has led to costly project delays or alterations and in some cases outright cancellation. Such resistance is not inevitable (Giordano et al. 2018) and can at times improve decision-making (Hager and Haddad 2015). But its prevalence and impact does constitute a risk to the transition to clean energy needed to avoid the worst effects of climate change. Renewable energy resistance is not a direct consequence of the movement to keep fossil fuels in the ground. As noted earlier, the academic literature on the social acceptance of renewable energy emerged before the climate movement made the strategic pivot to blocking infrastructure. The resistance dilemma is that the "keep it in the ground" movement builds the institutional, social, and cultural muscles that strengthen the capacity of groups intent on resistance to renewable energy.

Perhaps the most significant component of this dilemma is whether local governments should be granted veto power. Local control has frequently been demanded by the "keep it in the ground" movement, whether grounded in Indigenous rights or the idea that "only communities grant permission." If such rights are granted, it gives local authorities—Indigenous or not—the capacity to veto projects determined to be in the interests of the broader geographic political jurisdiction. Yet, if that power is taken away, local groups resent the disempowerment, and that can strengthen resistance. The engagement literature sees hope in giving communities a say but engaging them in meaningful processes that help community members see the broader public interest being promoted by projects that have impacts on treasured local values. Giving local communities a real governance role risks resistance, but shutting them out probably results in a much greater chance of impactful project opposition.

Overcoming place-based resistance is critical to decarbonization. If governments around the world can't get projects sited and built because of local resistance, fundamental human needs will not be met. How can we avoid shackling this transition with a process legitimacy crisis? Fortunately, the literature on public engagement contains a wealth of insights into how to gain greater acceptance for contested infrastructure processes. Presented in chapter 11, that literature demonstrates the importance of deep and meaningful engagement with stakeholders in ways that governments have traditionally been quite reluctant to do.

IV Can Innovative Processes Avoid Paralysis?

11 Overcoming Place-Based Resistance to Renewable Energy Infrastructure

While the climate movement benefits from stymied fossil fuel projects, addressing the climate crisis also requires the rapid deployment of a significant amount of renewable energy infrastructure. Yet our energy infrastructure decision-making process struggles to gain social legitimacy. Conflicts over energy facility siting, whether for fossil fuels or renewables, have proven to be so challenging in large part because of the tensions between place-based interests and the preferences of broader political collectives. This chapter addresses the book's third question: is there hope in more innovative processes of energy infrastructure decision-making that can promote social acceptance of the rapid transition to a clean energy system but avoid the confrontational politics that have characterized fossil fuel resistance?

This question will be addressed by exploring processes that have explicitly sought to overcome place-based resistance. The first section examines the theory and arguments behind proposals to move beyond site-specific decision-making to more promising strategic approaches and reviews the lessons learned from processes that have been used in the specific domain of energy planning and facility siting. The chapter will also explore innovative processes that have been applied in related areas such as environmental assessment and land-use planning to address competing demands on the same land base.

Criteria for Evaluating Energy Infrastructure Decision-Making Processes

Research on and practice with designing assessment and review processes contains significant insights into the kinds of processes that foster legitimacy and social acceptance.[1] There are four strands of the "good process" literature: (1) public participation and stakeholder engagement, (2) strategic and cumulative environmental assessment, (3) social acceptance of renewable energy, and (4) business-government relations. These literature strands form distinct but overlapping approaches to promoting good processes, quality decisions, and sustainable outcomes. The different strands of literature outline a variety of practices that can be considered criteria for a "good process."

Evaluating a process involves consideration of the process itself as well as the outcomes associated with it. Process-oriented criteria focus on the procedural aspects of decision-making that are associated with common features of democratic governance, such as transparency, accountability, fairness, and representation (Cleland et al. 2016; Kasperson 2006; Peterson St-Laurent et al. 2020; see also "acceptance criteria" in Rowe and Frewer 2000). Process-oriented criteria also focus on the types of communication and interactions that should occur between participants in decision-making processes (Renn 2006; Gregory 2017; Noble, Skwaruk, and Patrick 2014). This involves the inclusion of diverse perspectives and types of knowledge, respectful communication aimed at facilitating a shared understanding of information and values, and systematic, transparent characterization and evaluation of alternatives and trade-offs.

Outcome criteria focus on the quality of the decision itself and secondary outcomes associated with how the process affects participants and society apart from the implementation of the decision at hand. Overall, the literature on consideration of outcomes shows more variation, in part because the outcomes desired vary so much by context. Examples of outcome criteria include whether the decision increases joint gains, whether the process approaches the issue from a more holistic and integrated manner, whether the process is responsive and competent in the eyes of the public, and whether the decision is robust, sustainable, and fair (Sheppard 2005; Jami and Walsh 2016; Renn 2006). Secondary outcomes of good processes include enhanced participatory skills and social, cognitive, and normative learning outcomes. Trust in and satisfaction with the process are also desirable outcomes (Kasperson 2006; Peterson St-Laurent et al. 2020; Jami and Walsh 2016; Chess and Purcell 1999).

Public Participation and Engagement

There is broad consensus in all the literature that public participation in decision-making is desirable, but there are significant differences in views about who should be engaged and how. The literature on public participation and engagement in technically complex domains, including energy and climate policy, contains a variety of well-developed typologies for evaluative criteria. A long history of democratic, political, and, more broadly, social theory provides normative justification for public participation in democratic societies. More recently, participation in the form of deliberative dialogue has been considered by some the "essence of democracy" (Dryzek 2002; Abelson et al. 2003). The particular focus on deliberation in participative processes derives from Habermas's theory of communicative action (Habermas 1984) and how theory and discourse can shape practice (Taylor 1985).

Other scholars emphasize the instrumental benefits of public participation in addition to the normative ones (Cleland et al. 2016; Stirling 2008). Beierle (2002) conducted a comprehensive review of literature on stakeholder engagement and demonstrated that, contrary to the fears of many skeptics, "the majority of cases contain evidence of stakeholders improving decisions over the status quo; adding new information, ideas, and analysis; and having adequate access to technical and scientific resources" (Beierle 2002, 739).

Renn, Webler, and Wiedemann (1995) made a significant contribution to the establishment of evaluative criteria for deliberative public participation processes. They built on Habermas's work, proposing a framework that consists of competency and fairness. Competence refers to the content of discussion and the quality of input contributed by participants. Fairness refers to the reduction of power imbalances and equal opportunity to contribute to the process by all involved.

Subsequent conceptualizations of evaluative criteria involved a similar focus on both content and process. Petts (2001) proposed 10 effectiveness criteria that revolve around getting the "right science" and the "right participation" (Stern and Fineberg 1996) and combing the principles of publicity and accountability (Gutman and Thompson 1996) and fairness and competence. Rowe and Frewer (2000) assessed a variety of public participation methods and forums along two lines of criteria that they drew from the literature: acceptance criteria and process criteria. Acceptance criteria consisted of representativeness, independence, early involvement, influence on outcome, and transparency. Process criteria included resource accessibility, definition of purpose, presence of structured decision-making methods, and cost-effectiveness.

Abelson et al. (2003) draw on these works and the seminal work of others, including Beierle (2002), to suggest four categories for representing valued components of such processes: representation, procedural rules, the information that is used, and outcomes or decisions. Peterson St-Laurent et al. (2020) adopted a similar model but with an emphasis on analytic methods in integrating science and values, saying, "Recurrent insights point to the need for assessments of analytic-deliberative processes that reflect on the relationship between technical and scientific knowledge (analytic) as well as the interactions between participants (deliberative)." Their four-part framework includes representation, deliberation, knowledge and analysis, and outcome. They draw from seven evaluative works: Abelson et al. (2003); Papadopoulos and Warin (2007); Renn (2004); Stern and Fineberg (1996); Rauschmayer and Wittmer (2006); Chilvers (2007); and Sheppard (2005).

Analytic-deliberative processes, first introduced by the National Research Council (Stern and Fineberg 1996), hold considerable promise in bridging the knowledge gap between experts and lay members of the public through early engagement and partnerships between decision-makers, domain experts, stakeholders, and other members of the public (Chilvers 2007).

Integration with Policies and Higher-Level Plans

The public participation and engagement literature strongly demonstrates the importance of developing appropriate approaches to participation and deliberation, including getting the right kind of knowledge, information, and analysis in order for the process to be well informed. However, this literature tends to focus on only one scale of decision-making. The strategic environmental and cumulative effects assessment literature provides more emphasis on cross-scale linkages. In particular, it is concerned with how the goals of the project review process are situated in relation to higher-level policies and plans that establish the context for project decision-making (Pidgeon et al. 2014; Wilsdon and Willis 2004). Strategic environmental assessment (SEA) is defined as "a strategic framework instrument that helps to create a development context toward sustainability, by integrating environment and sustainability issues in decision-making, assessing strategic development options and issuing guidelines to assist implementation" (Partidário 2012). The promise and challenges of strategic/cumulative assessment will be discussed further.

Social Acceptance of Renewable Energy: The Importance of the Local

Widespread resistance has spawned a substantial amount of literature on the social acceptance of renewable energy (e.g., Cleland et al. 2016; Batel et al. 2013; Devine-Wright 2009; Fast 2013; Wustenhagen et al. 2007), discussed in chapter 10. Some of the literature resonates with the themes of the public engagement and strategic/cumulative assessment literature. For example, the social acceptance literature shares an emphasis with the public engagement literature on engaging host communities early and meaningfully in the process and on demonstrating how community input influences project design (Devine-Wright, Devine-Wright, and Cowell 2016; Fast et al. 2016). Like the strategic/cumulative assessment literature, the social acceptance literature stresses the importance of integration and coordination between different levels of government (Mulvilhill, Winfield, and Etcheverry 2013; Pidgeon et al. 2014; Devine-Wright, Devine-Wright, and Cowell 2016).

The distinctive element introduced by the social acceptance literature is the importance of local values. It emphasizes respecting the attachment of people to place and providing economic benefits to those impacted by the project, and the meaningful involvement of local communities (Devine-Wright, Devine-Wright, and Cowell 2016; Fast et al. 2016; Sovacool and Ratan 2012; Hyland and Bertsch 2018). One systematic review of the literature concludes that "local communities may be more willing to accept projects if developers site and design them in ways that work with, rather than against, local identities and people's attachment to specific places" (Devine-Wright, Devine-Wright, and Cowell 2016, 5). Some studies emphasize the importance of community ownership or shared ownership in fostering public acceptance (Cleland et al. 2016; Devine-Wright, Devine-Wright, and Cowell 2016). Others find that local economic benefits are more important than actual ownership (Hyland and Bertsch 2018). Regardless, there is a general consensus that some form of substantial community benefit is essential.

The Business Lens and Process Efficiency

The final strand of criteria comes not from the academic literature but from business group submissions related to improving energy infrastructure decision-making. In general, business groups are concerned with certainty, timeliness, and process costs. These issues are well illustrated by the

Canadian Association of Petroleum Producers (CAPP) in its submission to the review of environmental assessment processes in Canada: "To achieve certainty and consistency, the EA process needs to have predictable costs, timelines and a well-defined scope. Proponents need to have a very active role in assessment preparations so they can leverage the full benefit of assessment as a planning tool and adjust their designs and execution plans as new information becomes available. In addition, the EA decision making process must be transparent and timely. Without these elements, investment in Canada's resources will continue to diminish" (Canadian Association of Petroleum Producers 2017).

The Canadian Energy Pipeline Association's (CEPA) submission to the same review demonstrated overlapping concerns: "In particular CEPA recommended that processes should avoid duplication, outline clear accountabilities, be based on transparent rules and processes, ensure procedural certainty for project proponents, allow meaningful participation and balance the need for timeliness and inclusiveness" (CEPA 2017).

These process efficiency concerns of timeliness, certainty, and cost receive surprisingly little attention in the scholarly literature on public engagement and strategic environmental assessment. In fact, given the number of different typologies of process criteria, it is striking that none of them specifically address these concerns. Some might be tempted to dismiss these process efficiency concerns as self-interested lobbying by profit-oriented proponents, but they are important for three reasons. First, governments frequently share their concerns. The Canadian government's environmental assessment reforms of 2012 reflected these values: "What is needed is a system that provides predictable, certain and timely reviews, reduced duplication, strengthened environmental protection and enhanced Aboriginal consultations" (Government of Canada 2012). Third, even if the government is less business oriented, the power of business in a market-oriented democracy means they need to be taken seriously. Finally, given the urgency of the decarbonization imperative, timeliness takes on a new importance even among those who are less likely to share the values of the business community.

Criteria for Sustainable Energy Policy Decision-Making

Based on these four strands of literature, processes for energy infrastructure decision-making can be evaluated with the following nine criteria:

1. *Representative* is about "getting the right participation" (Stern and Feinberg 1996): ensuring the appropriate affected interests are involved and that the selection process is fair and legitimate (Abelson et al. 2003).
2. *Deliberative* is about "getting the participation right": legitimate and responsive procedures, including participant engagement in the design of procedures, respectful and inclusive dialogue, and effective facilitation (Peterson St-Laurent et al. 2020).
3. *Transparent, impartial, and accountable* ensures that the process is well understood and respected by direct participants and others with an interest in the decision, and that the ultimate decision is demonstrably influenced by the process.
4. *Well informed* is about "getting the right knowledge, and getting the knowledge right" (Stern and Feinberg 1996): the information presented and developed through the process is accessible, readable, digestible, and reflects a diversity of expertise and knowledge sources (Abelson et al. 2003).
5. *Integrated* means across levels of plans and policies, to ensure consistency between project-level outcomes and broader social and political goals.
6. *Efficient* ensures the process is affordable, feasible, and timely.
7. *Legitimate* means the outcome is socially acceptable to politically relevant constituencies.
8. *Equitable* ensures that benefits and risks of the project are seen as fairly distributed, particularly to host communities.
9. *Sustainable* refers to protecting environmental values and in this context ensuring that project decisions contribute to decarbonization.

The first six criteria address the review process. The remaining three address attributes of the outcome of the decision-making process.

Tensions among Criteria

Like any multicriteria exercise, conflicts among criteria will inevitably arise. Some of the most challenging tensions involve the efficiency criteria and a number of the other process criteria. Getting the desirable level of informed interaction between experts and process participants can be a time-consuming process, and advocates of deliberative processes rarely favor the types of strict process deadlines that industry proponents frequently feel are essential.

The public engagement literature also highlights a more surprising tension between representative on the one hand and deliberative and well informed on the other. This tension is particularly acute when processes are designed to be broadly inclusive and open to mass participation, which has become more commonplace over the past several decades—the recent Keystone XL hearing process, for example (Gregory 2017; Rossi 1997). This trade-off emerges because of a lack of government capacity or motivation to facilitate mass deliberation as well as a concurrent desire to enhance legitimacy in democratic societies through inclusivity. The result is that governments request public input without creating the forum or capacity for responsiveness, dialogue, and learning that is often required and desired by citizens in decision-making. Gregory recommends shifting the emphasis away from inclusivity as a guiding criterion and more carefully identifying a smaller group of the most appropriate representative participants. This could free up scarce time and resources, which could be dedicated to more engaged, deliberative processes. This type of process could be more successful in clarifying stakeholder priorities and incorporating quality information and as a result wield greater influence over decision-making (Gregory 2017, 161).

The Promise of More Strategic Assessments and Plans

There will probably always be a place for project-based assessments and reviews, but experience has demonstrated that if they proceed in the absence of coherent higher-level policies and plans, they are much more likely to attract strong opposition. The reviews of Canadian oil sands pipelines are a case in point. Some of the opposition to the pipelines was focused specifically on environmental concerns about pipeline or tanker spills, but the reviews also became venues where opponents sought to express grievances about climate policies or the role of Indigenous groups in decision-making, issues that the review panels had no jurisdiction to address (Ministerial Panel 2016).

Because of the inherent limits in how effectively project-based reviews can address the cumulative effects of multiple projects that affect the same values, these types of conflicts are not unusual. Cumulative effects can be described as "progressive nibbling—the accumulation of effects that occurs through many, often small-scale activities" (Noble, Skwaruk, and Patrick 2014, 317). For example, greenhouse gas emissions come from a variety of sources in any jurisdiction. Assessing the significance of one proposed facility is not meaningful in the absence of both a jurisdictional target for emissions and an understanding of existing and potential future sources of emissions. Similarly, several wind turbines may not fundamentally alter an agricultural region's sense of place, but a large number of turbines easily could.

These shortcomings of project-based reviews can be addressed by adopting more strategic processes. Two notable related models are strategic environmental assessments (SEAs) and collaborative land-use planning.

Strategic Environment and Cumulative Impact Assessment

After the early development of project-focused environmental impact assessment, concerns emerged about unnecessary conflict and uncertainty resulting from the lack of coherent policies or plans. One of the most stubborn challenges for impact assessment has been the consideration of cumulative effects. Cumulative effects of a development project consist of the consideration of its additional impacts in conjunction with projects that already exist, that are under consideration, or that could occur in the future (Seitz et al. 2011). This evaluation is typically carried out through cumulative effects assessment, a particular type of strategic environmental assessment.

In recent years, SEAs have drawn a lot of attention from academics, practitioners, and governments. Notable attempts at implementing SEAs have been observed worldwide, with recent examples in Australia (Coffey et al. 2011), Canada (Acharibasam and Noble 2014), and Europe (Polido, João, and Ramos 2016; Partidário 2012; SEPA 2011). However, notwithstanding the widespread attention directed at SEAs, theoreticians and practitioners alike do not concur on their interpretation of SEAs, and there is still no agreed understanding of what SEAs should look like (Baresi, Vella, and Sipe 2017; Noble and Nwanekezie 2016). For instance, Noble and Nwanekezie (2016) identify two main categories of SEAs. SEAs based on impact assessment are similar to traditional environmental impact assessment in that they aim to assess potential projects and their impacts. However, they focus their assessment on broader policies and plans instead of projects. In contrast, strategy-based SEAs go beyond traditional impact assessment to focus on the strategic directions a particular region's development should take (i.e., the design of policies and plans and/or of future alternatives).

A good SEA should be fully integrated into policies and plans at an early stage, be based on appropriate scientific information, engage the public well, engage in fulsome generation and evaluation of strategic directions, reflect ongoing evolution, communicate the SEA results in a timely and effective manner, deal with uncertainty and adverse effects, be coordinated by a lead agency, and be sufficiently resourced (SEPA 2011; Noble, Skwaruk, and Patrick 2014).

SEAs, when well implemented, provide numerous advantages over other approaches to cumulative effects assessment (SEPA 2011; Partidário 2012). For instance, they can provide an understanding of the challenges and opportunities associated with sustainability, which can then be incorporated early in the decision-making process. SEAs can also facilitate the identification of development options and alternatives, inform planners and decision-makers on the sustainability (or lack thereof) of different development options, ensure a democratic decision process by allowing the participation of different stakeholders, thereby increasing the credibility of policy decisions, and potentially change political mentalities by encouraging principles of strategic decision-making.

However, while these principles are attractive in theory, many initiatives have failed to live up to expectations over the years (Seitz et al. 2011; Noble, Skwaruk, and Patrick 2014; Acharibasam and Noble 2014). In Ontario, SEAs were used for several prominent assessments in the 1980s and then were abandoned in favor of project-level environmental impact assessments (Mulvihill, Winfield, and Etcheverry 2013). Various barriers to the successful implementation of SEAs have being identified in the literature. For instance, some authors draw attention to institutional issues in terms of capacity and available resources as well as a shared lack of vision about what an SEA should look like (which is also related to the lack of a clear definition) (Acharibasam and Noble 2014). Others note challenges associated with data availability (particularly baseline data about existing conditions) and long-term monitoring (Cronmiller and Noble 2018; Noble, Skwaruk, and Patrick 2014). However, most importantly, the literature identifies an uncoupling between theories and conceptual methodologies and actual implementation by practitioners (Lobos and Partidário 2014; Partidário 2015; Baresi et al. 2017; Noble and Nwanekezie 2016). In other words, there is a great deal of knowledge about how SEAs should be implemented but very little institutional capacity to carry them out successfully.

Strategic Land-Use Planning in British Columbia

Strategic environmental assessments can be applied to policies or a technology like nuclear power, but it can also be applied to a defined geographical region to establish land-use plans. One successful strategic land-use planning initiative occurred in British Columbia. The initiative arose in the early 1990s through a combination of political and legal events. In the 1980s, conflicts erupted over forestry in the province, especially the practice of clear-cutting old growth forests. Indigenous groups were also beginning to have success at using the courts to block road building and timber harvesting in areas of importance to them. In fall 1991, the New Democratic Party of Mike Harcourt was elected, and his social democratic party replaced a conservative party that had dominated provincial politics for decades. The Harcourt government had a strong environmental agenda, much of it focused on improving the protection of environmental values in the forests (Pralle 2006a; Shaw 2004).

Shortly after coming into office, the new government decided to address the "valley-by-valley" conflicts that continued to erupt with a strategic land-use planning process. It started in four of the most divisive regions in the province and had expanded to include virtually the entire provincial land base by the time their government was defeated in 2001 (Cashore et al. 2001). Thomas Gunton calls it "the most comprehensive application of collaborative planning to date" (Gunton 2017). Each of the regions had a planning table consisting of a range of stakeholders in the forest and resource sectors, including forest and mining companies, labor, environmentalists, community groups, and Indigenous groups. After a year or two of negotiations with the help of professional facilitation and government technical experts, they agreed to a list of consensus recommendations on zoning the forested land base for different uses. Those reports were then submitted to the government for formal decision-making.

When the New Democratic Party was swept out of office in 2001, it was replaced by leader Gordon Campbell's BC Liberals, a more conservative, business-oriented party. They were not keen on continuing strategic land-use planning, but they were committed to completing the intensely controversial land-use planning process in the north and central coast regions, which environmentalists had reframed as the "Great Bear Rainforest." Completion of this process was given priority by the Campbell government. The multistakeholder consultation process was completed with an agreement in 2004. The government then engaged in unprecedented government-to-government negotiations with First Nations in the area, culminating in the 2006 agreement. The agreement set aside one-third of the region as protected areas, and the remaining two-thirds would be subject to new ecosystem-based management principles (Cullen et al. 2011).

By 2008, land-use plans had been completed for 86% of the province. When the process was launched in 1992, the government committed to doubling protected areas in the province from 6% to 12%, but the province exceeded that objective. The new protected areas in the Great Bear Rainforest, along with other areas of the province, increased the amount of protected areas in the province to 14% of the land area, a significant accomplishment. It also contributed to a dramatic decline in conflict over forestry operations (Hoberg 2017).

While the BC Liberal government did succeed in completing the Great Bear Rainforest plan, the strategic land-use process was terminated in 2006. Any new planning would be undertaken only under a specific set of conditions: new legislation, accommodating First Nations interests, or a major environmental change. Despite the havoc wrought on the forestland base by the mountain pine beetle epidemic, the government has not yet reopened any of the existing land-use plans. The innovative collaborative planning model introduced by the New Democrats in the 1990s would also be abandoned (Hoberg 2017).

The reluctance to continue strategic planning, despite its demonstrable success at reducing conflict, undermined the BC Liberals' own energy agenda in the new millennium's first decade. Its 2002 Energy Plan moved the province away from relying on the government-owned utility, BC Hydro, to generate new power to meet growing demand. As a result, a number of private "independent power producers" proposed new renewable energy projects. While some wind projects were developed, the most common new projects were run-of-the-river hydro projects. For a variety of reasons, many of these projects were strongly opposed by local and provincial environmental groups, resulting in a number of project delays and cancellations. The established regional land-use plans focused mostly on forest resources and did not address siting energy facilities. As a result, there was no agreement on which areas of the province were appropriate for energy development and which were not. Despite pressure from a variety of interests to do more strategic planning around energy, the government declined to do so (Jaccard, Melton, and Nyboer 2011; Shaw 2011; Hoberg 2010).

Strategic Land-Use Planning in the Lower Athabasca Region

Next to British Columbia's land-use planning processes, the most prominent Canadian initiative in strategic land-use planning has been for the oil sands region in northern Alberta. Relatively early in the oil sands development process, there was a growing awareness of the need to move beyond facility-by-facility assessment and regulation to consider the regionwide cumulative effects of oil sands development. In 2000, the Cumulative Environmental Management Association (CEMA) was created as a voluntary stakeholder partnership among government, industry, environmentalists, and others. CEMA was slow to deliver, even as oil sands development was accelerating. Concerned by the pace of development in the absence of regional habitat protection plans, in 2008 the CEMA group working on habitat protection recommended a moratorium on new oil sands leases, but the government refused. The group then recommended the protection of up to 40% of the region (Hoberg and Phillips 2011). Rather than adopting that recommendation, the government of Alberta introduced a new land-use framework in 2008, creating a new planning process. Priority was given to the region where oil sands development had been the most intense, the Lower Athabasca region in the northeastern part of the province.

The Lower Athabasca Regional Plan (LARP) was completed in 2012. It provided regional standards for surface water quality and air quality, and most importantly created new conservation areas that increased protected areas from 6% to 22% of the region. It also committed to establishing specific targets for other regional ecosystem values (Government of Alberta 2012). The plan's commitment to address cumulative effects and its establishment of more protected areas were welcome improvements. However, the plan only covers part of the area of oil sands development, and highly valued species, such as boreal caribou, remain at serious risk. As a panel struck to review the plan's implementation concluded, "Despite the LARP's new conservation areas, the cumulative impacts on wildlife have exceeded or are reaching thresholds in significant adverse effects on biodiversity, some of which are likely permanent" (Lower Athabasca Regional Plan Review Panel 2015).

Conclusion

If humanity has any hope of limiting climate change to manageable levels, a massive and rapid transformation of energy system infrastructure is required. Governments need to have processes with sufficient legitimacy to comprehensively transform the energy system away from carbon-emitting sources. Because there is an inevitable difference in the intensity of preferences regarding impacts to local place values, it is no surprise that energy infrastructure siting processes are challenging. No process can guarantee that local opposition can be prevented, but the hope is that more engaged processes can both ensure meaningful local input and benefit and help local opponents see the collective benefits of the project.

Deeper engagement processes have been so uncommon not because they are ineffective or impractical but because they are perceived to run afoul of the ideology of some governments in power. Experience with strategic environmental assessment, from British Columbia's and Alberta's land-use planning, suggests that innovative processes with deeper engagement can improve decision quality and reduce conflict. But in some areas the practice of SEA has never been tried, or where it has, as in Ontario, was abandoned as the values of governing parties changed. British Columbia's process was terminated because the new business-oriented government associated it too much with the ideology and culture of the social democratic party that created it (Hoberg 2017). In Alberta, it was persistent resistance from powerful resource interests that prevented this process from achieving clearer success.

Despite the promise of deeper engagement processes to improve public acceptance of new energy projects, policymakers (particularly in North America) have shown great reluctance to engage in the sorts of processes that reflect best practices. As chapter 12 shows, this is because such processes are perceived to conflict with policymakers' core incentives to control process costs, duration, and outcomes.

12 Conclusion

Thus far, this book has addressed three questions: (1) Has the strategy of place-based resistance to fossil fuel development been effective at promoting climate action and the reduction of global warming emissions? (2) Does the strategy risk the unintended consequence of feeding place-based resistance to the clean energy transformation? (3) Is there hope in more innovative processes of regulatory review and facility siting that can promote social acceptance of the rapid transition to the clean energy system but avoid the confrontational politics that have characterized fossil fuel resistance? After a brief summary of the answers to these three questions, this concluding chapter addresses the book's fourth and final question: if innovative approaches have been demonstrated to reduce conflict, why are they so rarely used?

Place-based resistance was effective at delaying and in some cases leading to outright cancellations of pipeline projects. The anti-pipeline movement inflicted economic pain on the oil sands industry. When the Alberta and Canadian governments were run by conservative parties, they retrenched and defied environmental critics. That was changed in 2015 by the provincial election in Alberta and by Canada's federal election, and led to significant advancement in provincial and federal climate policy. The resistance strategy reduced oil sands production and emissions below what they would otherwise have been and eventually forced Canadian governments to adopt stronger climate policies than they would have absent the effects of that resistance strategy. The most important thing the pipeline movement did was force the federal government to get involved in climate policy in ways that it never had. It was not the only thing or even the most important reason why this happened, but it was pivotal.

If humanity has any hope of limiting climate change to manageable levels, a massive and rapid transformation of energy system infrastructure is required. The strategic decision by the climate movement to focus on blocking new fossil fuel infrastructure may facilitate the transition to clean energy, but it also risks creating a "resistance dilemma" by legitimizing place-based resistance that can then be mobilized to thwart needed clean energy infrastructure. As stated in chapter 10, the purpose of highlighting this resistance dilemma is not to question the legitimacy, sincerity, or efficacy of place-based resistance or to cast blame on the climate movement or challenge the wisdom or merits of its place-based resistance to fossil fuel infrastructure. Place-based resistance has a long tradition predating its adoption by climate activists. Rather, the reason for highlighting this resistance dilemma is that there is an urgent need to build massive amounts of clean and renewable energy infrastructure, and historically our institutions have not been effective at resolving the tensions between local desires to minimize impacts and the broader public interest in establishing necessary infrastructure. In short, we have a process legitimacy crisis, and we need to address that in order to effectively address the climate crisis.

Overcoming place-based resistance is critical to decarbonization. If governments around the world can't get projects sited and built because of local resistance, fundamental human needs will not be met. Fortunately, the literature on public engagement contains a wealth of insights into how to gain greater acceptance for contested infrastructure processes. Since the 1980s, public engagement practitioners have developed processes that can improve decision quality, conflict resolution, and public acceptance of technically complex policy decisions. These innovative processes show promise in reducing the risk and impact of place-based resistance to renewable energy, yet policymakers (particularly in North America) have shown great reluctance to engage in such processes.

The Strategic Calculus behind Process Choice

Place-based resistance can indeed be shown to have resulted in climate action. Strengthening the institutional muscle for place-based groups to block infrastructure does create the risk that needed clean energy projects will be delayed or canceled, but there is hope that more innovative processes can overcome place-based resistance.

This book's final question is: if innovative approaches have been demonstrated to reduce conflict, why are they so rarely used? To address this question, it is necessary to examine the literature on the political incentives underlying the choice of administrative and planning procedures. Insights from this literature can be used to develop a strategic calculus to guide the design of planning and siting processes. Examining the incentive structure of government decision-makers and project proponents will help explain this dilemma and help inform the development of process reforms to promote the social acceptance of projects that facilitate the transition to clean energy.

The actor-centered analytical frameworks that guide this book's analysis also help explain this dilemma. Strategic actors, the central agents of policy, interact within a context of institutional rules. However, they also work to change institutional rules through venue shifting or other means (Baumgartner and Jones 2010; Hoberg 2001; Pralle 2006a). Institutional rules can be pivotal because when the location or form of authority changes, the balance of policy preferences guiding policy decisions could also change significantly. As our case studies demonstrate, many conflicts between energy and environmental policy have been about "the politics of structure," or the struggle over defining the rules of the game (Moe and Wilson 1994).

Scholars have differentiated between two types of institutional strategies: procedural strategies that require agencies to follow specific processes (e.g., performing an environmental assessment or consulting with affected interests) and structural strategies that influence the organizational design of agencies (e.g., the choice between an independent regulatory commission and a more traditional line agency) (Shapiro 2017). Depending on the circumstances, information resulting from complying with procedural requirements influences decisions, and the organizational structure can shape what information flows to decision-makers (Shapiro 2017). Others have explored how different organizational structures "might shape learning about problems and solutions, policy choices, and conflict resolution in quite predictable ways" (Egeberg 1999).

Much of the literature focuses on how legislators, acting as principals, use requirements for procedure or structure to influence the outcomes from administrative agents (McCubbins, Noll, and Weingast 1987). However, strategic actors outside government also have large stakes in structure and procedure. According to Moe and Wilson, "all political actors know that structure is the means by which policies are carried out or subverted, and that different structures can have enormously different consequences. As a result, there is inevitably a "politics of structural choice" (Moe and Caldwell

1994). In this structural politics, strategic actors in and out of government will advocate for rules and venues that give them the greatest likelihood of achieving policy outcomes that reflect their interests.

This politics of structure incentivizes various actors in energy-environment conflicts to promote different procedural and structural rules. These conflicts can be thought of as involving three major categories of actors: those in favor of a particular project, those opposed to the project, and the policymakers who decide both the process and ultimate fate of the project. Project proponents' interests are normally in a stable, certain process of manageable scale and duration controlled by a single decision-maker so they can minimize process costs. Generally, these interests create pressures for minimal process requirements. These interests are balanced by proponents' interest in gaining sufficient public legitimacy to minimize political risks to their projects. This combination of values is clearly revealed in the statements by the Canadian industry associations quoted earlier.

Project opponents obviously have different incentives. They prefer comprehensive information requirements, widespread public access to decision processes, demanding consultation procedures, lengthy proceedings, multiple veto points, and clear rights to appeal unfavorable decisions. Opponents actually have a strategic interest in increasing process costs and delays as a way to discourage proponents.

Politicians designing regulatory processes, in addition to needing to balance these competing demands, have their own policy, budgetary, and especially electoral interests to keep in mind. They can be expected to want strong control over decisions where there is an opportunity to claim credit for favorable outcomes and to keep an arm's length from decisions that are more likely to be politically unpopular (Harrison 1996; Weaver 1986). All else being equal, they would prefer to minimize process time and costs, but they also need to be attentive to political legitimacy. The political influence of interests opposed to or skeptical of new infrastructure projects leads both politicians and proponents to prefer regulatory processes that are more time consuming, elaborate, and costly than they would prefer.

In some circumstances, elected officials grant discretion to administrators. Research on the incentives of administrators to integrate public engagement into policy decisions suggests support for public participation in principle but that in practice administrators are very skeptical about the practical usefulness of public participation in improving decision quality or social acceptance (Liao and Schachter 2018). When the perceived benefits of public engagement are low, managerial concerns about process costs, time delays, and a loss of control over the agenda discourage them from initiating engagement processes (Moynihan 2003).

Since the 1980s, public engagement practitioners have developed processes that can improve decision quality, conflict resolution, and public acceptance of technically complex policy decisions. These innovative processes show promise in reducing the risk and impact of place-based resistance to renewable energy, yet policymakers (particularly in North America) have shown great reluctance to engage in the sorts of processes that reflect best practices because they are perceived to conflict with core incentives to control process costs, duration, and outcomes. A gap between political incentives and best practices is hardly novel, but in the case of planning for the transition to clean energy, they take on new importance. Perhaps a clearer understanding of how more sophisticated engagement processes can improve decision quality and public legitimacy while also being conscious of process costs and timeliness will promote a willingness to adopt approaches that are more effective.

Addressing the climate crisis is an urgent imperative for humanity, but to make significant progress, the process crisis also needs to be addressed. The institutions and practices for energy planning and project approval require urgent reform to build sufficient social and political legitimacy in the coming infrastructure transformation. The experience with processes that engage the public more deeply and do so not at the project level but at a more strategic level gives cause for hope. But the political incentives of those who design institutions for planning and project approval help explain the reluctance of policymakers to adopt the types of processes that analysts recommend.

The most significant component of the institutional dilemma for designing better processes is whether local governments should be granted veto power. If they are, it gives local authorities—Indigenous or not—the capacity to veto projects determined to be in the interests of the broader geographic political jurisdiction. But if that power is taken away, local groups resent the disempowerment, which can strengthen resistance. The engagement literature sees hope in giving communities a say but engaging them in meaningful processes that help community members see the broader public interests being promoted by projects that have impacts on treasured local values. Giving local communities a real governance role risks resistance, but shutting them out probably results in a much greater chance of impactful project opposition.

Notes

Chapter 1

1. "Pathways reflecting current nationally stated mitigation ambition until 2030 are broadly consistent with cost-effective pathways that result in a global warming of about 3°C by 2100, with warming continuing afterwards" (IPCC 2018, 20).

2. All the theories described have their roots in political science, but there's also considerable commonality with the sociological concept of "strategic action fields" (Fligstein and McAdam 2011).

3. The term *policy regime* is used differently by some policy scholars to refer to contents of policy design or approaches more broadly rather than the governing arrangement (actors, institutions, and ideas) from which policy content arises (e.g., Howlett 2009; Eisner 2000; Harris and Milkis 1989).

4. Many other approaches prefer the category of "interests" rather than actor to create an alliterative triad of "interests, institutions, and ideas" (e.g., Hall 1993; Harrison and Sundstrom 2000; May and Joachim 2013). The regime approach used here prefers the term "strategic actors" because it better encompasses actors both within and outside the government. Many other approaches either have separate categories of state or government actors or, ironically, given the subject matter, don't really conceptualize government actors effectively in their framework. Government actors share many of the same incentives, resources, and strategies as nongovernment actors. What distinguishes them is that they have the resource of government authority.

5. Janzwood (2020) applies to several pipeline conflicts an integration of the McAdam and Boudet framework with an early version of the one presented in this book (Hoberg 2013).

6. This is similar to McAdam and Boudet's (2012) "political opportunity" variable.

7. McAdam and Boudet (2012) find that when a community has experience with similar projects, opposition to new projects is much less likely.

8. "In process tracing, the researcher examines histories, archival documents, interview transcripts, and other sources to see whether the causal process a theory hypothesizes or implies in a case is in fact evident in the sequence and values of the intervening variables in the case" (George and Bennett 2005, 6).

Chapter 2

1. That 1.3 million barrels per day increase by 2035 is only a fraction of the 6 million barrels per day capacity that have received approval for construction (Canadian Association of Petroleum Producers 2017a, 8).

2. Calculated from NEB data series Estimated Production of Canadian Crude Oil and Equivalent.

3. NEB reports production data by oil product type in a series called Estimated Production of Canadian Crude Oil and Equivalent, with subtotals given from Alberta; 79% of Canadian heavy oil is bitumen from the oil sands, and 47% of Canadian light oil is upgraded bitumen. Trade data are reported in the 2016 Oil Exports Statistics Summary, divided by light and heavy categories. In 2016, 2.3 million barrels per day of heavy oil and 0.8 million barrels per day of light oil were exported from Canada. Assuming that the same proportion of light and heavy oil products from Alberta's oil sands and Canadian production as a whole are exported, 1.8 million barrels per day of heavy oil sands and 0.4 million barrels per day of oil sands upgraded to light would have been exported in 2016. That total of 2.2 million barrels per day would constitute 85% of the 2.6 million barrels per day exported from Alberta.

4. Throughout the book, all dollar figures are US$ unless stated otherwise.

5. These data come from government of Alberta annual reports and can be found at http://www.finance.alberta.ca/publications/annual_repts/govt/index.html.

6. This list of populations in the oil sands region comes from the Alberta Biodiversity Monitoring Network (2014, 38).

7. All these figures are from Government of Alberta (n.d.c), http://economicdashboard.alberta.ca/.

8. According to an April 2019 IHS Markit analysis, these costs have fallen significantly over the past several years because of reductions in both capital and operating costs: "All things being equal, the price of oil required to justify a new oil sands project—mining or SAGD—has fallen. . . . IHS Markit estimated that the lowest-cost oil sands project—an expansion of an existing SAGD facility—required a WTI price more than $65/bbl in 2014 to break even. In 2018, this price had fallen into the mid-$40s/bbl. A mine without an upgrader required a WTI price approaching $100/bbl in 2014 compared with nearly $65/bbl in 2018" (IHS Markit 2019).

9. One review of oil sands economics concludes that "this suggests that the greatest climate policy risks for oil sands are from the oil market impacts of global action on

climate change, not domestic climate change policies" (Heyes, Leach, and Mason 2018, 250).

10. The next two paragraphs come from Hoberg (2016).

11. The groups involved were Canadian Parks and Wilderness Society, Dogwood Initiative, Greenpeace Canada, Natural Resources Defense Council, Rainforest Action Network, Sierra Club of Canada, Sierra Legal Defence Fund, Pembina Institute, West Coast Environmental Law, and Wildsight.

12. The Mackenzie Valley Pipeline, which would ship natural gas from Canada's north to the oil sands region.

Chapter 3

1. In his thoughtful book *The Patch*, Chris Turner refers to these competing worldviews as High Modern and Anthropocene (Turner 2017).

2. The groups are Canadian Parks and Wilderness Society, David Suzuki Foundation, Dogwood Initiative, Ecojustice, Environmental Defence, Equiterre, Greenpeace Canada, Pembina Institute, Sierra Club BC, and West Coast Environmental Law.

3. Pierre Trudeau was prime minister of Canada from 1968 to 1984, with the exception of a nine-month period during 1979–1980.

4. The next two sections are adapted from Hoberg (2016).

5. This section is taken from Hoberg (2016).

Chapter 4

I'd like to thank Claire Allen and Xavier Deschênes-Philion for research assistance, and Geoff Salomons for early research on this controversy. Thanks to Elizabeth Bennett for comments on an earlier draft. An earlier version of this chapter was prepared for delivery at the Annual Meeting of the American Political Science Association, August 31–September 3, 2017, in San Francisco, California.

1. Using LexisNexis Academic, total mentions and keywords were searched for each year (January 1 to December 31) beginning January 1, 2008, and ending August 31, 2019. The publication sources included were limited to the *Washington Post* and the *New York Times* (US edition) and their associated blog posts. Keyword searches included "Keystone XL" for the total number of mentions, "Keystone XL AND climate OR greenhouse gas OR global warming," "Keystone XL AND jobs OR job OR economic OR economics," "Keystone XL AND energy security," "Keystone XL AND accident OR spill OR disaster OR damage OR leak OR Lac Megantic," and "Keystone XL AND Native OR First Nation OR aboriginal OR indigenous."

2. For an overview of the relationship between activist discourse, media reporting, and policy influence, see Howe, Stoddart, and Tindall (2020).

3. The applicable law at the time was Executive Order 13337, Issuance of Permits with Respect to Certain Energy-Related Facilities and Land Transportation Crossings on the International Boundaries of the United States, dated April 30, 2004, https://www.federalregister.gov/documents/2004/05/05/04-10378/issuance-of-permits-with-respect-to-certain-energy-related-facilities-and-land-transportation.

4. In 2013, the Sierra Club went so far as to reverse its century-old policy against civil disobedience to allow its leaders to risk arrest (Hodges and Stocking 2016).

5. Prior to the 2016 election, TransCanada responded to Obama's rejection by challenging the action in federal court. In addition, the company filed a $15 billion claim under chapter 14 of NAFTA. Given Trump's reversal, however, these two challenges were withdrawn.

Chapter 5

1. The report was preceded by five reports by the Pembina Institute that focused on salmon risks, upstream impacts in the oil sands, and the economic justification for the pipeline.

2. An alliance of nine First Nations from British Columbia's central coast, north coast, and Haida Gwaii.

3. The data come from Canadian Newsstream. The pipeline was identified with the terms "Northern Gateway" or "Northern Gateway pipeline." Issue mentions were captured with the following search terms: "greenhouse gas" OR "climate" OR "global warming"; "job" OR "jobs"; "accident" OR "accidents" OR "spill" OR "spills" OR "disaster" OR "disasters" OR "damage" OR "damages" OR "leak" OR "leaks"; "First Nation" OR "First Nations" OR "indigenous" OR "aboriginal" OR "aboriginals."

Chapter 6

1. Annual total mentions and keywords were searched on Canadian Newsstream for each year from 2012 to 2017. The total number of sources was crossed with four categories: climate change, jobs, accidents/spills, and First Nations. Search terms for each category included pipeline name "AND climate OR greenhouse gas OR global warming"; pipeline name "AND job OR economic OR economics"; pipeline name "AND accident OR spill OR disaster OR damage OR leak OR Lac Megantic"; and pipeline name "AND native OR first nation OR aboriginal OR indigenous."

2. This section relies heavily on Hoberg (2018).

3. The statement was not in prepared remarks but in response to a reporter's question about his reaction to the government of Quebec seeking an injunction against the Energy East pipeline in March 2016:

> I think there is a desire by provinces across the country, understandably, that they want to ensure that they're acquiring the kind of social license that hasn't been acquired in the past. And that's where we're looking at working constructively and collaboratively with jurisdictions across the country for projects in the national interest in a way that understands that even though governments grant permits, ultimately only communities grant permission. And drawing in voices from a range of perspectives is going to lead us to a better number of and better kinds of solutions, and better outcomes for everyone across the country. (CBC 2016c)

4. Prime minister of Canada's news page: https://pm.gc.ca/eng/news.

5. The Canadian House of Commons' parliamentary web page enables a keyword search of Hansard publications by parliament, session, and speaker, among other categorical search tags. Searches in English and French for "grant permission" returned zero related results for members of the Trudeau government. It was used three times by two different Liberal backbenchers but never by a member of the cabinet.

6. British Columbia's minister of environment and climate change strategy, George Heyman, stated that the premier told him that "stopping the project was beyond the jurisdiction of B.C., and to talk about it or frame our actions around doing that, as opposed to defending B.C.'s coast through a variety of measures that were within our jurisdiction, would be inappropriate and unlawful" (Pearson 2018).

7. British Columbia's statement of claim challenging the Alberta law contains a number of quotations by Alberta government officials explicitly referring to inflicting economic pain on British Columbia to justify the legislation.

8. Tsleil-Waututh Nation; Squamish Nation; Musqueam Indian Band; Coldwater Indian Band; Aitchelitz, Skowkale Shxwa:y Village, Soow Ahlie, Squiala First Nation, Tzeachten, Yakweakwioose, Skwah, Kwaw-Kwaw-Aplit & Ts'elxweyeqw Tribe et al (Sto:lo); Upper Nicola Band; and Stk'emlupsemc Te Secwepemc (West Coast Environmental Law 2017).

Chapter 7

1. Because the review process was interrupted, a comprehensive evaluation of these projected economic benefits stemming from Energy East was never conducted, but regional analyses that have been done suggest that TransCanada's estimates may have been overstated, as argued by the Ontario Energy Board and the Montreal Metropolitan Community, for instance (Carlson et al. 2015; Communauté Métropolitaine de Montréal 2016).

2. Using Canadian Newsstream, the total mentions for each pipeline in English Canadian media were searched for each year (January 1 to December 31) from 2012 to 2016. Keyword searches for the total number of mentions for each pipeline included "Northern Gateway OR Northern Gateway Pipeline"; "Trans Mountain Pipeline OR TransMountain OR Trans Mountain"; "Keystone XL"; and "Energy East OR Energy East Pipeline."

3. Using Canadian Newsstream and Eureka.cc to count English and French Canadian media, respectively, the total mentions and keywords were searched for each year (January 1 to December 31) from 2012 to 2016. The total number of sources was crossed with four categories: climate change, jobs, accidents/spills, and First Nations. Search terms for each category included "Energy East Pipeline OR Energy East AND climate OR greenhouse gas OR global warming" / "Pipeline Énergie Est OR Oléoduc Énergie Est OR Projet Énergie Est AND changement climatique OR gaz à effet de serre OR réchauffement climatique OR climatique OR climat"; "Energy East Pipeline OR Energy East AND job" / "Pipeline Énergie Est OR Oléoduc Énergie Est OR Projet Énergie Est AND emploi"; "Energy East Pipeline OR Energy East AND accident OR spill OR disaster OR damage OR lead OR Lac Megantic" / "Pipeline Énergie Est OR Oléoduc Énergie Est OR Projet Énergie Est AND accident OR déversement OR Lac-Mégantic OR dommage OR désastre"; and "Energy East Pipeline OR Energy East AND Native OR First Nation OR aboriginal OR indigenous" / "Pipeline Énergie Est OR Oléoduc Énergie Est OR Projet Énergie Est AND première nation OR premières nations OR autochtones."

4. The first one was initiated in Quebec on August 11 by Stratégies Énergétiques and the Association québécoise de lutte contre la pollution atmosphérique. It was followed by a second on August 22, by Ecojustice on behalf of an Ontario group opposed to Energy East.

5. Poitras (2018, epilogue) also provides an overview of reasons for the cancellation.

Chapter 8

1. This statement is based on multiple interviews with respondents who chose to be off the record on this issue.

Chapter 9

1. The environmental impacts of hydropower projects, especially large dams, are both complex and controversial. The energy resource is renewable, but there are a variety of environmental impacts that can occur. Because of their impacts on habitat, land, and fish and wildlife, large dams have traditionally been strongly opposed by environmentalists. As a result of these environmental impacts, many definitions of "renewable" in American state electricity regulations exclude dams above a certain size, even though they are technically renewable (Frey and Linke 2002).

2. Cox (2018) gives a thorough account of the resistance coalition.

3. "While the Project is not being proposed for exporting energy, the energy surplus in its early years would allow BC Hydro to assist other jurisdictions, such as California, in managing an increasing level of intermittent resources such as solar or wind. This assistance could be provided irrespective of the net import/export position of

BC Hydro compared to external jurisdictions. The dynamic capacity and storage would allow these external jurisdictions to integrate additional wind, solar, and run-of-river hydro, in turn lowering their GHG emissions and footprint of supply resources. BC Hydro's ratepayers would further benefit from the revenues associated with providing such a service" (Joint Review Panel 2014, 273).

4. See also Shaw and Zussman (2018) for a sense of the personal implications of Horgan's decision. Cox (2018, 252–253) also discusses the Horgan decision and its impact on Peace River area activists.

Chapter 10

1. The feed-in tariff included domestic content provisions, or "buy local" rules, in an effort to tie renewable energy production to provincial economic growth in green manufacturing (Walker 2010). For wind projects, this was initially stipulated at 25%. The "buy local" provisions of the Green Energy Act were changed in 2012 after a World Trade Organization dispute was launched by Japan, the United States, and the European Union. The government attempted to appeal but was denied, causing the "buy local" provisions to be rescinded (Hill 2017).

2. NSTAR was a utility company in Massachusetts. In 2015, NSTAR and other subsidiaries were merged to become one large company, Eversource Energy, which is also the project developer of the Northern Pass Transmission project.

Chapter 11

1. This section is derived from Hunsberger, Froese, and Hoberg (2020).

References

Abacus Data. 2016b. *Public Opinion about Site C.* June 2016. http://abacusdata.ca/wp-content/uploads/2016/06/Abacus-Site-C-Public-Opinion-Survey-June-2016_FINAL.pdf.

Abelson, J., P.-G. Forest, J. Eyles, P. Smith, E. Martin, and F.-P. Gauvin. 2003. "Deliberations about Deliberative Methods: Issues in the Design and Evaluation of Public Participation Processes." *Social Science & Medicine* 57 (2): 239–251. https://doi.org/10.1016/S0277-9536(02)00343-X.

Aboriginal Equity Partners. 2016. "Read Our Op-ed in the Vancouver Sun." January 25, 2016. http://www.aepowners.ca/read_our_op_ed_in_the_vancouver_sun.

Acharibasam, John, and Bram F. Noble. 2014. "Assessing the Impact of Strategic Environmental Assessment." *Impact Assessment and Project Appraisal* 32 (3): 177–187. https://doi.org/10.1080/14615517.2014.927557.

Adkin, Laurie, ed. 2016. *First World Petro-Politics: The Political Ecology and Governance of Alberta*. Toronto: University of Toronto Press.

Aguilera, Roberto. 2014. "Production Costs of Global Conventional and Unconventional Petroleum." *Energy Policy* 64:134–140.

Alberta Biodiversity Monitoring Network. 2014. *The Status of Biodiversity in the Oil Sands Region of Alberta Preliminary Assessment 2014.* https://abmi.ca/home/publications/1-50/40.

Alberta Ministry of Energy. 2018. "Bill 12: Preserving Canada's Economic Prosperity Act." May 18, 2018. https://www.qp.alberta.ca/documents/Acts/p21p5.pdf.

Allen, Myles, et al. 2009. "Warming Caused by Cumulative Carbon Emissions towards the Trillionth Tonne." *Nature* 458 (7242): 1163–1166. https://doi.org/10.1038/nature08019.

Angus Reid Institute. 2016. "Rocky Mountain Rumble: Alberta, B.C. Residents at Odds over TransMountain Pipeline." June 16, 2016. http://angusreid.org/transmountain-pipeline/.

Angus Reid Institute. 2017. "British Columbians Call on New Provincial Government to Deliver Electoral Reform, Energy Projects." September 26, 2017. http://angusreid.org/horgan-ndp-electoral-reform-energy-projects/.

Angus Reid Institute. 2018a. "Site C: By Margin of Two-to-One, BC Residents Say Province Made the Right Call to Finish Dam Project." January 12, 2018. http://angusreid.org/site-c-horgan/.

Angus Reid Institute. 2018b. "Pipeline Problems? Try Tanker Troubles: BC Kinder Morgan Opponents Want Spill Response Assurances." April 18, 2018. http://angusreid.org/kinder-morgan-transmountain/.

Assembly of First Nations of Quebec and Labrador. 2016. "First Nations of Quebec Officially Oppose Energy East Pipeline." *Cision,* June 15, 2016. https://www.newswire.ca/news-releases/first-nations-of-quebec-officially-oppose-energy-east-pipeline-583165411.html.

Attorney General of British Columbia. 2018. Statement of Claim, no. 1801 (ABCQB), https://news.gov.bc.ca/files/Statement_of_Claim_Final.pdf.

Bailey, Ian. 2017. "B.C. Greens Won't Topple NDP Government over Site C Decision, Weaver Says." *Globe and Mail,* December 11, 2017. https://www.theglobeandmail.com/news/british-columbia/bc-greens-wont-topple-ndp-government-over-site-c-decision-weaver-says/article37300031/.

Bailey, Ian, and Josh Wingrove. 2012. "B.C.'s Northern Gateway Demands Trigger Showdown with Alberta." *Globe and Mail,* July 23, 2012. https://www.theglobeandmail.com/news/british-columbia/bcs-northern-gateway-demands-trigger-showdown-with-alberta/article4435977/.

Baldrey, Keith. 2018. "Analysis: The Kinder Morgan Pipeline Row Is about to Get Real." *Global News,* April 11, 2018. https://globalnews.ca/news/4139323/analysis-kinder-morgan-pipeline-row/.

Ballingall, Alex. 2017. "TransCanada Ends Bid to Build Energy East Pipeline after 'Careful Review of Changed Circumstances.'" *Toronto Star,* October 5, 2017. https://www.thestar.com/business/2017/10/05/transcanada-ends-bid-to-build-energy-east-pipeline-after-careful-review-of-changed-circumstances.html.

Bankes, Nigel. 2015. "Pipelines, the National Energy Board and the Federal Court." *Energy Regulation Quarterly* 3 (2). http://www.energyregulationquarterly.ca/case-comments/pipelines-the-national-energy-board-and-the-federal-court#sthash.trjoNzYg.dpuf.

Bankes, Nigel, and Alastair R. Lucas. 2004. "Kyoto, Constitutional Law and Alberta's Proposals." *Alberta Law Review* 42:355–398.

Bankes, Nigel, and Martin Olszynski. 2018. "TMX v Burnaby: When Do Delays by a Municipal (or Provincial) Permitting Authority Trigger Paramountcy and

Interjurisdictional Immunity?" *ABlawg: The University of Calgary Faculty of Law Blog*. January 24, 2018. https://ablawg.ca/2018/01/24/tmx-v-burnaby-when-do-delays-by-a-municipal-or-provincial-permitting-authority-trigger-paramountcy-and-interjurisdictional-immunity/.

BAPE (Bureau d'audiences publiques sur l'environnement Québec). 2016a. "Documents sur les émissions de gaz à effet de serre et sur les changements climatiques." http://www.bape.gouv.qc.ca/sections/mandats/oleoduc_energie-est/documents/ges.htm.

BAPE (Bureau d'audiences publiques sur l'environnement Québec). 2016b. "La commission d'enquête et d'audience publique débute ses travaux" Press release. January 26, 2016. http://www.bape.gouv.qc.ca/sections/mandats/oleoduc_energie-est/communiques/16-01-26-oleoduc_energie-est.htm.

Baresi, U., K. J. Vella, and N. G. Sipe. 2017. "Bridging the Divide between Theory and Guidance in Strategic Environmental Assessment: A Path for Italian Regions." *Environmental Impact Assessment Review* 62:14–24.

Baril, Daniel. 2015. "Les Canadiens sont peu préoccupés par le réchauffement climatique." *Universite de Montreal Nouvelles*, November 27, 2015. http://nouvelles.umontreal.ca/article/2015/11/27/les-canadiens-sont-peu-preoccupes-par-le-rechauffement-climatique/.

Batel, Susana, Patrick Devine-Wright, Torvald Tangeland. 2013. "Social Acceptance of Low Carbon Energy and Associated Infrastructures: A Critical Discussion." *Energy Policy* 58: 1–5.

Baumgartner, Frank, and Bryan Jones. 1993. *Agendas and Instability in American Politics*. Chicago: University of Chicago Press; reprinted 2010.

BC Coalition for Sustainable Forest Solutions 2001. *Getting Beyond the Softwood Lumber Dispute: Solutions in BC's Interest*. http://www.wcel.org/sites/default/files/publications/Getting%20Beyond%20the%20Softwood%20Lumber%20Dispute%20Solutions%20in%20BC's%20Interest%20-%20Preliminary%20Recommendations.pdf.

BC Green Caucus and the BC New Democrat Caucus. 2017. *2017 Confidence and Supply Agreement between the BC Green Caucus and the BC New Democrat Caucus*. May 29, 2017. http://bcndpcaucus.ca/wp-content/uploads/sites/5/2017/05/BC-Green-BC-NDP-Agreement_vf-May-29th-2017.pdf.

BC Hydro. 2013. *Site C Clean Energy Project—Environmental Impact Statement Executive Summary*. http://www.ceaa-acee.gc.ca/050/documents_staticpost/63919/85328/Executive_Summary.pdf.

BC Hydro. 2015. "BC Hydro and Saulteau First Nations Reach Agreement on Site C." Info bulletin. July 13, 2015. https://www.bchydro.com/news/press_centre/news_releases/2015/saulteau-first-nations-agreement.html.

BC Hydro. 2016a. "BC Hydro and McLeod Lake Indian Band Reach Agreements on Site C." Info bulletin. July 5, 2016. https://www.bchydro.com/news/press_centre/news_releases/2016/mlib-agrmnt.html.

BC Hydro. 2016b. "BC Hydro and Dene Tha' First Nation Come to Agreement on Site C." Info bulletin. July 20, 2016. https://www.bchydro.com/news/press_centre/news_releases/2016/dene-tha-site-c-agreement.html.

BC Hydro. 2016c. "BC Hydro Responds to Public Opinion Poll by Insights West." Info bulletin. November 16. https://www.bchydro.com/news/press_centre/news_releases/2016/site-c-poll-response.html.

BC Hydro. 2017. "BC Hydro and Halfway River First Nation Reach Agreements on Site C." Info bulletin. March 27, 2017. https://www.bchydro.com/news/press_centre/news_releases/2017/agreement-halfway-river-first-nation.html.

BC Ministry of Environment and Climate Change Strategy. 2017. "Government Takes Action to Protect B.C. over Kinder Morgan Pipeline and Tanker Traffic Expansion." News release. August 10, 2017. https://news.gov.bc.ca/releases/2017ENV0046-001417.

BC Ministry of Environment and Climate Change Strategy. 2018. "Additional Measures Being Developed to Protect B.C.'s Environment from Spills." News release. January 30, 2018. https://news.gov.bc.ca/releases/2018ENV0003-000115.

BC New Democratic Party. 2013. *Change for the Better: Practical Steps*. https://www.poltext.org/sites/poltext.org/files/plateformes/bcndp2013_plt.pdf.

BC New Democratic Party. 2017. *2017 BC NDP Platform*. https://action.bcndp.ca/page/-/bcndp/docs/BC-NDP-Platform-2017.pdf.

Behn, Caleb, and Karen Bakker. 2019. "Rendering Technical, Rendering Sacred: The Politics of Hydroelectric Development on British Columbia's Saaghii Naachii/Peace River." *Global Environmental Politics* 19:33–54.

Beierle, Thomas C. 2002. "The Quality of Stakeholder-Based Decisions." *Risk Analysis* 22 (4).

Bennett, Dean. 2016. "Alberta Premier Rachel Notley Rethinking Stance on Northern Gateway Pipeline." *Global News*, April 21, 2016. http://globalnews.ca/news/2654945/alberta-premier-rachel-notley-rethinking-stance-on-northern-gateway-pipeline/.

Bergeron, Patrice 2016. "L'UPA sort de sa réserve et s'oppose à Énergie Est." *Le Devoir*, May 9, 2016. http://www.ledevoir.com/environnement/actualites-sur-l-environnement/470323/transcanada-l-upa-prend-position-contre-energie-est.

Berman, Tzeporah. 2016. Personal interview with the author. September 16, 2016.

Berman, Tzeporah, with Mark Leiren-Young. 2011. *This Crazy Time: Living Our Environmental Challenges*. Toronto: Alfred Knopf Canada.

Bishop, Bradford H. 2014. "Focusing Events and Public Opinion: Evidence from the Deepwater Horizon Disaster." *Political Behavior* 36 (1): 1–22. https://link.springer.com/article/10.1007/s11109-013-9223-7.

Blakes, Cassels & Graydon LLP. 2017. "A83658–1 Letter to NEB re Comment Process on Lists of Issues—A5L8U1." Filed at the National Energy Board, May 17, 2017. https://apps.neb-one.gc.ca/REGDOCS/Item/View/3266020.

Blewett, Taylor. 2019. "United We Roll Protest Convoy Set to Reach Parliament Hill on Tuesday." *Ottawa Citizen*, February 19, 2019. https://ottawacitizen.com/news/local-news/united-we-roll-protest-convoy-set-to-reach-parliament-hill-on-tuesday.

Bloc Québécois. 2016. "Dire non à Énergie Est." http://www.blocquebecois.org/dire-non-a-energie-est/.

Bloomberg-Nanos. 2014. *Bloomberg-Nanos BC Northern Gateway Pipeline Survey Part 2 of 2, June 2014 (Submission 2014–525).* Nanos Research. http://www.nanosresearch.com/sites/default/files/POLNAT-S14-T609.pdf.

BOEM (Bureau of Ocean Energy Management, US Department of the Interior). n.d. *Cape Wind.* https://www.boem.gov/Massachusetts-Cape-Wind/.

Bold Nebraska. n.d. "About—We Mobilize Unlikely Alliances to Protect the Land and Water." http://boldnebraska.org/about/.

Boothby, Lauren. 2018. "B.C. Government Takes Pipeline Question to Court." *Burnaby Now*, April 26, 2018. http://www.burnabynow.com/news/b-c-government-takes-pipeline-question-to-court-1.23281968.

Borenstein, Severin, and Ryan Kellogg. 2014. *Energy Journal* 35 (1): 15–33.

Boudet, Hilary S., Dylan Bugden, Chad Zanocco, and Edward Maibach. 2016. "The Effect of Industry Activities on Public Support for 'Fracking.'" *Environmental Politics* 25 (4): 593–612. https://doi.org/ 10.1080/09644016.2016.1153771.

Boudet, Hilary S., Chad M. Zanocco, Peter D. Howe, and Christopher E. Clarke. 2018. "The Effect of Geographic Proximity to Unconventional Oil and Gas Development on Public Support for Hydraulic Fracturing." *Risk Analysis* 38 (9): 1871–1890. https://doi.org/ 10.1111/risa.12989.

Bowles, Paul, and Fiona MacPhail. 2017. "The Town That Said "No" to the Enbridge Northern Gateway Pipeline: The Kitimat Plebiscite of 2014." *Extractive Industries and Society* 4 (1): 15–23.

Braid, Don. 2018. "Coming Soon, Alberta's Bill to Choke Off Fuel to B.C." April 11. https://nationalpost.com/news/politics/braid-coming-soon-the-ndp-bill-to-choke-off-fuel-to-b-c/wcm/969e09b1-c02f-477d-86c6-3a02857a0c7a.

Bratt, Duane, Duane Bratt, Keith Brownsey, Richard Sutherland, and David Taras. 2019. *Orange Chinook: Politics in the New Alberta*. Calgary: University of Calgary Press.

BrightSource Energy. 2013. *FAQs about Desert Tortoise Care and Protection at the Ivanpah Solar Project.* May 2013. http://www.brightsourceenergy.com/stuff/contentmgr/files/0/044130f70ec2977f6389387b679dd815/files/ivanpah_tortoise_care___may_2013_final.pdf.

BrightSource Energy. n.d. *Ivanpah Project Overview.* http://www.brightsourceenergy.com/ivanpah-solar-project#.XDIQeC3Myu4.

British Columbia Environmental Assessment Office. 2016. *Trans Mountain Expansion Project.* British Columbia Office of the Premier. 2017. "Five Conditions Secure Coastal Protection and Economic Benefits for All British Columbians." *BC Government News,* January 11, 2017. https://news.gov.bc.ca/releases/2017PREM0002-000050.

British Columbia Office of the Premier. 2018. "B.C. Government Moves Forward on Action to Protect Coast." *BC Government News,* February 22, 2018. https://news.gov.bc.ca/releases/2018PREM0002-000252.

British Columbia Utilities Commission. 1983. *Site C Report: Report & Recommendations to the Lieutenant Governor-in-Council.* https://www.ordersdecisions.bcuc.com/bcuc/decisions/en/112107/1/document.do.

British Columbia Utilities Commission. 2017. *Inquiry Respecting Site C: Executive Summary of the Final Report to the Government of British Columbia.* November 1, 2017. https://www.ordersdecisions.bcuc.com/bcuc/decisions/en/236682/1/document.do.

Brulle, Robert J. 2014. "Institutionalizing Delay: Foundation Funding and the Creation of US Climate Change Counter-Movement Organizations." *Climatic Change* 122 (4): 681–694.

Bruno, Kenny. 2017. Personal interview with the author. November 7, 2017.

Burbank, R. 2012. "Appalachian Mountain Club Study Identified Significant Visual Impacts." *Appalachian Mountain Club Outdoors.* September 26, 2012. https://www.outdoors.org/articles/newsroom/northern-pass-project-in-n-h-appalachian-mountain-club.

Burck, Jan, Franziska Marten, and Christoph Bals. 2016. *Climate Change Performance Index—Results 2016.* Germanwatch and Climate Action Network Europe. https://www.climate-change-performance-index.org/the-climate-change-performance-index-2016.

Bureau of Land Management (BLM). 2011. "Service Issues Biological Opinion for Ivanpah Solar Electric Project; BLM Lifts Suspension of Activities Order." Press release. June 10, 2011. https://web.archive.org/web/20130312112703/http://www.fws.gov/cno/press/release.cfm?rid=239.

Burgman, Tamsyn. 2014. "Protesters Poke Fun at Oil Pipeline by Posting Snarling Selfies." *Toronto Star,* November 9, 2014. https://www.thestar.com/news/canada/2014/11/09/protesters_poke_fun_at_oil_pipeline_by_posting_snarling_selfies.html.

Burkett, E. 2003. "A Mighty Wind." *New York Times*, June 15, 2003. https://www.nytimes.com/2003/06/15/magazine/a-mighty-wind.html.

Burnaby (City) v. Trans Mountain Pipeline ULC, 2015 BCSC 2140.

Butts, Gerald. 2012. "Our Ecological Treasure Is the Issue with Northern Gateway." *Globe and Mail*, January 11, 2012. https://www.theglobeandmail.com/opinion/our-ecological-treasure-is-the-issue-with-northern-gateway/article554316/.

Cain, N. L., and H. T. Nelson. 2013. "What Drives Opposition to High-Voltage Transmission Lines?" *Land Use Policy* 33:204–213.

California Desert District, Bureau Land of Management. 2011. *Decision: Immediate Temporary Suspension of Activities Issued Notice*. https://www.scribd.com/doc/53386569/Ivanpah-Temporary-Suspension-Notice.

Canadian Association of Petroleum Producers (CAPP). 2017a. *Statistical Handbook for Canada's Upstream Petroleum Industry*. Calgary: Canadian Association of Petroleum Producers. October 2017.

Canadian Association of Petroleum Producers. 2017b. "Re: CEAA Expert Panel Report." https://www.letstalkea.ca/3524/documents/6527.

Canadian Association of Petroleum Producers (CAPP). 2019. *2019 Crude Oil Forecast, Markets and Transportation*. https://www.capp.ca/publications-and-statistics/crude-oil-forecast.

Canadian Association of Petroleum Producers (CAPP). n.d. *Statistics*. https://www.capp.ca/resources/statistics/.

Canadian Council of Academies, Expert Panel on the Potential for New and Emerging Technologies to Reduce the Environmental Impacts of Oil Sands Development. 2015. *Technological Prospects for Reducing the Environmental Footprint of Canadian Oil Sands*. Canadian Council of Academics. http://www.scienceadvice.ca/uploads/ENG/AssessmentsPublicationsNewsReleases/OilSands/OilSandsFullReportEn.pdf.

Canadian Environmental Assessment Agency. 2014. *Decision Statement Issued under Section 54 of the Canadian Environmental Assessment Act, 2012 to British Columbia Hydro and Power Authority for the Site C Clean Energy Project*. October 14, 2014. http://www.ceaa-acee.gc.ca/050/documents/p63919/100567E.pdf.

Canadian Intergovernmental Conference Secretariat. 2016a. *Vancouver Declaration on Clean Growth and Climate Change*. March 3. 2016. https://scics.ca/en/product-produit/vancouver-declaration-on-clean-growth-and-climate-change/#.

Canadian Intergovernmental Relations Conference Secretariat. 2016b. *Pan-Canadian Framework on Clean Growth and Climate Change*. December 9, 2016. https://scics.ca/en/product-produit/pan-canadian-framework-on-clean-growth-and-climate-change/.

Canadian Upstream Strategy Working Group. 2006. *The Canadian Upstream Strategy Working Group Oil and Gas Strategy*. Canadian Upstream Strategy Working Group. February 2006.

Canadian Wind Energy Association. n.d. *Wind Energy in Ontario*. https://canwea.ca/wind-energy/ontario/.

Carbon Tracker Initiative. 2011. *Unburnable Carbon—Are the World's Financial Markets Carrying a Carbon Bubble?* November 2011. http://www.carbontracker.org/wp-content/uploads/2014/09/Unburnable-Carbon-Full-rev2-1.pdf.

Carlson, Richard, Rob Dorling, Peter Spiro, and Mike Moffatt 2015. *A Review of the Economic Impact of Energy East on Ontario*. Toronto: Mowat Centre, Monk School of Public Policy and Governance, University of Toronto.

Carter, A., G. Fraser, and A. Zalik. 2017. "Environmental Policy Convergence in Canada's Fossil Fuel Provinces? Regulatory Streamlining, Impediments, and Drift." *Canadian Public Policy / Analyse de politiques* 43 (1): 61–76.

Carter, Angela. 2016a. "The Petro-Politics of Environmental Regulation in the Tar Sands." In *First World Petro-Politics: The Political Ecology and Governance of Alberta*, edited by Laurie Adkin, 152–189. Toronto: University of Toronto Press.

Carter, Angela. 2016b. "Environmental Policy and Politics: The Case of Oil." In *Canadian Environmental Policy and Politics: The Challenges of Austerity and Ambivalence*, 4th ed., edited by D. VanNijnatten, 292–306. Don Mills: Oxford University Press.

Casey, M. 2017. "First Nation Opposes Proposed Powerline Project in New Hampshire." *CTV News*, July 20, 2017. https://www.ctvnews.ca/business/first-nation-opposes-proposed-powerline-project-in-new-hampshire-1.3512757.

Casey, M. 2018. "New Hampshire Committee Rejects Northern Pass Appeal." *Financial Post*. May 24, 2018. https://business.financialpost.com/pmn/business-pmn/new-hampshire-committee-rejects-northern-pass-appeal.

Casey-Lefkowitz, Susan. 2009. "Migratory Bird Haven Joins NRDC BioGems." Natural Resources Defense Council. February 3, 2009. https://www.nrdc.org/experts/susan-casey-lefkowitz/migratory-bird-haven-joins-nrdc-biogems.

Cashore, Benjamin, George Hoberg, Michael Howlett, Jeremy Rayner, and Jeremy Wilson. 2001. *In Search of Sustainabilty: British Columbia Forest Policy in the 1990s*. Vancouver: UBC Press.

Cassell, B. 2016. "Cape Wind Project Suffers Loss at Federal Appeals Court." *Renewable Energy World*, July 2016. https://www.renewableenergyworld.com/articles/2016/07/cape-wind-project-suffers-loss-at-federal-appeals-court.html.

Castonguay, Alec. 2015. "Oléoduc Énergie Est: La moitié des Québécois disent non." *L'Actualité*, January 12, 2015. http://lactualite.com/politique/2015/01/12/oleoduc-energie-est-la-moitie-des-quebecois-disent-non/.

Cattaneo, Claudia. 2012. "Ottawa Dials Down Support for Northern Gateway Pipeline, Citing 'Huge Challenges.'" *Financial Post*, November 30, 2012. http://business.financialpost.com/commodities/energy/ottawa-dials-down-support-for-northern-gateway-pipeline-citing-huge-challenges/wcm/188e0053-c13d-4653-b72e-bd10211160ec.

Cattaneo, Claudia. 2015. "Conflict of Interest Charges Mar National Energy Board Panel." *Financial Post*, August 27, 2015. http://business.financialpost.com/commodities/energy/conflict-of-interest-charges-mar-national-energy-board-panel.

Cattaneo, Claudia. 2016. "Irving Oil's President Says It Would Keep Saudi Imports Even If Energy East Goes Ahead." *Financial Post*, April 12, 2016. http://business.financialpost.com/commodities/energy/irving-oils-president-says-it-would-keep-saudi-imports-even-if-energy-east-goes-ahead/wcm/a24d5455-87fb-48c7-a0b3-a8db1d18f228.

CBC (Canadian Broadcasting Corporation). 2006. "'Impossible' for Canada to Reach Kyoto Targets: Ambrose." *CBC News*, April 7, 2006. http://www.cbc.ca/news/canada/impossible-for-canada-to-reach-kyoto-targets-ambrose-1.583826.

CBC (Canadian Broadcasting Corporation). 2008. "PM: Dion's Carbon Tax Would 'Screw Everybody.'" *CBC News*, June 20, 2008. http://www.cbc.ca/news/canada/pm-dion-s-carbon-tax-would-screw-everybody-1.696762.

CBC (Canadian Broadcasting Corporation). 2009. "New Law Will Keep NIMBY-ism from Stopping Green Projects: Ont. Premier." *CBC News*, February 10, 2009. https://www.cbc.ca/news/technology/new-law-will-keep-nimby-ism-from-stopping-green-projects-ont-premier-1.805978.

CBC (Canadian Broadcasting Corporation). 2010. "B.C. Oil Pipeline, Tankers Opposed by UBCM." *CBC News*, October 1, 2010. http://www.cbc.ca/news/canada/british-columbia/b-c-oil-pipeline-tankers-opposed-by-ubcm-1.945213.

CBC (Canadian Broadcasting Corporation). 2012a. "Gateway Pipeline Hearings Resume in Bella Bella, B.C." *CBC News*, April 3, 2012. http://www.cbc.ca/news/canada/british-columbia/gateway-pipeline-hearings-resume-in-bella-bella-b-c-1.1204727.

CBC (Canadian Broadcasting Corporation). 2012b. "Environmental Charities 'Laundering' Foreign Funds." *CBC News*, May 1, 2012. http://www.cbc.ca/news/politics/story/2012/05/01/pol-peter-kent-environmental-charities-laundering.html.

CBC (Canadian Broadcasting Corporation). 2014. "New Brunswick Premier Backs Energy East Pipeline on Alberta Visit." *CBC News*, October 21, 2014. http://www.cbc.ca/news/canada/calgary/new-brunswick-premier-backs-energy-east-pipeline-on-alberta-visit-1.2806889.

CBC (Canadian Broadcasting Corporation). 2015. "Rachel Notley's NDP Win Puts Future of Enbridge Northern Gateway Pipeline in Question." *CBC News*, May 6, 2015.

http://www.cbc.ca/news/canada/british-columbia/rachel-notley-s-ndp-win-puts-future-of-enbridge-northern-gateway-pipeline-in-question-1.3063190.

CBC (Canadian Broadcasting Corporation). 2016a. "Brian Gallant Pushes Energy East Pipeline on Tout le monde en parle." *CBC News*, February 8, 2016. http://www.cbc.ca/news/canada/new-brunswick/brian-gallant-energy-east-pipeline-1.3438534.

CBC (Canadian Broadcasting Corporation). 2016b. "Montreal Mayor Denis Coderre Says Energy East Pipeline Too Risky." *CBC News*, May 1, 2016. http://www.cbc.ca/news/politics/story/2012/05/01/pol-peter-kent-environmental-charities-laundering.html.

CBC (Canadian Broadcasting Corporation). 2018a. "Hydro-Quebec Wins Major Massachusetts Energy Contract." *CBC News*, January 25, 2018. https://www.cbc.ca/news/canada/montreal/hydro-québec-energy-contract-1.4504156.

CBC (Canadian Broadcasting Corporation). 2018b. "Hydro-Quebec's Northern Pass Electricity Project Rejected by New Hampshire." *CBC News*, February 2, 2018. https://www.cbc.ca/news/canada/montreal/northern-pass-hydro-project-rejected-1.4515681.

CEPA (Canadian Energy Pipeline Association). 2017. "Response to the Expert Panel Review of Environmental Assessment Processes Final Report, Building Common Ground: A New Vision for Impact Assessment in Canada." https://businessdocbox.com/Government/66478644-5-may-2017-prepared-by-canadian-energy-pipeline-association.html.

Champagne, Sarah R. 2016. "Coderre réclame la suspension des audiences sur Énergie Est." *Le Devoir*, August 25, 2016. http://www.ledevoir.com/politique/montreal/478566/coderre-demande-la-suspension-des-audiences-sur-energie-est.

Cheon, Andrew, and Johannes Urpelainen. 2018. *Activism and the Fossil Fuel Industry*. London: Routledge.

Chess, Caron, and Kristen Purcell. 1999. "Public Participation and the Environment: Do We Know What Works?" *Environmental Science & Technology* 3 (16): 2685–2692. https://doi.org/10.1021/es980500g.

Chilvers, J. 2007. "Deliberating Competence: Theoretical and Practitioner Perspectives on Effective Participatory Appraisal Practice." *Science, Technology & Human Values* 33 (2): 155–185. https://doi.org/10.1177/0162243907307594.

Christidis, T., and J. Law. 2012. "Annoyance, Health Effects, and Wind Turbines: Exploring Ontario's Planning Processes." *Canadian Journal of Urban Research* 21 (1): 81–105. http://search.ebscohost.com/login.aspx?direct=true&db=aph&AN=86001355&site=ehost-live%5Cnhttp://content.ebscohost.com/ContentServer.asp?T=P&P=AN&K=86001355&S=R&D=aph&EbscoContent=dGJyMNLe80Sep7E4y9fwOLCmr0yeprSr6m4TbaWxWXS&ContentCustomer=dGJyMPGqtkm1rbV.

Christie, Gordon. 2006. "Developing Case Law: The Future of Consultation and Accommodation." U.B.C. L. Rev. 39 (1): 139–184.

City of Burnaby. 2016. "Mayor Derek Corrigan Statement in Response to Federal Government Approval of Kinder Morgan Pipeline Proposal." News release. November 29, 2016. https://www.burnaby.ca/About-Burnaby/News-and-Media/Newsroom/Mayor-Derek-Corrigan-Statement-in-Response-to-Federal-Government-Approval-of-Kinder-Morgan-Pipeline-Proposal_s2_p5957.html.

City of Burnaby. 2018. "Burnaby to Appeal NEB Decision on City Bylaws to the Supreme Court of Canada." News release. March 27, 2018. https://www.burnaby.ca/About-Burnaby/News-and-Media/Newsroom/Burnaby-to-Appeal-NEB-Decision-on-City-Bylaws-to-the-Supreme-Court-of-Canada_s2_p6446.html.

City of Vancouver. n.d. *It's Not Worth the Risk.* https://notworththerisk.vancouver.ca/.

Cleland, Michael, Stephen Bird, Stewart Fast, Shafak Sajid, and Louis Simard. 2016. *A Matter of Trust: The Role of Communities in Energy Decision-Making.* Ottawa: Canada West Foundation and the University of Ottawa. November 24, 2016. https://cwf.ca/research/publications/a-matter-of-trust-the-role-of-communities-in-energy-decision-making/.

Clogg, Jessica. 2017. Personal interview with the author. September 29, 2017.

CNW. 2009. "Polls Show Overwhelming Support for Ontario Green Energy Act." April 26, 2009. https://www.newswire.ca/news-releases/polls-show-overwhelming-support-for-ontario-green-energy-act-537496541.html.

Coastal First Nations. 2010. *Coastal First Nations Declaration.* https://www.wcel.org/sites/default/files/old/files/file-downloads/Coastal%20First%20Nations%20Tanker%20Ban%20Declaration.pdf.

Coastal First Nations v. British Columbia (Environment), 2016, BCSC 34.

Coates, Kenneth, and Dwight Newman. 2014. *The End Is Not Nigh: Reason over Alarmism in Analysing the Tsilhqot'in Decision.* MacDonald-Laurier Institute. September 2014. http://www.macdonaldlaurier.ca/files/pdf/MLITheEndIsNotNigh.pdf.

Coffey, B., J. A. Fitzsimons, and R. Gormly. 2011. "Strategic Public Land Use Assessment and Planning in Victoria, Australia: Four Decades of Trailblazing but Where to from Here? *Land Use Policy* 28:306–313.

Coldwater et al. v. Canada (Attorney General) et al., 2020 FCA 34.

Collier, David. 2011. "Understanding Process Tracing." *Political Science and Politics* 44 (4): 823–830.

Commission on Environmental Cooperation. 2018. *Alberta Tailings Pond II.* http://www.cec.org/sem-submissions/alberta-tailings-ponds-ii.

Communauté Métropolitaine de Montréal. 2016. *Projet d'oléoduc Énergie Est Trans-Canada. Développement Économique*. Technical report. March 2016, revised August 2017. http://cmm.qc.ca/fileadmin/user_upload/documents/20170818_transCanada_rapport Technique_devEconomique.pdf.

Connolly, Joel. 2016. "U.S. Tribes Mobilize against Giant Proposed Canadian Pipeline, Oil Export Terminal." *Seattle Post-Intelligencer*, May 18, 2016. http://www.seattlepi.com/local/politics/article/U-S-tribes-mobilize-against-giant-Canadian-7680625.php.

COSEWIC (Committee on the Status of Endangered Wildlife in Canada). 2016. *Designatable Units for Beluga Whales (Delphinapterus leucas) in Canada*. Ottawa: Committee on the Status of Endangered Wildlife in Canada.

Cotton, M., and P. Devine-Wright. 2012. "Making Electricity Networks 'Visible': Industry Actor Representations of 'Publics' and Public Engagement in Infrastructure Planning." *Public Understanding of Science* 21 (1): 17–35.

Cotton, M., and P. Devine-Wright. 2013. "Putting Pylons into Place: A UK Case Study of Public Perspectives on the Impacts of High Voltage Overhead Transmission Lines." *Journal of Environmental Planning* 56 (8): 1225–1245.

Coule pas chez nous. 2016a. "La Fondation." http://www.coulepascheznous.com/lafondation.

Coule pas chez nous. 2016b. "Les projets financés." http://www.coulepascheznous.com/projets#project.

Council of the Federation. 2015. *Canadian Energy Strategy*. July 2015. http://canadaspremiers.ca/wp-content/uploads/2013/03/canadian_energy_strategy_eng_fnl.pdf.

Cour Supérieure du Québec. 2014a. *Centre québécois du droit de l'environnement c. Oléoduc Énergie Est ltée*. 2014 QCCS 4147, no. 500-17-082462-147. September 1, 2014.

Cour Supérieure du Québec. 2014b. *Centre québécois du droit de l'environnement c. Oléoduc Énergie Est ltée*. 2014 QCCS 4398, no. 500-17-082462-147. September 23, 2014.

Cox, Sarah. 2018. *Breaching the Peace: The Site C Dam and a Valley's Stand against Big Hydro*. Vancouver: On Point Press.

Cox, Sarah. 2020a. "To Understand B.C.'s Push for the Coastal GasLink Pipeline, Think Fracking, LNG Canada and the Site C Dam." *The Narwhal*, March 3, 2020. https://thenarwhal.ca/to-understand-b-c-s-push-for-the-coastal-gaslink-pipeline-think-fracking-lng-canada-and-the-site-c-dam/.

Cox, Sarah. 2020b. "Top B.C. Government Officials Knew Site C Dam Was in Serious Trouble over a Year Ago: FOI Docs." October 21, 2020. https://thenarwhal.ca/site-c-dam-geotechnical-problems-bc-government-foi-docs/.

CPUC (California Public Utility Commission). 2011. "CPUC Orders Edison to Stop Transmission Line Work in Chino Hills and to Submit Options." News release. November 10, 2011. http://docs.cpuc.ca.gov/published/News_release/151359.htm.

CPUC (California Public Utility Commission). 2013. *Decision 13-07-018. Decision Granting the City of Chino Hills's Petition for Modification of Decision 09-12-044 and Requiring Undergrounding of Segment 8A of the Tehachapi Renewable Transmission Project.* http://docs.cpuc.ca.gov/PublishedDocs/Published/G000/M072/K175/72175064.PDF.

Cronmiller, J. G., and B. F. Noble. 2018. "The Discontinuity of Environmental Effects Monitoring in the Lower Athabasca Region of Alberta, Canada: Institutional Challenges to Long-Term Monitoring and Cumulative Effects Management." *Environmental Reviews* 26:169–180.

Croteau, Martin. 2017. "Projet d'oléoduc: Québec suspend son évaluation d'Énergie Est." *La Presse*, September 15, 2017. http://www.lapresse.ca/environnement/economie/201709/14/01-5133480-projet-doleoduc-quebec-suspend-son-evaluation-denergie-est.php.

Cryderman, Kelly. 2017. "Kinder Morgan Deal with British Columbia Sets Payment Precedent." *Globe and Mail*, January 12, 2017. https://www.theglobeandmail.com/report-on-business/industry-news/energy-and-resources/kinder-morgan-deal-with-british-columbia-sets-payment-precedent/article33609752/#c-image-0.

Cryderman, Kelly, Carrie Tait, and Mike Hager. 2018. "Notley Threatens to Turn Off Oil Taps in Dispute with B.C. over Trans Mountain Pipeline." *Globe and Mail*, March 8, 2018. https://www.theglobeandmail.com/news/alberta/notley-threatens-to-broaden-dispute-with-bc-over-trans-mountain-pipeline/article38253632/.

Cullen, Drea, et al. 2010. "Collaborative Planning in Complex Stakeholder Environments: An Evaluation of a Two-Tiered Collaborative Planning Model." *Society and Natural Resources* 23 (4): 332–350. https://doi.org/10.1080/08941920903002552.

Danelski, D. 2017. "Ivanpah Solar Plant, Built to Limit Greenhouse Gases, Is Burning More Gasses." *Press-Enterprise*, January 23, 2017. https://www.pe.com/2017/01/23/ivanpah-solar-plant-built-to-limit-greenhouse-gases-is-burning-more-natural-gas/.

Daniel, Patrick. 2011. "Canada's Role as a Leader in the World's Energy Market." Speech delivered at the Empire Club, Toronto, March 31, 2011. https://www.ragan.com/Resource.ashx?sn=Enbridge1.

Department of Finance, Canada. 2018. "Backgrounder: Details of Agreement for the Completion of the Trans Mountain Expansion Project." May 29, 2018. https://www.canada.ca/en/department-finance/news/2018/05/backgrounder-details-of-agreement-for-the-completion-of-the-trans-mountain-expansion-project.html.

De Souza, Mike. 2016a. "Brad Wall Provokes Anti-Quebec Insults over Pipeline Opposition." *National Observer*, January 22, 2016. http://www.nationalobserver.com/2016/01/22/news/brad-wall-provokes-anti-quebec-insults-over-pipeline-opposition.

De Souza, Mike. 2016b. "Quebec's Jean Charest Had Private Meeting with Pipeline Watchdog after TransCanada Hired Him." *National Observer*, July 7, 2016. http://www.nationalobserver.com/2016/07/07/news/quebecs-jean-charest-had-secret-meeting-pipeline-watchdog-after-transcanada-hired.

De Souza, Mike. 2016c. "Canada Pipeline Panel Apologizes, Releases Records on Meeting with Charest." *National Observer*, August 4, 2016. http://www.nationalobserver.com/2016/08/04/news/canada-pipeline-panel-apologizes-releases-records-meeting-charest.

De Souza, Mike. 2016d. "New Allegations of Bias over Charest Meeting Shake TransCanada Pipeline Hearings." *National Observer*, August 22, 2016. http://www.nationalobserver.com/2016/08/22/news/new-allegations-bias-over-charest-meeting-shake-transcanada-pipeline-hearings.

Devine-Wright, P. 2009. "Rethinking NIMBYism: The Role of Place Attachment and Place Identity in Explaining Place-Protective Action." *Journal of Community & Applied Social Psychology* 19 (6): 426–441.

Devine-Wright, Patrick, Hannah Devine-Wright, and Richard Cowell. 2016. "What Do We Know about Overcoming Barriers to Siting Energy Infrastructure in Local Areas?" Placewise. https://orca.cf.ac.uk/93905/1/DECC_Infrastructure_PlacewiseLtd.pdf.

Difley, J. 2011. "Trees, Not Towers: Why the Forest Society Opposes Northern Pass and Why You Should Too." *New Hampshire Business Review*, March 11, 2011. https://www.nhbr.com/March-11–2011/Trees-not-towers/.

Difley, J., and W. Webb. 2016. "Northern Pass Behaving Badly: Tying the Project to the Balsams Redevelopment Is Divisive and Coercive." *New Hampshire Business Review*, April 29, 2016. https://www.nhbr.com/April-29-2016/Northern-Pass-behaving-badly/.

Dion, Jason, Dave Sawyer, and Phil Gass. 2014. *A Climate Gift or a Lump of Coal? The Emission Impacts of Canadian and U.S. Greenhouse Gas Regulations in the Electricity Sector*. International Institute for Sustainable Development. September 2014. http://www.iisd.org/sites/default/files/publications/climate-gift-or-lump-of-coal.pdf.

Diotalevi, Robert, and Susan Burhoe. 2016. "Native American Lands the Keystone Pipeline Expansion: A Legal Analysis." *Indigenous Policy Journal* 27 (2): 1–12.

Doern, Bruce G., and Glen Toner. 1985. *The Politics of Energy: The Development and Implementation of the NEP*. Toronto: Methuen.

Dogwood BC. n.d. "Let BC Vote." http://letbcvote.dogwoodbc.ca/.

Dogwood Initiative. 2011. "Stepping Up to the Mic." October 14, 2011. https://dogwoodbc.ca/news/mob-the-mic-success/.

Dogwood Initiative. 2013. *2013 Annual Report: Powered by You.* https://dogwoodinitiative.org/aboutus/annual-reports/2012-13.

Dombek, C. 2011. "California Supreme Court Refuses to Hear Challenge to SCE Tehachapi Line." *Transmission Hub,* December. https://www.transmissionhub.com/articles/2011/12/california-supreme-court-refuses-to-hear-challenge-to-sce-tehachapi-line.html.

Dombek, C. 2012. "SCE: Undergrounding Portion of Tehachapi Project Could Cost Up To 5x More Than Overhead Lines." *Transmission Hub,* December 23, 2011. https://www.transmissionhub.com/articles/2012/12/sce-undergrounding-portion-of-tehachapi-project-could-cost-up-to-5x-more-than-overhead-lines.html.

Dryzek, John S. 2002. *Deliberative Democracy and Beyond: Liberals, Critics, Contestations.* Oxford: Oxford University Press. http://www.oupcanada.com/catalog/9780199250431.html.

Eckhouse, B., and J. Ryan. 2017. "What Was Once Hailed as First U.S. Offshore Wind Farm Is No More." *Bloomberg,* December 1, 2017. https://www.bloomberg.com/news/articles/2017-12-01/cape-wind-developer-terminates-project-opposed-by-kennedys-koch.

Egeberg, Morton. 1999. "The Impact of Bureaucratic Structure on Policy Making." *Public Administration* 77 (1): 155–170.

Eisner, Marc Allen. 2000. *Regulatory Politics in Transition.* 2nd ed. Baltimore: Johns Hopkins University Press.

Ejima, M., J. Gaskin, P. Maskulrath, and K. Singh. 2015. *Cape Wind: The Collapse of the United States' Inaugural Offshore Wind Farm Project.* University of British Columbia, Department of Geography. https://environment.geog.ubc.ca/cape-wind-the-collapse-of-the-united-states-inaugural-offshore-wind-farm-project/.

Ekos. 2016. "Canadian Attitudes toward Energy and Pipelines: Survey Findings." March 17, 2016. http://www.ekospolitics.com/wp-content/uploads/full_report_march_17_2016.pdf.

Energy Information Administration. 2017. *Annual Energy Outlook 2017.* January 5, 2017. https://www.eia.gov/outlooks/aeo/pdf/0383(2017).pdf.

Energy Information Administration. n.d. "Europe Brent Spot Price." https://www.eia.gov/dnav/pet/hist/RBRTED.htm.

Environics Institute. 2019. *Focus Canada—Spring 2019 Canadian Public Opinion about Immigration and Refugees.* April 30, 2019. https://www.environicsinstitute.org/docs/default-source/project-documents/focus-canada-spring-2019/environics-institute--focus-canada-spring-2019-survey-on-immigration-and-refugees--final-report.pdf?sfvrsn=8dd2597f_2.

Environmental Defence. 2008. *Canada's Toxic Tar Sands: The Most Destructive Project on Earth*. February. https://environmentaldefence.ca/report/report-canadas-toxic-tar-sands-the-most-destructive-project-on-earth/.

Environmental Defence and ForestEthics. 2012. *Our Nation, Their Interest: The Case against the Northern Gateway Pipeline*. Report. March 2012. http://environmentaldefence.ca/report/report-nation-interest/.

Environmental Defence and Greenpeace. 2014. *TransCanada Vastly Exaggerating Energy East's Ability to Reduce Overseas Oil Imports. Report*. October 2014. https://environmentaldefence.ca/report/report-transcanada-vastly-exaggerating-energy-easts-ability-to-reduce-overseas-oil-imports/.

Environment and Climate Change Canada. 2016. *Trans Mountain Pipeline ULC—Trans Mountain Expansion Project Review of Related Upstream Greenhouse Gas Emissions Estimates*. November 2016. https://ceaa-acee.gc.ca/050/documents/p80061/116524E.pdf.

Environment and Climate Change Canada. 2017a. *Greenhouse Gas Emissions—Canadian Environmental Sustainability Indicators*. April 2017. https://www.canada.ca/content/dam/eccc/migration/main/indicateurs-indicators/18f3bb9c-43a1-491e-9835-76c8db9ddfa3/ghgemissions_en.pdf.

Environment and Climate Change Canada. 2017b. *Canada's 2016 Greenhouse Gas Emissions Reference Case*. https://www.canada.ca/en/environment-climate-change/services/climate-change/publications/2016-greenhouse-gas-emissions-case.html.

Environment and Climate Change Canada. 2017c. *Report on the Progress of Recovery Strategy Implementation for the Woodland Caribou (Rangifer tarandus caribou), Boreal Population in Canada for the Period 2012–2017*. Species at Risk Act Recovery Strategy Series. Ottawa: Environment and Climate Change Canada. http://registrelep-sararegistry.gc.ca/virtual_sara/files/Rs-ReportOnImplementationBorealCaribou-v00-2017Oct31-Eng.pdf.

Environment and Climate Change Canada. 2019. *Facility Greenhouse Gas Reporting: Overview of Reported Emissions 2017*. https://www.canada.ca/en/environment-climate-change/services/climate-change/greenhouse-gas-emissions/facility-reporting/overview-2017.html.

Environment and Climate Change Canada. 2020. *A Healthy Environment and a Healthy Economy: Canada's Strengthened Climate Plan to Create Jobs and Support People, Communities and the Planet*. https://www.canada.ca/en/services/environment/weather/climatechange/climate-plan/climate-plan-overview.html.

Équiterre. n.d. *Signez la Petition—Non Aux Sables Bitumeaux*. https://equiterre.org/sites/fichiers/e_petition_to_print.pdf.

Evans-Brown, S. 2014. "Understanding Northern Pass: Is Northern Pass Inevitable? Would We Ever Get Used To It? The Headlines Focus on Power and Towers, but It's

Really a Story about People and Land." *New Hampshire Magazine*, January 2014. https://www.nhmagazine.com/January-2014/Understanding-Northern-Pass/.

Fast, S. 2013. "Social Acceptance of Renewable Energy: Trends, Concepts, and Geographies." *Geography Compass* 7 (12): 293–317.

Fast, S. 2016. "Assessing Public Participation Tools during Wind Energy Siting." *Journal of Environmental Studies and Sciences* 7 (3): 386–393.

Fast, S., and W. Mabee. 2015. "Place-Making and Trust-Building: The Influence of Policy on Host Community Responses to Wind Farms." *Energy Policy* 81: 27–37. https://doi.org/10.1016/j.enpol.2015.02.008.

Fast, S., W. Mabee, J. Baxter, T. Christidis, L. Driver, S. Hill, and M. Tomkow. 2016. "Lessons Learned from Ontario Wind Energy Disputes." *Nature Energy* 1 (2): 15028. https://doi.org/10.1038/nenergy.2015.28.

Fast, S., W. Mabee, and J. Blair. 2015. "The Changing Cultural and Economic Values of Wind Energy Landscapes." *Canadian Geographer* 59 (2): 181–193. https://doi.org/10.1111/cag.12145.

Fekete, Jason. 2010. "Prentice Tells Oil Sands to Clean Up Its Act." *National Post*, February 1, 2010. http://www.nationalpost.com/rss/Prentice+tells+sands+clean/2509815/story.html.

Fekete, Jason. 2016. "Will Battle over Energy East Pipeline Threaten National Unity?" *Ottawa Citizen*, January 25, 2016. http://ottawacitizen.com/news/national/will-battle-over-energy-east-pipeline-threaten-national-unity.

Fenton, Cam. 2017a. Personal interview with the author. November 10, 2017.

Fenton, Cam. 2017b. "Climate Concerns Killed Energy East: And That's a Good Thing." *Huffington Post*, October 13, 2017. http://www.huffingtonpost.ca/cameron-fenton/climate-concerns-killed-energy-east-and-thats-a-good-thing_a_23241672/.

Fiorina, Morris. 1996. *Divided Government*. Boston: Allyn and Bacon.

Fligstein, Neil, and Doug McAdam, 2011. "Toward a General Theory of Strategic Action Fields." *Sociological Theory* 29 (1): 1–26. https://doi.org/10.1111/j.1467-9558.2010.01385.x.

ForestEthics Advocacy Association v. Canada (National Energy Board), 2014 FCA 245.

Forum Research. 2016. "Majority Approve of All Three Controversial Pipelines." March 16, 2016. http://poll.forumresearch.com/post/2480/most-see-pipelines-as-safer-than-rail-for-moving-crude/.

Fowlie, Jonathan. 2012a. "War of Words Heats Up between B.C.'s Christy Clark, Alberta's Alison Redford." *Vancouver Sun*, July 25, 2012. http://www.theprovince.com/news/politics/alison+redford+turns+heat+northern+gateway+royalty+words/6982233/story.html.

Fowlie, Jonathan. 2012b. "Pipeline Impasse Remains after Clark, Redford Hold 'Frosty' Meeting." *Vancouver Sun*, October 2, 2012. http://www.vancouversun.com/business/pipeline+impasse+remains+after+clark+redford+hold+frosty+meeting+with+video/7325098/story.html.

Franklin, Michael. 2017. "TransCanada Cancels Plans for Energy East Pipeline." *CTV News Calgary*, October 5, 2017. http://calgary.ctvnews.ca/transcanada-cancels-plans-for-energy-east-pipeline-1.3620224.

Frey, Gary, and Deborah Linke. 2002. "Hydropower as a Renewable and Sustainable Energy Resource Meeting Global Energy Challenges in a Reasonable Way." *Energy Policy* 30: 1261–1265. https://doi.org/10.1016/S0301-4215(02)00086-1.

Front commun pour la transition énergétique. 2019. "Accueil." https://www.pourlatransitionenergetique.org/.

Gardiner, Beth. 2012. "We're All Climate-Change Idiots." *New York Times*, July 21, 2012. http://www.nytimes.com/2012/07/22/opinion/sunday/were-all-climate-change-idiots.html?_r=0.

Garthwaite, J. 2013. "Mojave Mirrors: World's Largest Solar Plant Ready to Shine." *National Geographic*. https://news.nationalgeographic.com/news/energy/2013/07/130725-ivanpah-solar-energy-mojave-desert/.

Gattinger, Monica. 2012. "A National Energy Strategy for Canada: Golden Age or Golden Cage of Energy Federalism?" In *Canada: The State of the Federation 2012–2013*, edited by Loleen Berdahl and André Juneau, 36–69. Montreal and Kingston: McGill-Queen's University Press.

Gattinger, Monica. 2016. "The Harper Government Approach to Energy: Shooting Itself in the Foot." In *The Harper Era in Canadian Foreign Policy: Parliament, Politics, and Canada's Global Posture*, edited by Adam Chapnick and Christopher J. Kukucha, 151–156. Vancouver: UBC Press.

George, Alexander, and Andrew Bennett. 2005. *Case Studies and Theory Development in the Social Sciences*. Cambridge, MA: MIT Press.

Giordono, Leanne S., Hilary S. Boudet, Anna Karmazina, Casey L. Taylor, and Brent S. Steel. 2018. "Opposition 'overblown'? Community Response to Wind Energy Siting in the Western United States." *Energy Research and Social Science* 43:119–131.

Gitxaala Nation et al. v. Attorney General of Canada et al., 2016 FCA 187.

Global News. 2016. "Alberta Premier Rachel Notley Vows to Keep Promoting Energy East Pipeline." September 13, 2016. http://globalnews.ca/news/2937974/alberta-premier-rachel-notley-vows-to-keep-promoting-energy-east-pipeline/.

Goldenberg, Suzanne. 2014. "Revealed: Keystone Company's PR Blitz to Safeguard Its Backup Plan." *The Guardian*, November 18, 2014. https://www.theguardian.com

/environment/2014/nov/18/revealed-keystone-companys-pr-blitz-to-safeguard-its-backup-plan.

Goldstein, Judith, and Robert O. Keohane, eds. 1993. *Ideas and Foreign Policy: Beliefs, Institutions, and Political Change*. Ithaca, NY: Cornell University Press.

Goujard, Clothilde. 2017. "Trudeau Government Expresses Confidence in NEB after Its Report Says Opposite." *National Observer*, June 30, 2017. http://www.nationalobserver.com/2017/06/30/news/trudeau-government-expresses-confidence-neb-after-its-report-says-opposite.

Government of Alberta. 2011. *Annual Report*. http://www.finance.alberta.ca/publications/annual_repts/govt/index.html.

Government of Alberta. 2012. *Annual Report*. http://www.finance.alberta.ca/publications/annual_repts/govt/index.html.

Government of Alberta. 2013. *Annual Report*. http://www.finance.alberta.ca/publications/annual_repts/govt/index.html.

Government of Alberta. 2014. *Annual Report*. http://www.finance.alberta.ca/publications/annual_repts/govt/index.html.

Government of Alberta. 2015. *Annual Report*. http://www.finance.alberta.ca/publications/annual_repts/govt/index.html.

Government of Alberta. 2016. *Annual Report*. http://www.finance.alberta.ca/publications/annual_repts/govt/index.html.

Government of Alberta. 2017. *Annual Report*. http://www.finance.alberta.ca/publications/annual_repts/govt/index.html.

Government of Alberta. 2018. *Annual Report*. http://www.finance.alberta.ca/publications/annual_repts/govt/index.html.

Government of Alberta. 2012. *Lower Athabasca Regional Plan, 2012–2022*. https://landuse.alberta.ca/LandUse%20Documents/Lower%20Athabasca%20Regional%20Plan%202012-2022%20Approved%202012-08.pdf.

Government of Alberta. 2013. "Energy East Announcement a Win for Canada: Redford." News release. August 1, 2013. https://www.alberta.ca/release.cfm?xID=3470838396743-D675-3385-1B3DDD7AB8B35654.

Government of Alberta. 2015. *Climate Leadership Plan*. November 22, 2015.

Government of Alberta. 2017. *Highlights of the Alberta Economy*. http://www.albertacanada.com/files/albertacanada/SP-EH_highlightsABEconomyPresentation.pdf.

Government of Alberta. 2019. *Public Inquiry into Anti-Alberta Energy Campaigns*. https://www.alberta.ca/public-inquiry-into-anti-alberta-energy-campaigns.aspx.

Government of Alberta. 2020. "Provincial Investment Kick-Starts KXL Pipeline." News release. March 31, 2020. https://www.alberta.ca/release.cfm?xID=69965D6D6EE7A-92F8-DD89-BBB9E1FE323BD2DD.

Government of Alberta. n.d.a. "Oil Prices." Economic Dashboard. https://economicdashboard.alberta.ca/OilPrice.

Government of Alberta. n.d.b. *Gross Domestic Product by Industry*. http://economicdashboard.alberta.ca/GrossDomesticProduct#type.

Government of Alberta. n.d.c. *Oil Sands Facts and Statistics*. Accessed August 25, 2019. https://www.alberta.ca/oil-sands-facts-and-statistics.aspx.

Government of Alberta. n.d.d. *Resource Revenue Collected*. http://www.energy.alberta.ca/About_Us/2564.asp.

Government of British Columbia. 2007. *The BC Energy Plan: A Vision for Clean Energy Leadership*. https://www2.gov.bc.ca/assets/gov/farming-natural-resources-and-industry/electricity-alternative-energy/bc_energy_plan_2007.pdf.

Government of British Columbia. 2010. "Province Announces Site C Clean Energy Project." News release. April 19, 2010. https://archive.news.gov.bc.ca/releases/news_releases_2009-2013/2010prem0083-000436.htm.

Government of British Columbia. 2012a. "Environment Minister Sets Out Government's Position on Heavy Oil Pipelines." News release. August 1, 2012. https://news.gov.bc.ca/stories/environment-minister-sets-out-governments-position-on-heavy-oil-pipelines.

Government of British Columbia. 2012b. *Requirements for British Columbia to Consider Support for Heavy Oil Pipelines*. July 20, 2012. http://www.env.gov.bc.ca/main/docs/2012/TechnicalAnalysis-HeavyOilPipeline_120723.pdf.

Government of British Columbia. 2013. *Enbridge Northern Gateway Project Application: Argument of the Province of British Columbia*. http://www.env.gov.bc.ca/main/docs/2013/BC-Submission-to-NGP-JointReviewPanel_130531.pdf.

Government of British Columbia. 2014. "Site C to Provide More than 100 Years of Affordable, Reliable Clean Power." News release. December 16, 2014. https://news.gov.bc.ca/stories/site-c-to-provide-more-than-100-years-of-affordable-reliable-clean-power.

Government of British Columbia. 2016. "Province Reaffirms Trans Mountain Pipeline Must Meet Five Conditions." News release. January 11, 2016.

Government of British Columbia. 2018. "Government Will Complete Site C Construction, Will Not Burden Taxpayers or BC Hydro Customers with Previous Government's Debt." News release. December 11, 2018. https://news.gov.bc.ca/releases/2017PREM0135-002039.

Government of Canada, C.E.A.A. 2012. Canadian Environmental Assessment Agency—Policy and Guidance—Canadian Environmental Assessment Act: An Overview.

Government of Canada. 2013. "Minister Oliver Highlights Important Milestone towards Canadian Energy Security." Natural Resources Canada. August 1, 2013. https://www.nrcan.gc.ca/media-room/news-release/2013/11396.

Government of Canada. 2016a. "Interim Measures for Pipeline Reviews." News release. January 27, 2016. https://www.canada.ca/en/natural-resources-canada/news/2016/01/interim-measures-for-pipeline-reviews.html.

Government of Canada. 2016b. "Government Appoints Temporary Members to National Energy Board." News release. October 20, 2016. https://www.canada.ca/en/natural-resources-canada/news/2016/10/government-appoints-temporary-members-national-energy-board.html?wbdisable=true.

Government of Canada. 2017a. "National Energy Board Seeks Public Input on List of Issues for Energy East Hearing." News release. National Energy Board. May 10, 2017. https://www.canada.ca/en/canada-energy-regulator/news/2017/05/national_energy_boardseekspublicinputonlistofissuesforenergyeast.html.

Government of Canada. 2017b. *Beluga Whale (St. Lawrence Estuary Population)*. Fisheries and Ocean Canada. http://www.dfo-mpo.gc.ca/species-especes/profiles-profils/belugaStLa-eng.html.

Government of Canada and Government of British Columbia. 2014. *Federal/Provincial Consultation and Accommodation Report—Site C Clean Energy Project*. September 7, 2014. https://projects.eao.gov.bc.ca/api/document/58868f49e036fb010576803a/fetch.

Government of New Brunswick. 2016. "Motion in Support of Energy East Passed." News release, Office of the Premier. December 13, 2016. http://www2.gnb.ca/content/gnb/en/departments/premier/news/news_release.2016.12.1204.html.

Government of Ontario. n.d. *The End of Coal*. https://www.ontario.ca/page/end-coal#:~:text=Ontario%20enshrined%20its%20commitment%20in,to%20generate%20electricity%20in%20Ontario.

Government of Quebec. 2016a. "Energy East Pipeline—Motion for an Injunction against TransCanada: The Government Takes Action to Ensure Compliance with Québec Law." Press release. March 1, 2016.

Government of Quebec. 2016b. *Projet Oléoduc Énergie Est de TransCanada*. Ministère du Développement Durable, Environnement et Lutte contre les changements climatiques. http://www.mddelcc.gouv.qc.ca/evaluations/transcanada/index.htm.

Government of Saskatchewan. 2014. "Debates and Proceedings." Legislative Assembly of Saskatchewan. Fourth Session—Twenty-Seventh Legislature. n.s. vol. 57, no.19A. November 26, 2014. Published under the authority of the Honourable Dan D'Autremont, Speaker.

Gralnick, Daniel. 2016. "Constitutional Implications of Quebec's Review of Energy East." *Energy Regulation Quarterly* 4 (3).

Gravelle, Timothy, and Erick Lachapelle. 2015. "Politics, Proximity, and Pipeline: Mapping Public Attitudes toward Keystone XL." *Energy Policy* 83:99–108.

Green, Fergus, and Richard Denniss. 2018. "Cutting with Both Arms of the Scissors: The Economic and Political Case for Restrictive Supply-Side Climate Policies." *Climatic Change* 150 (1–2): 73–87.

Greenpeace Canada. 2014. *Leaked Documents Show TransCanada Planning "Dirty Tricks" Campaign to Support Energy East Pipeline*. Greenpeace Canada. November 18, 2014. http://www.greenpeace.org/canada/en/recent/Leaked-documents-show-TransCanada-planning-dirty-tricks-campaign-to-support-Energy-East-pipeline/.

Greenpeace Canada. 2016. Press release. November 7, 2016. http://www.greenpeace.org/canada/Global/canada/pr/2016/11/2016-11-07_Pressrelease_sondage.pdf.

Gregory, Robin S. 2017. "The Troubling Logic of Inclusivity in Environmental Consultations." *Science, Technology, & Human Values* 42 (1): 144–165. https://doi.org/10.1177/0162243916664016.

Gruenspecht, Howard. 2019. "The U.S. Coal Sector: Recent and Continuing Challenges." January. Brookings Institution. https://www.brookings.edu/research/the-u-s-coal-sector/.

Grunwald, Michael. 2013. "I'm with the Tree Huggers." *Time*, February 28, 2013. http://swampland.time.com/2013/02/28/im-with-the-tree-huggers/.

Guilbeault, Steven. 2017. "The Death of Energy East Marks the End of a Long Battle." Équiterre. October 13, 2017. https://equiterre.org/en/news/the-death-of-energy-east-marks-the-end-of-a-long-battle.

Gunster, Shane. 2019. "Extractive Populism and the Future of Canada." *Monitor*, July 2, 2019. Canadian Centre for Policy Alternatives. https://www.policyalternatives.ca/publications/monitor/extractive-populism-and-future-canada.

Gunster, Shane, Robert Neubauer, John Bermingham, and Alicia Massie. 2021. "'Our Oil': Extractive Populism in Canadian Social Media." In *Regime of Obstruction: How Corporate Power Blocks Energy Democracy*, edited by William Carroll. Edmonton: Athabasca University Press.

Gunton, Thomas. 2017. "Collaborate Resource Management." In *The International Encyclopedia of Geography*, edited by Douglas Richards, et al. Chichester, UK: Wiley-Blackwell.

Gutman, A., and D. Thompson. 1996. *Democracy and Disagreement.* Cambridge, MA: Harvard University Press.

Habermas, Jurgen. 1984. *The Theory of Communicative Action.* Boston: Beacon Press.

Haddock, Mark. 2010. *Environmental Assessment in British Columbia.* Environmental Law Centre, University of Victoria. http://www.elc.uvic.ca/documents/ELC_EA-IN-BC_Nov2010.pdf.

Hager, Carol, and Mary Alice Haddad, eds. 2015. *NIMBY Is Beautiful: Cases of Local Activism and Environmental Innovation around the World.* New York: Berghahn.

Hager, Mike. 2012. "B.C. First Nation Says Pipeline Panel 'Disrespectful.'" *Vancouver Sun,* April 2, 2012. http://www.vancouversun.com/technology/first+nation+says+pipeline+panel+disrespectful/6398576/story.html.

Hall, Chris. 2018. "Does Trudeau Have a Trans Mountain Plan That Goes beyond Talk?" *CBC News,* April 9, 2018. https://www.cbc.ca/news/politics/kinder-morgan-pipeline-deadline-1.4611873.

Hall, Peter. 1993. "Policy Paradigms, Social Learning, and the State: The Case of Economic Policymaking in Britain." *Comparative Politics* 25 (3): 275–296.

Hammel, Paul. 2017. "Future of Keystone XL Pipeline Becomes More Muddled after PSC Denies Revisions." *Omaha World-Herald,* December 21, 2017. https://www.omaha.com/news/nebraska/decision-by-nebraska-psc-casts-more-uncertainty-into-whether-transcanada/article_63053116-e4e6-11e7-8a21-3b6e865ae887.html.

Hansen, James. 2012. "Game Over for the Climate." *New York Times,* May 9, 2012. https://www.nytimes.com/2012/05/10/opinion/game-over-for-the-climate.html.

Hare, Bill. 1997. *Fossil Fuels and Climate Protection: The Carbon Logic.* Greenpeace International. September 1997. http://www.greenpeace.org/international/Global/international/planet-2/report/2006/3/fossil-fuels-and-climate-prote.pdf.

Harris, Paul G. 2013. *What's Wrong with Climate Politics and How to Fix It.* Cambridge: Polity.

Harris, Richard A., and Sidney M. Milkis. 1989. *The Politics of Regulatory Change: A Tale of Two Agencies.* New York: Oxford University Press.

Harrison, Kathryn. 1996. *Passing the Buck: Federalism and Canadian Environmental Policy.* Vancouver: UBC Press.

Harrison, Kathryn. 2007. "The Road Not Taken: Climate Change Policy in Canada and the United States." *Global Environmental Politics* 72 (4): 92–117.

Harrison, Kathryn. 2012. "A Tale of Two Taxes: The Fate of Environmental Tax Reform in Canada." *Review of Policy Research* 29 (3): 383–407. https://doi.org/10.1111/j.1541–1338.2012.00565.x.

Harrison, Kathryn. 2013. "Federalism and Climate Policy Innovation: A Critical Reassessment." *Canadian Public Policy* 39:S95–S108.

Harrison, Kathryn, and Lisa McIntosh Sundstrom. 2010. "Introduction." In *Global Commons, Domestic Decisions: The Comparative Politics of Climate Change*, edited by Kathryn Harrison and Lisa McIntosh Sundstrom, 1–22. Cambridge, MA: MIT Press.

Henn, Jamie, and Daniel Kessler. 2011. "Nation's Largest Environmental Organizations Stand Together to Oppose Oil Pipeline." *CommonDreams*, August 24, 2011. https://www.commondreams.org/newswire/2011/08/24/nations-largest-environmental-organizations-stand-together-oppose-oil-pipeline?amp.

Heyes, Anthony, Andrew Leach, and Charles F. Mason. 2018. "The Economics of Canadian Oil Sands." *Review of Environmental Economics and Policy* 12 (2): 242–263. https://doi.org/10.1093/reep/rey006.

Hill, B. 2017. "Ontario Energy Minister Admits Mistake with Green Energy Program." *Global News*, February 24, 2017. https://globalnews.ca/news/3272095/ontario-energy-minister-admits-mistake-with-green-energy-program/.

Hislop, Markham. 2019. "Debunked: Vivian Krause's Tar Sands Campaign Conspiracy Narrative." EnergiMedia. May 14, 2019. https://energi.media/deep-dives/debunked-vivian-krauses-tar-sands-campaign-conspiracy-narrative/.

Hoberg, George. 2000. "How the Way We Make Policy Governs the Policy We Make." In *Forging Truces in the War in the Woods: Sustaining the Forests of the Pacific Northwest*, edited by Donald Alper and Debra Salazar, 26–53. Vancouver: UBC Press.

Hoberg, George. 2001. "Policy Cycles and Policy Regimes: A Framework for Studying Public Policy." In *In Search of Sustainability: British Columbia Forest Policy in the 1990s*, edited by Benjamin Cashore, George Hoberg, Michael Howlett, Jeremy Rayner, and Jeremy Wilson. Vancouver: UBC Press.

Hoberg, George. 2010. "Bringing the Market Back In: BC Natural Resource Policies during the Campbell Years." In *British Columbia Politics and Government*, edited by Micheal Howlett, Dennis Pilon, and Tracy Sommerville. Toronto: Edmond Montgomery.

Hoberg, George. 2013. "The Battle over Oil Sands Access to Tidewater: A Political Risk Analysis of Pipeline Alternatives." *Canadian Public Policy—Analyse de politiques* 39 (3): 371–391.

Hoberg, George. 2015. "Pipeline Resistance as Political Strategy: 'Blockadia' and the Future of Climate Politics." Presented at the Canadian Political Science Association annual meeting, Ottawa, June 2–4, 2015.

Hoberg, George. 2016. "Unsustainable Development: Energy and Environment in the Harper Decade." In *The Harper Factor: Assessing a Prime Minister's Policy Legacy*, edited by Jennifer Ditchburn and Graham Fox, 253–263. Montreal and Kingston: McGill-Queen's University Press.

Hoberg, George. 2017. "Ambition without Capacity: Environmental and Natural Resource Policy in the Campbell Era." In *The Campbell Revolution: Power and Politics in British Columbia from 2001 to 2011*, edited by Tracy Summerville and Jason Lacharite, 177–193. Montreal and Kingston: McGill-Queen's University Press.

Hoberg, George. 2018. "Pipelines and the Politics of Structure: Constitutional Conflicts in the Canadian Oil Sector." *Review of Constitutional Studies* 23 (1): 52–89.

Hoberg, George, and Jeffrey Phillips. 2011. "Playing Defence: Early Responses to Conflict Expansion in the Oil Sands Policy Subsystem." *Canadian Journal of Political Science* 44:507–527. http://dx.doi.org/10.1017/S0008423911000473.

Hochstetler, Kathryn. 2011. "The Politics of Environmental Licensing: Energy Projects of the Past and Future in Brazil." *Studies in Comparative International Development* 46:349–371. https://doi.org/10.1007/s12116-011-9092-.

Hodges, Heather, and Galen Stocking. 2016. "A Pipeline of Tweets: Environmental Movements' Use of Twitter in Response to the Keystone XL Pipeline." *Environmental Politics* 22 (2): 223–247.

Hoekstra, Gordon. 2013. "Kinder Morgan Files Formal Application for Trans Mountain Pipeline Expansion." *Vancouver Sun*, December 17, 2013. http://www.vancouversun.com/business/Kinder+Morgan+files+formal+application+Trans+Mountain+pipeline+expansion/9292403/story.html.

Horgan, John. 2017. "Honourable George Heyman Mandate Letter." Government of British Columbia. July 18, 2017. https://www2.gov.bc.ca/assets/gov/government/ministries-organizations/premier-cabinet-mlas/minister-letter/heyman-mandate.pdf.

Horgan, John. 2018. "John Horgan: 'It Doesn't Matter Who Owns the Kinder Morgan Pipeline, the Risks Remain.'" *Maclean's*, May 30, 2018. https://www.macleans.ca/opinion/john-horgan-kinder-morgan-op-ed/.

Horter, Will. 2016. Personal interview with the author. August 18, 2016.

Housty, Jess. 2017. Personal interview with the author. December 4, 2017.

Howe, Adam C., Mark C. J. Stoddart, and David B. Tindall. 2020. "Media Coverage and Perceived Policy Influence of Environmental Actors: Good Strategy or Pyrrhic Victory?" *Politics and Governance* 8 (2): 298–310. https//doi.org/10.17645/pag.v8i2.2595.

Howlett, Michael. 2009. "Governance Modes, Policy Regimes and Operational Plans: A Multi-level Nested Model of Policy Instrument Choice and Policy Design." *Policy Sciences* 42 (1): 73–89.

Hunold, C., and S. Leitner. 2011. "'Hasta la vista, Baby?' The Solar Grand Plan, Environmentalism, and Social Constructions of the Mojave Desert." *Environmental Politics* 20 (5): 687–704.

Hunsberger, Carol, Sarah Froese, and George Hoberg. 2020. "Toward 'Good Process' in Regulatory Reviews: Is Canada's New System Any Better Than the Old?" *Environmental Impact Assessment Review* 82: 106379. https://doi.org/10.1016/j.eiar.2020.106379.

Hunter, Justine. 2017a. "BC Liberals, NDP Spar over Site C Dam." *Globe and Mail*, May 5, 2017. https://www.theglobeandmail.com/news/british-columbia/everything-you-need-to-know-about-the-bc-electionplatforms/article34912667/.

Hunter, Justine. 2017b. "BC Green Leader Andrew Weaver Says Economics, Indigenous Rights Changed His Mind on Site C." *Globe and Mail*, June 20, 2017. https://www.theglobeandmail.com/news/british-columbia/site-c-the-bc-green-partys-big-dam-dilemma/article35386000/.

Hunter, Justine. 2018. "B.C. Prepares Court Challenge as Alberta Threatens to Cut Off Oil Shipments." *Globe and Mail*, May 17, 2018. https://www.theglobeandmail.com/canada/british-columbia/article-bc-prepares-court-challenge-as-alberta-threatens-to-cut-off-oil/.

Hunter, Justine. 2020. "BC Hydro's Site C Dam Project on Shaky Ground." *Globe and Mail*, November 21. https://www.theglobeandmail.com/canada/british-columbia/article-bc-hydros-site-c-dam-project-on-shaky-ground/.

Hunter, Justine, and Mark Hume. 2016. "NEB's Pipeline Approval Puts Christy Clark in the Hot Seat." *Globe and Mail*, May 19, 2016. http://www.theglobeandmail.com/news/british-columbia/nebs-pipeline-approval-puts-christy-clark-in-the-hot-seat/article30104153/.

Hussey, Ian, Eric Pineault, Emma Jackson, and Susan Cake. 2018. *Boom, Bust, and Consolidation: Corporate Restructuring in the Alberta Oil Sands*. Corporate Mapping Project. November 7, 2018. https://www.corporatemapping.ca/boom-bust-and-consolidation/.

Hyland, Marie, and Valentin Bertsch. 2018. "The Role of Community Involvement Mechanisms in Reducing Resistance to Energy Infrastructure Development." *Ecological Economics* 146:447–474.

IHS Markit. 2015. *Oil Sands Cost Competitiveness*. October 2015. https://ihsmarkit.com/products/energy-industry-oil-sands-dialogue.html.

IHS Markit. 2019. *Four Years of Change: Oil Sands Cost and Competitiveness in 2018*. April 2019. https://ihsmarkit.com/products/energy-industry-oil-sands-dialogue.html.

IISD (International Institute for Sustainable Development). 2015. *The End of Coal: Ontario's Coal Phase-out*. International Institute for Sustainable Development. https://www.iisd.org/sites/default/files/publications/end-of-coal-ontario-coal-phase-out.pdf.

Immergut, Ellen. 1990. "Institutions, Veto Points, and Policy Results: A Comparative Analysis of Health Care." *Journal of Public Policy* 10 (4): 391–416.

Independent Contractors and Business Association. 2016. Super bowl ad. https://vimeo.com/154212765.

Indigenous Environmental Network. 2013. *International Treaty to Protect the Sacred from Tar Sands Projects*. January 25, 2013. http://www.ienearth.org/international-treaty-to-protect-the-sacred-from-tar-sands-projects/.

Insights West. 2016a. "Half of British Columbians Remain Opposed to Northern Gateway." News release. September 6, 2016. https://insightswest.com/news/half-of-british-columbians-remain-opposed-to-northern-gateway/.

Insights West. 2016b. "Seventy per cent of British Columbians Support Pausing Site C Construction to Investigate Alternatives." News release. November 16, 2016. http://www.insightswest.com/news/seventy-per-cent-of-british-columbians-support-pausing-site-c-construction-to-investigate-alternatives/.

Insights West. 2018. "A Majority of British Columbians Oppose the Federal Government's Decision to Buy the Trans Mountain Pipeline Project, but Now Believe It Is More Likely to Get Built." News release. June 1, 2018. https://insightswest.com/news/a-majority-of-british-columbians-oppose-the-federal-governments-decision-to-buy-the-trans-mountain-pipeline-project-but-now-believe-it-is-more-likely-to-get-built/.

International Energy Agency. 2011. *World Energy Outlook 2011*. Paris: International Energy Agency. https://www.iea.org/publications/freepublications/publication/WEO2011_WEB.pdf.

IPCC (Intergovernmental Panel on Climate Change). 2014. *Climate Change 2014: Synthesis Report. Contribution of Working Groups I, II and III to the Fifth Assessment Report of the Intergovernmental Panel on Climate Change* (Core Writing Team, R. K. Pachauri and L. A. Meyer, eds.). Geneva, Switzerland: IPCC, 151 pp.

IPCC (Intergovernmental Panel on Climate Change). 2018. *Global Warming of 1.5 °C: Special Report*. https://www.ipcc.ch/sr15/.

IPSOS. 2018a. "Healthcare (54%), Economy (36%), Taxes (29%) and Energy Costs (28%) Remain Top Issues of Campaign." May 10, 2018. https://www.ipsos.com/en-ca/news-polls/Global-News-Ontario-Vote-Issues-May-10-2018.

IPSOS. 2018b. "Six in Ten (61%) Ontarians Say Hydro Prices Will Impact Their Vote." May 29, 2018. https://www.ipsos.com/en-ca/news-polls/Global-News-Ontario-Vote-Hydro-Poll-May-29-2018.

Israel, Benjamin. 2017. *The Real GHG Trend: Oilsands among the Most Carbon Intensive Crudes in North America. Oilsands at 50 Series—the Real Cost of Development, Part 1*. Pembina Institute. October 4, 2017. https://www.pembina.org/blog/real-ghg-trend-oilsands.

Jaccard, Mark. 2017. "Opinion: BCUC Must Consider 'Dispatchability' in Site C Review." *Vancouver Sun*, August 3, 2017. https://vancouversun.com/opinion/op-ed/opinion-bcuc-must-consider-dispatchability-in-site-c-review.

Jaccard, Mark. 2020. "The Die Has Been Cast on Canada's Carbon Tax. Now We Just Need the Courage to Implement It across the Country." *Globe and Mail*, December 14. https://www.theglobeandmail.com/opinion/article-the-die-has-been-cast-on-canadas-carbon-tax-now-we-just-need-the/.

Jaccard, Mark, James Hoffele, and Torsten Jaccard. 2018. "Global Carbon Budgets and the Viability of New Fossil Fuel Projects." *Climatic Change* 150 (1–2): 15–28.

Jaccard, Mark, Noel Melton, and John Nyboer. 2011. "Institutions and Processes for Scaling Up Renewables: Run-of-River Hydropower in British Columbia." *Energy Policy* 39 (7): 4042–4050.

James, Patrick, and Robert Michelin. 1989. "The Canadian National Energy Program and Its Aftermath: Perspective on an Era of Confrontation." *American Review of Canadian Studies* 19 (1): 59–81.

Jami, Anahita, and Philip Walsh. 2016. "Wind Power Deployment: The Role of Public Participation in the Decision-Making Process in Ontario, Canada." *Sustainability* 8 (713). https://doi.org/10.3390/su8080713.

Janzwood, Amy. 2020. "Explaining Variation in Oil Sands Pipeline Projects." *Canadian Journal of Political Science* 53 (3): 540–559. https://doi.org/10.1017/S0008423920000190.

Jenkins-Smith, Hank C., Daniel Nohrstedt, Christopher M. Weible, and Paul A. Sabatier. 2014. "The Advocacy Coalition Framework: Foundations, Evolution and Future Challenges." In *Theories of the Policy Process*, 3rd ed., edited by Paul Sabatier and Christopher Weible, 183–224. Boulder, CO: Westview Press.

Joint Review Panel. 2011. *Enbridge Northern Gateway Project Joint Review Panel—Panel Session Results and Decisions*. January 19, 2011. https://docs.neb-one.gc.ca/ll-eng/llisapi.dll/fetch/2000/90464/90552/384192/620327/624909/662325/A22-3_-_Panel_Session_Results_and_Decision_A1X2L8.pdf?nodeid=662117&vernum=-2.

Joint Review Panel. 2013. *Report of the Joint Review Panel for the Enbridge Northern Gateway Project*. http://publications.gc.ca/site/eng/456575/publication.html.

Joint Review Panel. 2014. *Report of the Joint Review Panel, Site C Clean Energy Project*. May 1, 2014. https://www.ceaa-acee.gc.ca/050/documents/p63919/99173E.pdf.

Judd, Amy. 2014. "'We Will Fight for Our Legacy': Chief Stewart Phillip on Northern Gateway Decision." *Global News*, June 17, 2014. http://globalnews.ca/news/1400501/we-will-fight-for-our-legacy-chief-stewart-phillip-on-northern-gateway-decision/.

Judd, Amy, and Richard Zussman. 2018. "B.C. Taking Legal Action against Alberta over Bill Allowing Province to Cut Off Gas." *Global News*, May 22, 2018. https://globalnews.ca/news/4224275/bc-legal-action-against-alberta-bill-cut-off-gas/.

Kagan, Robert. 2001. *Adversarial Legalism: The American Way of Law*. Cambridge, MA: Harvard University Press.

Karl, Terry Lynn. 1997. *The Paradox of Plenty: Oil Booms and Petro-States*. Berkeley: University of California Press.

Kasperson, Roger E. 2006. "Rerouting the Stakeholder Express." *Global Environmental Change* 16 (4): 320–322. https://doi.org/10.1016/j.gloenvcha.2006.08.002.

Keir, L. S., and S. H. Ali. 2014. "Conflict Assessment in Energy Infrastructure Siting: Prospects for Consensus Building in the Northern Pass Transmission Line Project." *Negotiation Journal* 30 (2): 169–189.

Keller, James. 2014. "Judge Throws Out Charges against Dozens of Activists Arrested at Anti-Kinder Morgan Pipeline Protests." *National Post*, January 24, 2014. http://news.nationalpost.com/news/canada/judge-throws-out-charges-against-dozens-of-activists-arrested-at-anti-kinder-morgan-pipeline-protests.

Kennedy, R. F., Jr. 2005. "An Ill Wind Off Cape Cod." *New York Times*, December 16, 2005. https://www.nytimes.com/2005/12/16/opinion/an-ill-wind-off-cape-cod.html.

Kenney, Jason. 2019. "Election Victory Speech." *National Post*, April 17, 2019. https://nationalpost.com/news/canada/read-jason-kenneys-prepared-victory-speech-in-full-after-ucp-wins-majority-in-alberta-election.

Kerlin, K. 2018. "Can Solar Energy and Wildlife Coexist?" *Washington Post*, September 9, 2018. https://www.washingtonpost.com/brand-studio/wp/2018/09/09/feature/can-solar-energy-and-wildlife-coexist/?utm_term=.2167864ae55f.

Kheraj, Sean. 2013. *Historical Background Report: Trans Mountain Pipeline, 1947–2013*. Prepared for the city of Vancouver. https://vancouver.ca/images/web/pipeline/Sean-Kheraj-history-of-TMP.pdf.

Kilbourn, William. 1970. *Pipeline: Transcanada and the Great Debate, a History of Business and Politics*. Toronto: Clarke, Irwin.

Kinder Morgan Canada. 2018. "Kinder Morgan Canada Limited Suspends Non-essential Spending on Trans Mountain Expansion Project." News release. April 8, 2018. https://ir.kindermorgancanadalimited.com/2018-04-08-Kinder-Morgan-Canada-Limited-Suspends-Non-Essential-Spending-on-Trans-Mountain-Expansion-Project.

Kingdon, J. 1995. *Agendas, Alternatives and Public Policies*. New York: Harper Collins.

Kinsella, William J. 2016. "A Question of Confidence: Nuclear Waste and Public Trust in the United States after Fukushima." In *The Fukushima Effect: A New Geopolitical Terrain*, edited by Richard Hindmarsh and Rebecca Priestly, 223–236. Routledge: New York.

Klein, Naomi. 2014. *This Changes Everything: Capitalism vs the Climate*. Toronto: Alfred Knopf Canada.

Klein, Naomi, and Bill McKibben. 2014. "Foreword." In *A Line in the Tar Sands: Struggles for Environmental Justice*, edited by Toban Black, Stephen D'Arcy, Tony Weis, and Joshua Kahn Russel, xvii–xviii. Oakland: PM Press.

Knoema. n.d. *Cost of Producing a Barrel of Crude Oil by Country*. https://knoema.com/rqaebad/cost-of-producing-a-barrel-of-crude-oil-by-country.

Knopper, L. L. D., and C. A. Ollson. 2011. "Health Effects and Wind Turbines: A Review of the Literature." *Environmental Health* 10 (1): 78–87. https://doi.org/10.1186/1476-069X-10-78.

Krugel, Lauren. 2013. "TransCanada Going Ahead with Energy East Line, an 'historic opportunity.'" *Maclean's*, August 1, 2013. http://www.macleans.ca/general/transcanada-going-ahead-with-energy-east-line-between-alberta-and-new-brunswick/.

Lachapelle, Erick, and Simon Kiss. 2019. "Opposition to Carbon Pricing and Rightwing Populism: Ontario's 2018 General Election." *Environmental Politics* 28 (5): 970–976. https://doi.org/10.1080/09644016.2019.1608659.

Larsen, Karin. 2018. "Kennedy Stewart Pleads Guilty to Criminal Contempt for Kinder Morgan Protest." *CBC News*, May 14, 2018. https://www.cbc.ca/news/canada/british-columbia/kennedy-stewart-pleads-guilty-to-criminal-contempt-for-kinder-morgan-protest-1.4662089.

Lassonde, Maryse. 2016. Letter to The Right Honourable Justin Trudeau from the President of the Royal Society of Canada. May 19, 2016.

Lavant, Ezra. 2010. *Ethical Oil: The Case for Canada's Oil Sands*. Toronto: McClelland and Stewart.

Lavant, Ezra. 2013. "Freedom Oil: Energy East Pipeline Appealing and Has a Politically Important Spinoff." *Toronto Sun*, August 4, 2013. https://torontosun.com/2013/08/04/freedom-oil-energy-east-pipeline-appealing-and-has-a-politically-important-spinoff.

Lavoie, Judith. 2014. "B.C. Business Community Slams 'Astronomical' Cost of Building Site C Dam." Desmog Canada. June 10, 2014. https://www.desmog.ca/2014/06/10/b-c-business-community-slams-astronomical-cost-building-site-c-dam.

Lawlor, Andrea, and Timothy B. Gravelle. 2018. "Framing Trans-Border Energy Transportation: The Case of Keystone XL." *Environmental Politics* 27 (4): 666–685. https://doi.org/10.1080/09644016.2018.1425106.

Laxer, Gordon. 2015. *After the Sands: Energy and Ecological Security for Canadians*. Vancouver: Douglas and McIntyre.

Lazarus, Richard J. 2009. "Super Wicked Problems and Climate Change: Restraining the Present to Liberate the Future." 94 Cornell L. Rev. 1153–1234.

Leach, Andrew. 2011. "On the Potential for Oilsands to Add 200ppm of CO2 to the Atmosphere." *Rescuing the Frog – Andrew Leach's Energy and Climate Blog*. June 4. http://andrewleach.ca/oilsands/on-the-potential-for-oilsands-to-add-200ppm-of-co2-to-the-atmosphere/.

Leach, Andrew. 2012. "Alberta's Specified Gas Emitters Regulation." *Canadian Tax Journal* 60 (4): 881–898.

Leach, Andrew. 2014. "New Language, Same Old Story on Keystone from Climate Scientists." *Maclean's*, May 8, 2014. https://www.macleans.ca/economy/economicanalysis/new-language-same-old-story-on-keystone-from-climate-scientists/.

Leach, Andrew. 2017. "How Donald Trump Killed the Energy East Pipeline." *Globe and Mail*, October 9, 2017. https://beta.theglobeandmail.com/report-on-business/rob-commentary/how-donald-trump-killed-the-energy-east-pipeline/article36527153/?ref=http://www.theglobeandmail.com&.

Lefebvre, Sarah-Maude. 2016. "La polémique sur Énergie Est, bonne nouvelle pour la souveraineté?" *Journal de Montréal*, January 27, 2016. http://www.journaldemontreal.com/2016/01/27/la-polemique-sur-energie-est-bonne-pour-la-souveraineté.

Lemphers, Nathan, and Dan Woynillowicz. 2012. *In the Shadow of the Boom: How Oilsands Development Is Reshaping Canada's Economy*. Pembina Institute. May 30, 2012. http://www.pembina.org/pub/shadow-of-boom.

Levi, Michael. 2012. "Five Myths about the Keystone XL Pipeline." *Washington Post*, January 18, 2012. https://www.washingtonpost.com/opinions/five-myths-about-the-keystone-xl-pipeline/2011/12/19/gIQApUAX8P_story.html.

Levi, Michael. 2013. *The Power Surge: Energy, Opportunity, and the Battle for America's Future*. Oxford: Oxford University Press.

Levin, Kelly, Benjamin Cashore, Steven Bernstein, and Graeme Auld. 2012. "Overcoming the Tragedy of Super Wicked Problems: Constraining Our Future Selves to Ameliorate Global Climate Change." *Policy Sciences* 45 (2): 123–152.

Liao, Yuguo, and Hindy L. Schachter. 2018. "Exploring the Antecedents of Municipal Managers' Attitudes towards Citizen Participation." *Public Management Review* 20 (9): 1287–1308. https://doi.org/10.1080/14719037.2017.1363903.

Liberal Party of Canada. 2013. "Liberal Party of Canada Leader Justin Trudeau's Speech to the Calgary Petroleum Club." October 30, 2013. https://www.liberal.ca/liberal-party-canada-leader-justin-trudeaus-speech-calgary-petroleum-club/.

Liberal Party of Canada. 2015a. *Environmental Assessments*. https://www.liberal.ca/realchange/environmental-assessments/.

Liberal Party of Canada. 2015b. *Real Change: A Plan for a Strong Middle Class*. https://liberal.ca/wp-content/uploads/sites/292/2020/09/New-plan-for-a-strong-middle-class.pdf.

Lindblom, Charles E. 1982. "The Market as Prison." *Journal of Politics* 44 (2): 324–336.

LINGO (Leave It in the Ground). n.d. "LINGO and the XYZ Agenda." http://leave-it-in-the-ground.org/why-lingo-2/.

Linnitt, Carol. 2016. "Premier Clark's Proposal to 'Electrify Oilsands' with Site C Dam Has an 'Air of Desperation': Panel Chair." *The Narwhal*, April 13, 2016. https://thenarwhal.ca/premier-clark-s-proposal-electrify-oilsands-site-c-dam-has-air-desperation-panel-chair/.

Lizza, Ryan. 2010. "As The World Burns." *New Yorker*, October 11, 2010.

Lizza, Ryan. 2013. "The President and the Pipeline." *New Yorker*, September 16, 2013.

Lobos, V., and M. Partidário. 2014. "Theory versus Practice in Strategic Environmental Assessment (SEA)." *Environmental Impact Assessment Review* 48:34–46.

London Economics International. 2014. *Cost-Effectiveness Evaluation of Clean Energy Projects in the Context of Site C*. Prepared for the Clean Energy Association of British Columbia. September 16, 2014. https://www.cleanenergybc.org/wp-content/uploads/2015/12/LondonEI_20141016.pdf.

Long, G. A. 2011. "It's about New Hampshire's Energy Future." *New Hampshire Business Review*, March 11, 2011. https://www.nhbr.com/March-11-2011/Its-about-New-Hampshires-energy-future/.

Loudermilk, M. S. 2016. "Renewable Energy Policy and Ontario Wind Turbine Development." Presentation at the Ontario Network for Sustainable Energy Policy, Ivey School of Business, University of Western Ontario, April 20–22, 2016. https://www.ivey.uwo.ca/cmsmedia/3775606/renewable-energy-policy-and-wind-generation-in-ontario.pdf.

Loudermilk, M. S. 2017. *Renewable Energy Policy and Wind Generation in Ontario*. Policy Brief. January 2017. Energy Policy and Management Centre, Ivey Business School. https://www.ivey.uwo.ca/cmsmedia/3775606/january-2017-renewable-energy-policy-and-wind-generation-in-ontario.pdf.

Love, C. 2014. "Case Study: Cape Wind Project—America's First Offshore Wind Farm." *National Geographic*. https://www.nationalgeographic.org/news/case-study-cape-wind-project/.

Lower Athabasca Regional Plan Review Panel. 2015. *Lower Athabasca Regional Plan Review Panel: Report 2015*. https://www.landuse.alberta.ca/LandUse%20Documents/Lower%20Athabasca%20Regional%20Plan%20Review%20Panel%20Recommendations%20-%202016-06.pdf.

MacDonald, Douglas. 2020. *Carbon Province, Hydro Province: The Challenge of Canadian Energy and Climate Federalism*. Toronto: University of Toronto Press.

Mahoney, James. 2010. "After KKV: The New Methodology for Qualitative Research." *World Politics* 61 (1): 120–147.

Major Projects Management Office. 2016. Ministerial Panel Examining the Proposed Trans Mountain Expansion Project. https://mpmo.gc.ca/measures/262.

Marandola, Sabrina. 2016. "NEB Panel Members Step Down after Flurry of Criticism." *CBC News*, September 9, 2016. http://www.cbc.ca/news/canada/montreal/neb-panel-steps-down-1.3755872.

Marshall, George. 2014. *Don't Even Think About It: Why Our Brains Are Wired to Ignore Climate Change*. New York: Bloomsbury USA.

Masnadi, Mohammad S., Hassan M. El-Houjeiri, Dominik Schunack, Yunpo Li, Jacob G. Englander, Alhassan Badahdah, Jean-Christophe Monfort, James E. Anderson, Timothy J. Wallington, Joule A. Bergerson, Deborah Gordon, Jonathan Koomey, Steven Przesmitzki, Inês L. Azevedo, Xiaotao T. Bi, James E. Duffy, Garvin A. Heath, Gregory A. Keoleian, Christophe McGlade, D. Nathan Meehan, Sonia Yeh, Fengqi You, Michael Wang, and Adam R. Brandt. 2018. "Global Carbon Intensity of Crude Oil Production." *Science* 31 (6405): 851–853.

May, P. J., and A. E. Jochim. 2013. "Policy Regime Perspectives: Policies, Politics, and Governing." *Policy Studies Journal* 4 (2): 426–452.

Mayer, Jane. 2011. "Taking It to the Streets." *New Yorker*, November 28, 2011. http://www.newyorker.com/magazine/2011/11/28/taking-it-to-the-streets.

McAdam, Doug, and Hilary Boudet. 2012. *Putting Social Movements in Their Place: Explaining Opposition to Energy Projects in the United States, 2000–2005*. Cambridge: Cambridge University Press.

McCarthy, Shawn. 2012. "Ottawa's New Anti-terrorism Strategy Lists Eco-extremists as Threats." *Globe and Mail*, February 10, 2012. http://www.theglobeandmail.com/news/politics/ottawas-new-anti-terrorism-strategy-lists-eco-extremists-as-threats/article533522/.

McCarthy, Shawn. 2014a. "Greenpeace Sees 'Dirty Tricks' in PR Firm's TransCanada Plan." *Globe and Mail*, November 17, 2014. https://beta.theglobeandmail.com/report-on-business/industry-news/energy-and-resources/greenpeace-sees-dirty-tricks-in-pr-firms-transcanada-plan/article21630761/?ref=http://www.theglobeandmail.com&.

McCarthy, Shawn. 2014b. "Harper Calls Climate Regulations on Oil and Gas Sector 'Crazy Economic Policy.'" *Globe and Mail*, December 9, 2014. http://www.theglobeandmail.com/news/politics/harper-it-would-be-crazy-to-impose-climate-regulations-on-oil-industry/article22014508/.

McCarthy, Shawn. 2015. "'Anti-petroleum' Movement a Growing Security Threat to Canada, RCMP Say." *Globe and Mail*, February 17, 2015. http://www.theglobeandmail.com/news/politics/anti-petroleum-movement-a-growing-security-threat-to-canada-rcmp-say/article23019252/.

McCarthy, Shawn, and Jeffrey Jones. 2013. "The Promise and the Perils of a Pipe to Saint John." *Globe and Mail*, August 1, 2013. https://www.theglobeandmail.com

/report-on-business/industry-news/energy-and-resources/transcanada-to-push-ahead-with-major-new-oil-pipeline-to-eastern-canada/article13545064/.

McConaghy, Dennis. 2017. *Dysfunction: Canada after Keystone XL*. Toronto: Dundern.

McCright, Aaron M., and Riley E. Dunlap. 2010. "Anti-reflexivity: The American Conservative Movement's Success in Undermining Climate Science and Policy." *Theory, Culture and Society* 27 (2–3): 100–133.

McCubbins, Mathew, Roger Noll, and Barry Weingast. 1987. "Administrative Procedures as Instruments of Political Control."*Journal of Law, Economics, and Organization* 3 (2): 243–277.

McFarlane, Allison M., and Rodney C. Ewing, eds. 2006. *Uncertainty Underground: Yucca Mountain and the Nation's High-Level Nuclear Waste*. Cambridge, MA: MIT Press.

McGlade, Christofe, and Paul Ekins. 2015. "The Geographical Distribution of Fossil Fuels Unused When Limiting Global Warming to 2 C. *Nature* 517 (7533): 187–190.

McGowan, Elizabeth. 2011. "NASA's Hansen Explains Decision to Join Keystone Pipeline Protests." *Inside Climate News*, August 21, 2011. https://insideclimatenews.org/news/20110826/james-hansen-nasa-climate-change-scientist-keystone-xl-oil-sands-pipeline-protests-mckibben-white-house.

McGowan, Katharine, and Nino Antadze. 2019. "Value Conflict, Normative Narratives, and Sustainability Transitions: The Confounding Case of the Canadian Oil Sands." Presentation at the 10th International Sustainability Transitions Conference, School of Public Policy and Administration at Carleton University, June 23–26, 2019.

McKenna, Frank. 2012. "Let's Build a Canadian Oil Pipeline from Coast to Coast." *Globe and Mail*, June 18, 2012. https://www.theglobeandmail.com/opinion/lets-build-a-canadian-oil-pipeline-from-coast-to-coast/article4299451/.

McKibben, Bill. 2012. "Global Warming's Terrifying New Math." *Rolling Stone*, July 19, 2012. http://www.rollingstone.com/politics/news/global-warmings-terrifying-new-math-20120719.

McKibben, Bill. 2013a. "The Fossil Fuel Resistance." *Rolling Stone*, April 11, 2013. http://www.rollingstone.com/politics/news/the-fossil-fuel-resistance-20130411.

McKibben, Bill. 2013b. *Oil and Honey: The Education of an Unlikely Activist*. New York: Times Books.

McKibben, Bill, et al. 2011. "Join Us in Civil Disobedience to Stop the Keystone XL Tar-Sands Pipeline." *Grist*, June 23, 2011. https://grist.org/climate-change/2011-06-23-join-us-in-civil-disobedience-to-stop-the-keystone-xl-tar-sands/.

McNeill, Jodi. 2017a. *Tailings Ponds: The Worst Is Yet to Come. Pembina's Oilsands at 50 Series—the Real Cost of Development, Part 2*. Pembina Institute. October 10, 2017. http://www.pembina.org/blog/tailings-ponds-worst-yet-come.

McNeill, Jodi. 2017b. "The Oilsands Sector's Toxic Liquid Legacy." *iPolitics*, November 7, 2017. https://ipolitics.ca/2017/11/07/the-oilsands-sectors-toxic-liquid-legacy/.

McNeill, Jodi, and Nina Lothian. 2017. *Review of Directive 085 Tailings Management Plans Backgrounder*. Pembina Institute. March 13, 2017. http://www.pembina.org/reports/tailings-whitepaper-d85.pdf.

McSheffrey, Elizabeth. 2018. "Trudeau Spills on Kinder Morgan Pipeline." *National Observer*, February 14, 2018. https://www.nationalobserver.com/2018/02/14/news/inside-interview-trudeau-spills-kinder-morgan-pipeline.

Meinshausen, Malte. 2009. "Greenhouse-Gas Emission Targets for Limiting Global Warming to 2 °C." *Nature* 458 (7242): 1158–1162. https://doi.org/10.1038/nature08017.

Mercer, Rick. 2016. "The Mercer Report." *CBC Television*, January 25, 2016.

Metcalfe, J. 2016. "Birds and Insects Sizzle at the World's Biggest Solar Facility." *City Lab*, July 28, 2016. https://www.citylab.com/life/2016/07/watch-insects-and-a-bird-get-vaporized-in-a-huge-solar-facility/493352/.

Meyer, Carl. 2019. "How Climate Policy Dominated Canada's Election." *Mother Jones*, October 22. https://www.motherjones.com/environment/2019/10/how-climate-policy-dominated-canadas-election/.

Meyer, David S., and Suzanne Staggenborg. 1996. "Movements, Countermovements, and the Structure of Political Opportunity." *American Journal of Sociology* 101 (6): 1628–1660. https://doi.org/10.1086/230869.

Ministerial Panel. 2016. *Report from the Ministerial Panel for the Trans Mountain Expansion Project*. November 1, 2016. https://www.nrcan.gc.ca/sites/www.nrcan.gc.ca/files/files/pdf/16-011_TMX%20Full%20Report-en_nov2-11-30am.pdf.

Minister of the Environment, Canada, and Minister of the Environment, British Columbia. 2012. *Agreement to Conduct a Cooperative Environmental Assessment, Including the Establishment of a Joint Review Panel of the Site C Clean Energy*. https://www.ceaa-acee.gc.ca/050/documents/54272/54272E.pdf.

Moe, Terry, and Michael Caldwell. 1994. "The Institutional Foundations of Democratic Government: A Comparison of Presidential and Parliamentary Systems." *Journal of Institutional and Theoretical Economics* 150 (1): 171–195.

Moe, Terry, and Scott Wilson. 1994. "Presidents and the Politics of Structure." *Law and Contemporary Problems* 57:1–44.

Monbiot, George. 2007. "The Real Answer to Climate Change Is to Leave Fossil Fuels in the Ground." *The Guardian*, December 11, 2007. https://www.theguardian.com/commentisfree/2007/dec/11/comment.greenpolitics.

Montpetit, Éric, Erick Lachapelle, and Alexandre Harvey. 2016. "Advocacy Coalitions, the Media and the Politics of Hydraulic Fracturing in the Canadian Provinces of British Columbia and Quebec." In *Comparing Coalition Politics: Policy Debates on Hydraulic Fracturing in North America and Western Europe*, edited by Christopher M. Weible, Karin Ingold, Manuel Fischer, and Tanya Heikkila, 53–79. London: Palgrave.

Montreal Gazette. 2017. "Group Seeks Changes to Hydro-Quebec's Northern Pass Project in the Eastern Townships." March 27, 2017. https://montrealgazette.com/news/local-news/group-opposed-to-hydro-quebecs-northern-pass-project-in-the-eastern-townships.

Moore, Dene. 2014. "First Nations Insist B.C. Choose between Site C Dam and LNG Projects in Peace River Region." *Globe and Mail*, September 24, 2014. https://www.theglobeandmail.com/news/british-columbia/first-nations-insist-bc-choose-between-site-c-dam-and-lng-projects-in-peace-river-region/article20771822/.

Moore, S., and E. Hackett. 2016. "The Construction of Technology and Place: Concentrating Solar Power Conflicts in the United States." *Energy Research and Social Science* 11:67–78.

Morgan, Geoffrey. 2015. "TransCanada Corp's Decision to Shelve Quebec Oil Terminal Plans May Delay Energy East Pipeline by Two Years." *Financial Post*, April 2, 2015. http://business.financialpost.com/commodities/energy/transcanada-corp-wont-be-building-quebec-energy-east-terminal-because-of-whales/wcm/da191d7d-9136-40f6-9950-87116a094199.

Morgan, Granger, Ahmed Abdulla, Michael J. Ford, and Michael Rath. 2018. "US Nuclear Power: The Vanishing Low-Carbon Wedge." *Proceedings of the National Academy of Sciences* 115 (28): 7184–7189. https://doi.org/10.1073/pnas.1804655115.

Morneau, Jennifer. 2015. "Burnaby Mayor Derek Corrigan Ready to End Career with Pipeline Arrest." *Vancouver Sun*, May 20, 2015. http://www.vancouversun.com/Burnaby+Mayor+Derek+Corrigan+ready+career+with+pipeline+arrest/11072064/story.html.

Morrow, Adrian. 2014. "Wynne Drops Main Climate Change Requirement in Considering Energy East Pipeline." *Globe and Mail*, December 3, 2014. https://www.theglobeandmail.com/news/politics/ontario-plays-down-climate-change-concerns-of-energy-east-pipeline/article21907743/.

Morton, Joseph. 2015. "Foes of Keystone XL Pipeline Call TransCanada's Latest Move a 'Hail Mary.'" *Omaha World Herald*, November 3, 2015. http://www.omaha.com/news/nebraska/foes-of-keystone-xl-pipeline-call-transcanada-s-latest-move/article_a6313cd2-81ba-11e5-8ea5-ef3d1c3df5ae.html.

Moynihan, D. P. 2003. "Normative and Instrumental Perspectives on Public Participation Citizen Summits in Washington, DC." *American Review of Public Administration* 33 (2): 164–188.

Mulvihill, P., M. Winfield, and J. Etcheverry. 2013. "Strategic Environmental Assessment and Advanced Renewable Energy in Ontario: Moving Forward or Blowing in the Wind?" *Journal of Environmental Assessment Policy and Management* 15 (2). https://doi.org/10.1142/S1464333213400061.

Nanos, Nik. 2020. "As Lockdowns Drag on and Stress Levels Increase, Canadians Turn to Soul-Searching." *Globe and Mail*, June 13, 2020. https://www.theglobeandmail.com/opinion/article-data-dive-with-nik-nanos-covid-19-triggers-soul-searching-among/.

Nanos Research. Data provided to the author through private communication.

National Energy Board (NEB). 2010. *Reason for Decision: TransCanada Keystone Pipeline GP Ltd.* OH-1–2009.

National Energy Board (NEB). 2014a. *Ruling on Participation.* Hearing Order OH-001–2014. April 2, 2014. https://docs.neb-one.gc.ca/ll-eng/llisapi.dll/fetch/2000/130635/2445932/Letter_-_Application_for_Trans_Mountain_Expansion_Project_-_Ruling_on_Participation_-_A3V6I5.pdf?nodeid=2445819&vernum=-2.

National Energy Board (NEB). 2014b. *Decision Statement Issued under Section 54 of the Canadian Environmental Assessment Act, 2012 and Paragraph 104 (4) (b) of the Jobs, Growth and Long-Term Prosperity Act.* June 17, 2014. http://www.ceaa.gc.ca/050/documents/p21799/99414E.pdf.

National Energy Board (NEB). 2016a. *Key Milestones for Trans Mountain Expansion Project Review.* http://www.neb-one.gc.ca/pplctnflng/mjrpp/trnsmntnxpnsn/mlstns-eng.html.

National Energy Board (NEB). 2016b. *Trans Mountain Expansion Report.* https://docs.neb-one.gc.ca/ll-eng/llisapi.dll?func=ll&objId=2969867&objAction=browse.

National Energy Board (NEB). 2016c. *Canada's Energy Future 2016: Energy Supply and Demand Projections to 2040.* January 2016. https://www.neb-one.gc.ca/nrg/ntgrtd/ftr/2016/index-eng.html#s2.

National Energy Board (NEB). 2016d. *Northern Gateway Pipelines Inc. (Northern Gateway).*

National Energy Board (NEB). 2017a. "NEB Names New Energy East Hearing Panel." News release. January 9, 2017.https://www.canada.ca/en/national-energy-board/news/2017/01/names-new-energy-east-hearing-panel.html.

National Energy Board (NEB). 2017b. *Market Snapshot: Canadian Crude Oil Imports from the U.S. Decline in 2016, Overseas Imports Increase.* February 21, 2017. https://www.neb-one.gc.ca/nrg/ntgrtd/mrkt/snpsht/2017/02-04cndncrlmprtsdcln-eng.html.

National Energy Board (NEB). 2017c. "Expanded Focus for Energy East Assessment." News release. August 23, 2017. https://www.canada.ca/en/canada-energy-regulator/news/2017/08/expanded_focus_forenergyeastassessment.html.

National Energy Board (NEB). 2017d. "NEB Suspends Energy East and Eastern Mainline Hearing." News release. September 8, 2017. https://www.nebenergyeast.ca/.

National Energy Board (NEB). 2017e. *Canada's Renewable Power Landscape—Energy Market Analysis 2017*. https://www.neb-one.gc.ca/nrg/sttstc/lctrct/rprt/2017cndrnwblpwr/2017cndrnwblpwr-eng.pdf.

National Energy Board (NEB). 2017f. *Reasons for Decision*. Order MO-057–2017. https://apps.neb-one.gc.ca/REGDOCS/File/Download/3436250.

National Energy Board (NEB). 2019. *Reconsideration Report—Trans Mountain Pipeline ULC*. MH-052–2018. February 2019. https://apps.neb-one.gc.ca/REGDOCS/Item/Filing/A98021.

National Energy Board and British Columbia Environmental Assessment Office. 2010. *Environmental Assessment Equivalency Agreement*. June 21, 2010. http://www.eao.gov.bc.ca/pdf/NEB-EAO_Equivalancy_Agreement_20100621.pdf.

National Transportation Safety Board. 2012. "Pipeline Rupture and Oil Spill Accident Caused by Organizational Failures and Weak Regulations." Press release. July 10, 2012. https://www.ntsb.gov/news/press-releases/Pages/PR20120710.aspx.

Natural Resources Canada. 2012. "Harper Government Announces Plan for Responsible Resource Development." News release. April 17, 2012. http://www.nrcan.gc.ca/media-room/news-release/2012/45/2001.

Natural Resources Canada. 2016. "Government of Canada Moves to Restore Trust in Environmental Assessment—Statement." January 27, 2016. http://news.gc.ca/web/article-en.do?nid=1029999.

New England Clean Energy Connect. 2020. "Army Corps of Engineers Grants Permit to AVANGRID'S New England Clean Energy Connect Clean Energy Corridor." News release. November 4, 2020. https://www.necleanenergyconnect.org/necec-milestones.

Newfoundland and Labrador Statistics Agency. 2017. *Annual Average Unemployment Rate: Canada and Provinces 1976–2016*. Economics and Statistics Branch. January 9, 2017. http://www.stats.gov.nl.ca/statistics/Labour/PDF/UnempRate.pdf.

Newman, Dwight. 2014. "Provinces Have No Right to Pipeline 'Conditions.'" *Globe and Mail*, December 3, 2014. https://www.theglobeandmail.com/opinion/provinces-have-no-right-to-pipeline-conditions/article21887449/.

Newman, Dwight. 2017. *Political Rhetoric Meets Legal Reality: How to Move Forward on Free, Prior, and Informed Consent in Canada*. MacDonald-Laurier Institute. August 2017. http://macdonaldlaurier.ca/files/pdf/MLIAboriginalResources13-NewmanWeb_F.pdf.

Nikiforek, Andrew. 2010. *Tar Sands: Dirty Oil and the Future of a Continent*. Vancouver: Greystone Books.

Nikiforuk, Andrew. 2013. "Oh, Canada: How America's Friendly Northern Neighbor Became a Rogue, Reckless Petrostate." *Foreign Affairs*, June 24, 2013. https://foreignpolicy.com/2013/06/24/oh-canada/.

Nisperos, N. 2016. "Edison about to Flip the Switch on Controversial Power Line Project through Chino Hills." *Daily Bulletin*, October 30, 2016. https://www.dailybulletin.com/2016/10/30/edison-about-to-flip-the-switch-on-controversial-power-line-project-through-chino-hills/.

Noble, Bram F., and Kelechi Nwanekezie. 2016. "Conceptualizing Strategic Environmental Assessment: Principles, Approaches and Research Directions." *Environmental Impact Assessment Review* 62:165–173. http://dx.doi.org/10.1016/j.eiar.2016.03.005.

Noble, Bram F., Jessie S. Skwaruk, and Robert J. Patrick. 2014. "Toward Cumulative Effects Assessment and Management in the Athabasca Watershed, Alberta, Canada." *Canadian Geographer* 58 (3): 315–328. https://doi.org/10.1111/cag.12063.

Northern Gateway Joint Review Panel. 2013. *Report of the Joint Review Panel for the Enbridge Northern Gateway Project.* http://publications.gc.ca/site/eng/456575/publication.html.

Northern Pass Transmission, LLC. n.d. *The Northern Pass: Project Overview.* http://www.northernpass.us/project-overview.htm.

Northern Plains Resource Council et al. v. Army Corps of Engineers et al., Case 4:19-cv-00044-BMM, April 15, 2020.

Notley, Rachel. 2015. Speech introducing Alberta's Climate Leadership Plan, November 22, 2015. https://www.youtube.com/watch?v=WkwlLWfVag8.

Notley, Rachel. 2016. "Premier Rachel Notley's Address to Albertans." News release. Government of Alberta. April 7, 2016. http://www.alberta.ca/release.cfm?xID=41528807CD312-F230-C30D-3FF1F0E8B998BBC2.

Notley, Rachel. 2018. "Premier Notley: Further Measures to Defend Alberta." News release. Government of Alberta. February 9, 2018. https://www.alberta.ca/release.cfm?xID=52389DF7A690D-0626-F431-10F8D00BBA6AE467.

Obama, Barack. 2013. "Remarks by the President on Climate Change." White House Office of the Press Secretary. June 25, 2013. https://obamawhitehouse.archives.gov/the-press-office/2013/06/25/remarks-president-climate-change.

Obama, Barack. 2015. "Statement by the President on the Keystone XL Pipeline." White House Office of the Press Secretary. November 6, 2015. https://obamawhitehouse.archives.gov/the-press-office/2015/11/06/statement-president-keystone-xl-pipeline.

OEB (Ontario Energy Board). 2015. *Giving a Voice to the Ontarians.* Report to the Minister. August 13, 2015. https://www.oeb.ca/sites/default/files/uploads/energyeast_finalreport_EN_20150813.pdf.

Oliver, Joe. 2012. "An Open Letter from Natural Resources Minister Joe Oliver." *Globe and Mail*, January 9, 2012. https://www.theglobeandmail.com/news/politics/an-open-letter-from-natural-resources-minister-joe-oliver/article4085663/.

Olson, Mancur. 1965. *The Logic of Collective Action*. Cambridge, MA: Harvard University Press.

Olszynski, Martin. 2015. "From 'Badly Wrong' to Worse: An Empirical Analysis of Canada's New Approach to Fish Habitat Protection Laws." *Journal of Environmental Law and Practice* 28 (1): 1–51.

Olszynski, Martin. 2016. "Provisions and Unwittingly Affirms Regressiveness of 2012 Budget Bills." *ABlawg: The University of Calgary Faculty of Law Blog*. July 2016. http://ablawg.ca/wp-content/uploads/2016/07/Blog_MO_NorthernGatewayFC_July2016.pdf.

Olszynski, Martin. 2018. "Testing the Jurisdictional Waters: The Provincial Regulation of Interprovincial Pipelines." *Review of Constitutional Studies* 23 (1): 91–128.

Omaha World Herald. 2017. "Timeline: Keystone XL pipeline." January 24, 2017. http://www.omaha.com/news/nebraska/timeline-keystone-xl-pipeline/article_f5827de2-e24d-11e6-bd42-6b28e1d7be97.html.

O'Neil, Peter. 2011a. "Protesters Won't Derail Gateway Pipeline." *Vancouver Sun*, December 7, 2011. http://www.vancouversun.com/business/Protests+stop+Northern+Gateway+pipeline+minister+says/5820908/story.html#ixzz4pCg6jYgd.

O'Neil, Peter. 2011b. "B.C. Premier Won't Take Stand on Northern Gateway Pipeline." *Vancouver Sun*, December 14, 2011.

Ontario Chief Medical Officer of Health. 2010. *The Potential Health Impact of Wind Turbines*. May 2010.

Ontario Office of the Premier. 2014. "Backgrounder: Agreements Reached at Québec-Ontario Joint Meeting of Cabinet Ministers." News release. November 21, 2014. http://news.ontario.ca/opo/en/2014/11/agreements-reached-at-quebec-ontario-joint-meeting-of-cabinet-ministers.html.

Ontario Progressive Conservatives. 2018. *A Plan for the People*. https://www.ontariopc.ca/plan_for_the_people.

Orsekes, Namoi, and Erik Conway. 2010. *Merchants of Doubt: How a Handful of Scientists Obscured the Truth on Issues from Tobacco Smoke to Global Warming*. New York: Bloomsbury Press.

Ostrom, Elinor, with Michael Cox and Edella Schlager. 2014. "An Assessment of the Institutional Analysis and Development Framework and Introduction of the Social-Ecological Systems Framework." In *Theories of the Policy Process*, 3rd ed., edited by Paul Sabatier and Christopher Weible. Boulder, CO: Westview Press.

Overton, T. W. 2014. "Ivanpah Solar Electric Generating System Earns Power's Highest Honor." *Power Magazine*. https://www.powermag.com/ivanpah-solar-electric-generating-system-earns-powers-highest-honor/.

Owen, Brenda. 2018. "We've Got New Trans Mountain Data and We're Sharing It." *APTN National News*, July 3, 2018. http://aptnnews.ca/2018/07/03/weve-got-new-trans-mountain-data-and-were-sharing-it/.

Palmer, V. 2012. "Oil Pipeline Looks Dead and Buried." *Vancouver Sun*, July 25, 2012.

Palmer, Vaughn. 2014. "Clean Energy Report Becomes a Political Football." *Vancouver Sun*, November 20, 2014. http://www.vancouversun.com/technology/Vaughn+Palmer+Clean+energy+report+becomes+political+football/10396576/story.html.

Palmer, Vaughn. 2017. "Getting Site C to Point of No Return a Damning Progress Report, So Far." *Vancouver Sun*, January 5, 2017. http://vancouversun.com/opinion/columnists/vaughn-palmer-getting-site-c-to-point-of-no-return-a-damning-progress-report-so-far.

Pammett, Jon, and Christopher Dornan. 2016. *The Canadian Election of 2015*. Toronto: Dundurn.

Papadopoulos, Y., and Warin, P. 2007. "Are Innovative, Participatory and Deliberative Procedures in Policy Making Democratic and Effective?" *European Journal of Political Research* 46 (4): 445–472. https://doi.org/10.1111/j.1475-6765.2007.00696.x.

Parfomak, Paul. 2019. *Keystone XL Pipeline: The Saga Continues*. CRS Insight 11131. Congressional Research Service. June 11. https://fas.org/sgp/crs/misc/IN11131.pdf.

Parfomak, Paul, et al. 2013. *Keystone XL Pipeline Project: Key Issues*. Congressional Research Service. December 2, 2013. https://fas.org/sgp/crs/misc/R41668.pdf.

Parfomak, Paul, Linda Luther, Richard K. Lattanzio, Jonathan L. Ramseur, Adam Vann, Robert Pirog, and Ian F. Fergusson. 2015. *Keystone XL Pipeline: Overview and Recent Developments*. Congressional Research Service 7–5700.

Partidário, M. R. 2012. Strategic Environmental Assessment Better Practice Guide: Methodological Guidance for Strategic Thinking in SEA. http://www.civil.ist.utl.pt/shrha-gdambiente/SEA_Guidance_Portugal.pdf.

Partidário, M. R. 2015. "A Strategic Advocacy Role in Sea for Sustainability." *Journal of Environmental Assessment Policy and Management* 17 (1): 1550015.

Pearson, Natalie. 2018. "B.C. Premier Knows He Has No Legal Power to Block Trans Mountain. But That's Not Stopping Him." *Financial Post*, April 13, 2018. https://business.financialpost.com/commodities/energy/b-c-premier-knows-he-has-no-legal-power-to-block-trans-mountain-but-thats-not-stopping-him.

Pembina Institute. 2005. *Oilsands Fever*. November 23, 2005. http://www.pembina.org/pub/203.

Penner, Derrick. 2014. "Site C Mega-project a Welcome Boost for Construction Industry." *Vancouver Sun*, December 18, 2014. http://www.vancouversun.com/business

/energy/Site+mega+project+welcome+boost+construction+industry/10661544/story.html?__lsa=daaa-345d.

Pentland, William. 2018. "New Hampshire Blocks Major Power Transmission Project." *Forbes*, February 4, 2018. https://www.forbes.com/sites/williampentland/2018/02/04/new-hampshire-blocks-major-power-transmission-project/#11dc1fdc7fdb.

Penty, Rebecca. 2015. "TransCanada Confirms It Won't Build Cacouna Marine Terminal, Exploring other Quebec Sites." *Calgary Herald*, April 2, 2015. http://calgaryherald.com/business/energy/transcanada-confirms-it-wont-build-cacouna-marine-terminal-exploring-other-quebec-sites.

Peterson St-Laurent, Guillaume, George Hoberg, Stephen R. J. Sheppard, and Shannon Hagerman. 2020. "Designing and Evaluating Analytic-Deliberative Engagement Processes for Natural Resources Management." *Elementa: Science of the Anthropocene* 8 (8): 1–17. https://doi.org/10.1525/elementa.402.

Petts, Judith. 2001. "Evaluating the Effectiveness of Deliberative Processes: Waste Management Case-Studies." *Journal of Environmental Planning and Management* 44 (2): 207–226. https://doi.org/10.1080/09640560120033713.

Pew Research Center. 2017. "Public Divided over Keystone XL, Dakota Pipelines; Democrats Turn Decisively against Keystone." February 21, 2017. http://www.pewresearch.org/fact-tank/2017/02/21/public-divided-over-keystone-xl-dakota-pipelines-democrats-turn-decisively-against-keystone/.

Pidgeon, Nick, Christina Demski, Catherine Butler, Karen Parkhill, and Alexa Spence. 2014. "Creating a National Citizen Engagement Process for Energy Policy." *Proceedings of the National Academy of Sciences* 111 (Supplement 4): 13606–13613. https://doi.org/10.1073/pnas.1317512111.

Pierson, Paul. 2000. "Not Just What, but When: Timing and Sequence in Political Processes." *Studies in American Political Development* 14 (1): 72–92. https://doi.org/10.1017/S0898588X00003011.

Piggot, Georgia. 2018. "The Influence of Social Movements on Policies That Constrain Fossil Fuel Supply." *Climate Policy* 18 (7): 942–954. https://doi.org/10.1080/14693062.2017.1394255.

Poitras, Jacques. 2016. "Edmundston 'Strongly Opposes' Current Energy East Pipeline Route." *CBC News*, April 27, 2016. http://www.cbc.ca/news/canada/new-brunswick/edmundston-energy-east-opposition-1.3555076.

Poitras, Jacques. 2018. *Pipe Dreams: The First for Canada's Energy Future*. Toronto: Viking/Penguin Canada.

Polido, A., E. João, and T. B. Ramos. 2016. "Strategic Environmental Assessment practices in European Small Islands: Insights from Azores and Orkney Islands." *Environmental Impact Assessment Review* 57:18–30.

Postmedia News. 2016. "Kathleen Wynne Gives Tentative Backing to Energy East Pipeline as Rachel Notley Faces Criticism over Project." *Financial Post,* January 22, 2016. http://business.financialpost.com/news/economy/kathleen-wynne-gives-tentative-backing-to-energy-east-pipeline-as-rachel-notley-faces-criticism-over-project.

Pralle, Sarah. 2006a. *Branching In, Digging Out: Environmental Advocacy and Agenda Setting*. Washington, DC: Georgetown University Press.

Pralle, Sarah. 2006b. "Timing and Sequence in Agenda-Setting and Policy Change: A Comparative Study of Lawn Care Pesticide Politics in Canada and the US." *Journal of European Public Policy* 13 (7): 987–1005. https://doi.org/10.1080/13501760600923904.

Press-Enterprise. 2016. "Environment: Soda Mountain Solar Plant Approved, Despite Objections." April 6, 2016. https://www.pe.com/2016/04/06/environment-soda-mountain-solar-plant-approved-despite-objections/.

Price, Matt. 2017. Personal interview. September 21, 2017.

Prophet River First Nation v. British Columbia (Environment), 2017 BCCA 58.

Prystupa, Mychaylo. 2014. "Burnaby Mountain Battle: Our Notes from the Courts, the Woods and 100 Arrests." *Vancouver Observer,* November 30, 2014. http://www.vancouverobserver.com/news/burnaby-mountain-battle-our-notes-courts-woods-and-100-arrests.

Rabson, Mia, and the Canadian Press. 2018. "Canada Will Do What It Must to Keep B.C. from Blocking Trans Mountain: Carr." *CBC News,* February 12, 2018. https://www.cbc.ca/news/politics/carr-trans-mountain-bc-1.4531962.

Ramana, M. V. 2018. "Technical and Social Problems of Nuclear Waste." *WIREs Energy and Environment* 7 (4): 1–2. https://doi.org/10.1002/wene.289.

Rauschmayer, F., and H. Wittmer. 2006. "Evaluating Deliberative and Analytical Methods for the Resolution of Environmental Conflicts." *Land Use Policy* 23 (1): 108–122. https://doi.org/10.1016/j.landusepol.2004.08.011.

Read, Andrew. 2014. *Climate Change Policy in Alberta: Backgrounder.* Pembina Institute. July 2014. http://www.pembina.org/reports/sger-climate-policy-backgrounder.pdf.

Reference re Environmental Management Act (British Columbia), 2019 BCCA 181.

Reject and Protect. n.d. "About." http://rejectandprotect.org/about-2/.

Renn, Ortwin. 2004. "The Challenge of Integrating Deliberation a Expertise: Participation and Discourse in Risk Management." In *Risk Analysis and Society: An Interdisciplinary Characterization of the Field,* edited by T. McDaniels and M. Small, 289–366. Cambridge: Cambridge University Press. https://doi.org/10.1017/CBO9780511814662.009.

Renn, Ortwin. 2006. "Participatory Processes for Designing Environmental Policies." *Land Use Policy, Resolving Environmental Conflicts: Combining Participation and Multi-Criteria Analysis,* 23 (1): 34–43. https://doi.org/10.1016/j.landusepol.2004.08.005.

Renn, Ortwin, Thomas Webler, and Peter Wiedemann. 1995. "The Pursuit of Fair and Competent Citizen Participation." In *Fairness and Competence in Citizen Participation*, 339–367. Dordrecht: Springer. https://doi.org/10.1007/978-94-011-0131-8_20.

Resource Works. n.d. "About." http://www.resourceworks.com/about.

Revkin, Andrew. 2011. "Can Obama Escape the Alberta Tar Pit?" *New York Times*, September 5, 2011. https://dotearth.blogs.nytimes.com/2011/09/05/can-obama-escape-the-alberta-tar-pit/?src=tp.

Rivard, Christine, Denis Lavoie, René Lefebvre, Stephan Séjourné, Charles Lamontagne, and Mathieu Duchesne. 2014. "An Overview of Canadian Shale Gas Production and Environmental Concerns." *International Journal of Coal Geology* 126:64–76.

Robitaille, David, Jean Baril, Ghislain Otis, Benoît Pelletier, Sophie Thériault, and Pierre Thibault. 2014. "Le Canada n'est pas un pays unitaire." *Le Devoir*, December 11, 2014. http://www.ledevoir.com/politique/canada/426316/la-replique-juridictions-sur-energie-est-le-canada-n-est-pas-un-etat-unitaire.

Ross, Michael. 2012. *The Oil Curse: How Petroleum Wealth Shapes the Development of Nations*. Princeton, NJ: Princeton University Press.

Rossi, Jim. 1997. "Participation Run Amok: The Costs of Mass Participation for Deliberative Agency Decisionmaking." *Northwestern University Law Review* 97 (1): 173–250.

Rowe, Gene, and Lynn J. Frewer. 2000. "Public Participation Methods: A Framework for Evaluation." *Science, Technology, & Human Values* 25 (1): 3–29.

Rowland, Robin. 2014. "Kitimat residents vote 'no' in pipeline plebiscite." *Globe and Mail*, April 12. https://www.theglobeandmail.com/news/british-columbia/kitimat-residents-vote-in-northern-gateway-oil-pipeline-plebiscite/article17949815/.

Royal Society of Canada. 2010. *Environmental and Health Impacts of Canada's Oil Sands Industry*. https://rsc-src.ca/en/environmental-and-health-impacts-canadas-oil-sands-industry.

Royal Society of Canada. n.d. *Over 200 Leading Scholars Call on Government to Suspend Site C Dam*. http://www.rsc.ca/en/about-us/our-people/our-priorities/over-200-leading-scholars-call-government-to-suspend-site-c-dam.

Rusbridger, Alan. 2015. "Climate Change: Why the Guardian Is Putting Threat to Earth Front and Centre." *The Guardian*, March 6, 2015. https://www.theguardian.com/environment/2015/mar/06/climate-change-guardian-threat-to-earth-alan-rusbridger.

Sabatier, Paul, and Hank Jenkins-Smith, eds. 1993. *Policy Change and Learning: An Advocacy Coalition Approach*. Boulder, CO: Westview Press.

Sahagun, L. 2015. "L.A. Won't Buy Power from Mojave Desert Solar Plant, after All." *Los Angeles Times*, June 12, 2015.https://www.latimes.com/local/california/la-me-solar-20150612-story.html.

Sahagun, L. 2016a. "San Bernardino County Rejects a Controversial Solar Power Plant Proposed for the Mojave Desert." *Los Angeles Times*, August 24, 2016. https://www.latimes.com/local/california/la-me-soda-mountain-solar-20160824-snap-story.html.

Sahagun, L. 2016b. "This Mojave Desert Solar Plant Kills 6,000 Birds a Year. Here's Why That Won't Change Any Time Soon." *Los Angeles Times*, August 31, 2016. https://www.latimes.com/local/california/la-me-solar-bird-deaths-20160831-snap-story.html.

Salomons, Geoffrey, and George Hoberg. 2014. "Setting Boundaries of Participation in Environmental Impact Assessment." *Environmental Impact Assessment Review* 45:69–75.

San Bernardino Sun. 2014. "Emerging Solar Plants in Mojave Desert Scorch Birds in Mid-air. August 18, 2014. https://www.sbsun.com/2014/08/18/emerging-solar-plants-in-mojave-desert-scorch-birds-in-mid-air/.

Save the Fraser Gathering of First Nations. 2013. "Save the Fraser Declaration." May 2013. http://savethefraser.ca/images/Fraser-Declaration-May2013.jpg.

Savoie, Donald. 1999. *Governing from the Centre: The Concentration of Power in Canadian Politics*. Toronto: University of Toronto Press.

Scott, Adam. 2016. Personal interview with the author. October 4, 2016.

Scotti, Monique. 2017. "Notley Responds after Rona Ambrose Says: 'I Don't See Energy East Getting through Montreal.'" *AM730*, March 27, 2017. http://www.am730.ca/syn/98/201214/rona-ambrose-i-dont-see-energy-east-getting-through-montreal.

Seelye, K. Q. 2017. "After 16 Years, Hopes for Cape Cod Wind Farm Float Away." *New York Times*, December 19, 2017. https://www.nytimes.com/2017/12/19/us/offshore-cape-wind-farm.html.

Seitz, N. E., C. J. Westbrook, and B. F. Noble. 2011. "Bringing Science into River Systems Cumulative Effects Assessment Practice." *Environmental Impact Assessment Review* 31: 172–179.

Semeniuk, Ivan. 2012. "Scientists March on Canadian Parliament." *Nature Newsblog*, July 10, 2012. http://blogs.nature.com/news/2012/07/scientists-march-on-canadian-parliament.html.

SEPA. 2011. "The Scottish Strategic Environmental Assessment Review—Summary." 1–30. https://www.sepa.org.uk/media/27556/sea-review_summary.pdf.

Shapiro, Stuart. 2017. "Structure and Process: Examining the Interaction between Bureaucratic Organization and Analytical Requirements." *Review of Policy Research* 34 (3): 682–699.

Sharpe, Sydney, and Don Braid. 2016. *Notley Nation: How Alberta's Political Upheaval Swept the Country*. Toronto: Dundurn.

Shaw, Karena. 2004. "The Global/Local Politics of the Great Bear Rainforest." *Environmental Politics* 13 (2) 373–392. https://doi.org/10.1080/0964401042000209621.

Shaw, Karena. 2011. "Climate Deadlocks: The Environmental Politics of Energy Systems." *Environmental Politics* 20 (5): 743–763. https://doi.org/10.1080/09644016.2011.608538.

Shaw, Rob, and Richard Zussman. 2018. *A Matter of Confidence: The Inside Story of the Political Battle for BC*. Victoria: Heritage.

Shearon, Kimberly. 2013. "That's a Wrap! Northern Gateway Hearings Come to a Close." *Ecojustice*, June 10, 2013. http://www.ecojustice.ca/blog/thats-a-wrap-northern-gateway-hearings-come-to-a-close.

Sheppard, Stephen R. J. 2005. "Participatory Decision Support for Sustainable Forest Management: A Framework for Planning with Local Communities at the Landscape Level in Canada." *Canadian Journal of Forest Research* 35 (7): 1515–1526.

Shields, Alexandre. 2014. "Pipeline Énergie Est—L'ONE fera fi de l'impact du pétrole." *Le Devoir*, November 1, 2014.

Shields, Alexandre. 2016. "La FTQ s'oppose malgré les dissensions." *Le Devoir*, August 11, 2016. http://www.ledevoir.com/environnement/actualites-sur-l-environnement/477470/oleoduc-energie-est-la-ftq-s-oppose-malgre-les-dissensions.

Shingler, Benjamin, and Stephen Smith. 2016. "NEB Cancels 2 Days of Energy East Hearings in Montreal after 'Violent Disruption.'" *CBC News*, August 29, 2016. http://www.cbc.ca/news/canada/montreal/neb-hearings-energy-east-protest-quebec-2016-1.3739215.

Shogren, E. 2012. "Firm Blamed in the Costliest Oil Spill Ever." National Public Radio, July 10, 2012. http://www.npr.org/2012/07/10/156561319/oilcompany-knew-michigan-pipeline-was-cracked.

Shrivastava, Meenal, and Lorna Stefanik, eds. 2015. *Alberta Oil and the Decline of Democracy in Canada*. Edmonton: Athabasca University Press.

Simpson, Jeffrey, Mark Jaccard, and Nic Rivers. 2008. *Hot Air: Meeting Canada's Climate Change Challenge*. Toronto: McClelland & Stewart.

Sinoski, Kelly. 2016. "Burnaby Mayor to Rally the Public to Fight Kinder Morgan Expansion." *Vancouver Sun*, May 20, 2016. http://vancouversun.com/news/local-news/burnaby-mayor-to-rally-the-public-to-fight-kinder-morgan-expansion.

Smith, Joanna. 2016. "Trudeau Wants Canada to Play Key Role in Fighting Climate Change." *Toronto Star*, March 2. https://www.thestar.com/news/canada/2016/03/02/canada-will-play-leading-role-in-new-economy-trudeau-says.html.

Smith, Merran, and Art Sterritt. 2016. *From Conflict to Collaboration: The Story of the Great Bear Rainforest.* http://coastfunds.ca/wp-content/uploads/2016/02/StoryoftheGBR.pdf.

Snyder, Jesse. 2017. "NEB Decision Could Set Back TransCanada's Energy East Pipeline as Keystone XL Moves Ahead." *Financial Post,* January 27, 2017. http://business.financialpost.com/commodities/energy/ottawas-review-of-energy-east-pipeline-project-is-scrapping-work-already-done-and-starting-from-beginning/wcm/7e8adc5a-64cb-424c-a10a-767e6b58a24f.

Society for the Protection of New Hampshire Forests (the Forest Society). n.d. *Trees Not Towers: Bury Northern Pass!* https://forestsociety.org/project/trees-not-towers-bury-northern-pass.

Southern California Edison. n.d. *Tehachapi Renewable Transmission Project: Delivering Renewable Power Affordably throughout Southern California.* https://www.sce.com/about-us/reliability/upgrading-transmission/TRTP-4-11.

Sovacool, Benjamin, and Pushkala Lakshmi Ratan. 2012. "Conceptualizing the Acceptance of Wind and Solar Electricity." *Renewable and Sustainable Energy Review* 16 (7): 5268–5279.

State Impact New Hampshire. n.d. *Why the Northern Pass Project Matters.* https://stateimpact.npr.org/new-hampshire/tag/northern-pass/.

Statement of Concerned Scholars on the Site C Dam Project. 2016. https://sitecstatement.org/.

Statistics Canada. 2018. *Distribution of Gross Domestic Product of Alberta, Canada in 2018, by Industry.* https://www.statista.com/statistics/608354/gdp-distribution-of-alberta-canada-by-industry/.

Steinberg, J. 2016. "Supervisors Reject Soda Mountain Solar Plant; Developers Vow to Continue Efforts." *San Bernardino Sun,* August 24, 2016. https://www.sbsun.com/2016/08/24/supervisors-reject-soda-mountain-solar-plant-developers-vow-to-continue-efforts/.

Stern, P. C., and H. V. Fineberg, eds. 1996. *Understanding Risk: Informing Decisions in a Democratic Society.* Washington, DC: National Academy Press.

Sterritt, Art. 2017. Personal interview with the author. November 10, 2017.

Stewart, Keith. 2016. Personal interview with the author. October 3, 2016.

Stirling, Andy. 2008. "'Opening Up' and 'Closing Down': Power, Participation, and Pluralism in the Social Appraisal of Technology." *Science, Technology, & Human Values* 33 (2): 262–294. https://doi.org/10.1177/0162243907311265.

Stokes, L. C. 2013. "The Politics of Renewable Energy Policies: The Case of Feed-in Tariffs in Ontario, Canada." *Energy Policy* 56:490–500. https://doi.org/10.1016/j.enpol.2013.01.009.

Stokes, Leak. 2016. "Electoral Backlash against Climate Policy: A Natural Experiment on Retrospective Voting and Local Resistance to Public Policy." *American Journal of Political Science* 60 (4): 958–974.

Supreme Court of Canada. 2020. "Judgement in Leave Applications." July 2. https://decisions.scc-csc.ca/scc-csc/news/en/item/6899/index.do.

Sweet, C. 2015. "High-Tech Solar Projects Fail to Deliver." *Wall Street Journal*, 2015. https://www.wsj.com/articles/high-tech-solar-projects-fail-to-deliver-1434138485.

Swift, Anthony, Nathan Lemphers, Susan Casey-Lefkowitz, Katie Terhune, Danielle Droitsch. 2011. *Pipeline and Tanker Trouble: The Impact to British Columbia's Communities, Rivers, and Pacific Coastline from Tar Sands Oil Transport.* Natural Resources Defense Council, Pembina Institute, and Living Oceans Society. November 2011. https://www.nrdc.org/sites/default/files/PipelineandTankerTrouble.pdf.

Taft, Kevin. 2017. *Oil's Deep State.* Toronto: Lorimer.

Tasci, C. 2013. "Edison Notifies Chino Hills Resident of Undergrounding Power Lines." *Daily Bulletin*, August 29, 2013. https://www.dailybulletin.com/2013/08/29/edison-notifies-chino-hills-residents-of-undergrounding-power-lines/.

Taylor, Charles. 1985. *Social Theory as Practice.* Cambridge: Cambridge University Press.

Tedesco, Theresa. 2013. "Why New Brunswick's Premier Has Become the Public Face of the West-East Pipeline." *Financial Post*, April 18, 2013. http://business.financialpost.com/commodities/energy/new-brunswick-eastern-pipeline/wcm/865e261e-86d5-4bbf-bf86-72aea94b7776.

Ternes, Brock, James Ordner, and David Heath Cooper. 2020. "Grassroots Resistance to Energy Project Encroachment: Analyzing Environmental Mobilization against the Keystone XL Pipeline." *Journal of Civil Society* 16 (1): 44–60. https://doi.org/10.1080/17448689.2020.1717151.

Thomson v. Heineman. 2015. Nebraska Supreme Court 289 Neb. 798. https://www.nebraska.gov/apps-courts-epub/public/viewCertified?docId=N00004799PUB.

350.org. n.d. "Give Energy East a People's Intervention." http://act.350.org/letter/energy_east.

Tierney, S. F., and P. G. Darling. 2017. *The Proposed Northern Pass Transmission Project: Assessing Its Impacts on New Hampshire.* Analysis Group, Inc. http://www.analysisgroup.com/uploadedfiles/content/insights/publishing/pgs_final_northern_pass_report_2017-2-8.pdf.

Toensing, G. C. 2011. "Aquinnah Wampanoag Sues Feds over Cape Wind." *Indian Country Today*, July 14, 2011. https://newsmaven.io/indiancountrytoday/archive/aquinnah-wampanoag-sues-feds-over-cape-wind-OAkvpCXRH06s8ZIAF4K6zg/.

Tombe, Trevor. 2016. "Policy, Not Pipelines, Will Determine If We Meet Our Goals." *Maclean's*, December 2. https://www.macleans.ca/economy/economicanalysis/policy-not-pipelines-will-determine-if-we-meet-our-goals/.

Toner, Glen, and Jennifer McKee. 2014. "Harper's Partisan Wedge Politics: Bad Environmental Policy and Bad Energy Policy." In *How Ottawa Spends, 2014–2015*, edited by G. Bruce Doern and Christopher Stoney. Montreal and Kingston: McGill-Queen's University Press.

Town of Barnstable, Massachusetts, et al. v. Ann G. Berwick, et al. 2014. United States District Court—District of Massachusetts.

Tracy, P. 2016. "Balsams Developer Support Northern Pass." WMUR9, March 9, 2016. https://www.wmur.com/article/balsams-developer-supports-northern-pass/5209281.

TransCanada. 2015. *Project Scope of the Energy East Pipeline Project.* http://www.energyeastpipeline.com/wp-content/uploads/2015/12/Energy-East-Pipeline-Project-Backgrounder.pdf.

TransCanada. 2016a. "TransCanada Reports Fourth Quarter and Year-End 2015 Financial Results." News release. February 11, 2016. https://www.transcanada.com/fr/announcements/2016-02-11transcanada-reports-fourth-quarter-and-year-end-2015-financial-results/.

TransCanada. 2016b. *TransCanada PipeLines Limited. Consolidated Application. Executive Summary.* https://apps.neb-one.gc.ca/REGDOCS/File/Download/2957481.

TransCanada. 2017a. "TransCanada Seeks 30-Day Suspension of Energy East Pipeline and Eastern Mainline Project Applications." TransCanada Corporation, News release, September 7, 2017. http://www.marketwired.com/press-release/transcanada-seeks-30-day-suspension-energy-east-pipeline-eastern-mainline-project-applications-tsx-trp-2232994.htm.

TransCanada. 2017b. "TransCanada Announces Termination of Energy East Pipeline and Eastern Mainline Projects." News release. October 5, 2017. https://www.globenewswire.com/news-release/2017/10/05/1300421/0/en/TransCanada-Announces-Termination-of-Energy-East-Pipeline-and-Eastern-Mainline-Projects.html.

Trans Mountain. 2013. *Trans Mountain Expansion Project.* vol. 2, 2–27. https://apps.neb-one.gc.ca/REGDOCS/Item/Open/2484763.

Trans Mountain. 2018. "43 Aboriginal Groups Have Signed Agreements in Support of the Trans Mountain Expansion Project." News release. April 19, 2018. https://www

.transmountain.com/news/2018/43-aboriginal-groups-have-signed-agreements-in-support-of-the-trans-mountain-expansion-project.

Trans Mountain. n.d. *Proposed Expansion.* https://www.transmountain.com/proposed-expansion.

Treaty Alliance. 2016a. *Treaty Alliance against Tar Sands.* http://www.treatyalliance.org/.

Treaty Alliance. 2016b. *The Story of the Treaty.* http://www.treatyalliance.org/treaty#storyContent.

Treaty 8 Tribal Association. 1899. *Treaty 8 Agreement between Nations of Alberta, Saskatchewan, and Northwest Territories.* June 21, 1899. http://treaty8.bc.ca/wp-content/uploads/2015/07/Treaty-No-8-Easy-Read-Version.pdf.

Treaty 8 Tribal Association. 2011. *Treaty 8 First Nations Declaration on the Site C Dam Proposal.* March 9, 2011. https://fathertheo.wordpress.com/2011/03/09/treaty-8-first-nations-declaration-on-the-site-c-dam-proposal/.

Trudeau, Justin. 2015. "Remarks by Liberal Party of Canada Leader Justin Trudeau at the Canadian Club of Calgary on February 6, 2015." Liberal Party of Canada. https://www.liberal.ca/justin-trudeau-pitches-medicare-approach-to-fight-climate-change-in-canada/.

Trudeau, Justin. 2016a. "Prime Minister Justin Trudeau Delivers a Speech on Pricing Carbon Pollution." October 3, 2016. https://pm.gc.ca/en/news/speeches/2016/10/03/prime-minister-justin-trudeau-delivers-speech-pricing-carbon-pollution.

Trudeau, Justin. 2016b. "Prime Minister Justin Trudeau's Pipeline Announcement." News release. November 30, 2016. http://pm.gc.ca/eng/news/2016/11/30/prime-minister-justin-trudeaus-pipeline-announcement.

Trudeau, Justin. 2019. "Trans Mountain Expansion Will Fund Canada's Future Clean Economy." News release. June 18, 2019. https://pm.gc.ca/en/news/news-releases/2019/06/18/trans-mountain-expansion-will-fund-canadas-future-clean-economy.

Tsebelis, George. 2000. "Veto Points and Institutional Analysis." *Governance* 13:441–474.

Tsleil-Waututh First Nation. 2015. *Resolution in Respect of Tsleil-Waututh Nation's Stewardship Policy Decision about the Trans Mountain Pipeline and Tanker Expansion Proposal.* May 21, 2015. http://twnsacredtrust.ca/wp-content/uploads/2015/05/2015-05-21-TWN-BCR-re-TMEX.pdf.

Tsleil-Waututh Nation v. Canada (Attorney General), 2018 FCA 153.

Tucker, Bronwen. 2017. "Social Movements Played a Huge Part in Derailing Energy East." *CBC News*, Opinion, October 12, 2017. http://www.cbc.ca/news/opinion/social-movements-energy-east-1.4344080.

Turner, Chris. 2017. *The Patch: The People, Pipelines, and Politics of the Oil Sands*. New York: Simon and Schuster.

Tweed, K. 2010. "Tehachapi Renewable Transmission Project Completes Phase One." *Green Tech Media*, 2010. https://www.greentechmedia.com/articles/read/phase-one-completed-in-tehachapi-renewable-transmission-project#gs.4yepea.

Uechi, Jenny. 2013. "Tar Sands Solutions Network Launches Oil Sands News Aggregation Site." *Vancouver Observer*, May 14, 2013. https://www.vancouverobserver.com/environment/tar-sands-solutions-network-launches-oil-sands-news-aggregation-site.

UNESCO. 2017. *Report of the Joint WHC/IUCN Reactive Monitoring Mission to Wood Buffalo National Park, Canada*. http://whc.unesco.org/en/documents/156893.

UN General Assembly. 2007. *United Nations Declaration on the Rights of Indigenous Peoples: Resolution/Adopted by the General Assembly*, October 2, 2007, A/RES/61/295. https://www.refworld.org/docid/471355a82.html.

Union of BC Indian Chiefs (UBCIC). 2018. "As Trudeau Ramps Up Pressure to Build, First Nations from across Canada Stand in Solidarity against Kinder Morgan Pipeline." February 8, 2018. https://www.ubcic.bc.ca/nokm2018.

United Conservative Party. 2019. *Alberta Strong and Free—Getting Alberta Back to Work*. Election platform. March 30, 2019. https://www.albertastrongandfree.ca/wp-content/uploads/2019/04/Alberta-Strong-and-Free-Platform-1.pdf.

Urquhart, Ian. 2018. *Costly Fix: Power, Politics, and Nature in the Tar Sands*. Toronto: University of Toronto Press.

US Bureau of Reclamation. n.d. *The Story of Hoover Dam—Essays: What Is the Biggest Dam in the World?* https://www.usbr.gov/lc/hooverdam/history/essays/biggest.htm.

US Department of Energy. 2017. "Department of Energy Approves Presidential Permit for Northern Pass Transmission Line Project." https://www.energy.gov/articles/department-energy-approves-presidential-permit-northern-pass-transmission-line-project.

US Department of State. 2011a. *Final Environmental Impact Statement for the Keystone XL Project*. August 26, 2011. https://2012-keystonepipeline-xl.state.gov/archive/dos_docs/feis/index.htm.

US Department of State. 2011b. *Keystone XL Pipeline Project Review Process: Decision to Seek Additional Information*. November 10, 2011. https://2009-2017.state.gov/r/pa/prs/ps/2011/11/176964.htm.

US Department of State. 2012. *Denial of the Keystone XL Pipeline Application*. January 18, 2012. https://2009-2017.state.gov/r/pa/prs/ps/2012/01/181473.htm.

US Department of State. 2014. *Final Supplemental Environmental Impact Statement for the Keystone XL Project*. January. https://www.keystonepipeline-xl.state.gov/final seis/.

US Department of State. 2015. *Record of Decision and National Interest Determination—TransCanada Keystone Pipeline LP Application for a Presidential Permit*. November 6, 2015. https://2012-keystonepipeline-xl.state.gov/documents/organization/249450 .pdf.

US Department of State. 2017. *Record of Decision and National Interest Determination—TransCanada Keystone Pipeline LP Application for a Presidential Permit*. March 23, 2017. https://keystonepipeline-xl.state.gov/documents/organization/269323.pdf.

Valiante, Giuseppe. 2016. "TransCanada's Energy East Pipeline Rouses Quebec Nationalism." *Huffington Post*, February 11, 2016. http://www.huffingtonpost.ca/2016 /02/11/transcanada-must-overcome-quebec-identity-politics-in-quest-to-build-energy -east_n_9207070.html.

Vancouver Sun. 2018. "Trudeau Tells B.C. That Pipeline Wouldn't Get Approval without Environmental Confidence." April 5, 2018. https://vancouversun.com/news/local -news/trudeau-tells-b-c-that-pipeline-wouldnt-get-approval-without-environmental -confidence.

Vanderklippe, Nathan. 2011. "Ottawa Energy Strategy Targets Diverse Marketplace." *Globe and Mail*, July 18, 2011. https://beta.theglobeandmail.com/report-on -business/industry-news/energy-and-resources/ottawa-energy-strategy-targets-diverse -marketplace/article590827/?ref=http://www.theglobeandmail.com&.

Van Praet, Nicolas, and Les Perreaux. 2014. "Energy East Port Must Be Away from Belugas, Quebec and Alberta Premiers Say." *Globe and Mail*, December 2, 2014. https://www .theglobeandmail.com/news/politics/energy-east-port-must-be-away-from-belugas -quebec-and-alberta-premiers-say/article21878827/.

Victor, David. 2011. *Global Warming Gridlock*. Cambridge: Cambridge University Press.

Vis, Matt. 2017. "Council Supports Proposed Energy East Pipeline." *tbnewswatch*, May 30, 2017. https://www.tbnewswatch.com/local-news/council-supports-proposed -energy-east-pipeline-628597.

Walker, Chad, and Jamie Baxter. 2017. "Procedural Justice in Canadian Wind Energy Development: A Comparison of Community-Based and Technocratic Siting Processes." *Energy Research & Social Sciences* 29:160–169.

Walker, Sven. 2010. *Recent Developments Arising under Ontario's Green Energy Act*. Dale and Lessmann LLP, October 19, 2010. https://www.dalelessmann.com/sites/default /files/Green%20Energy%20Act%20%28English%20-October%2018%202010%29%20 final.pdf.

Walls, W. D., and Xiaoli Zheng. 2020. "Pipeline Capacity Rationing and Crude Oil Price Differentials: The Case of Western Canada." *Energy Journal* 41 (1): 241–258. https://www.iaee.org/en/Publications/ejarticle.aspx?id=3458.

Watson, Greg, and Farah Courtney. 2004. "Nantucket Sound Offshore Wind Stakeholder Process." *Boston College Environmental Affairs Law Review* 31:263284.

Washington Post Editorial Board. 2019. "Canada's Election Results Are a Victory for the Planet." *Washington Post*, October 22. https://www.washingtonpost.com/opinions/global-opinions/canadas-election-results-are-a-victory-for-the-planet/2019/10/22/d4dbf440-f4fd-11e9-a285-882a8e386a96_story.html.

Weaver, R. Kent. 1986. "The Politics of Blame Avoidance." *Journal of Public Policy* 6 (4): 371–398.

Weber, Bob. 2012. "PM Concerned 'Foreign Money' Could Hold Up Pipeline Hearing Process." *Winnipeg Free Press*, January 7, 2012. https://www.winnipegfreepress.com/business/pm-fears-pipeline-hearings-to-be-hijacked-136867423.html.

West, Ben. 2016. Personal interview with the author. September 28, 2016.

West Coast Environmental Law. 2017. "See You in Court." September 14, 2017. https://www.wcel.org/blog/see-you-in-court-kinder-morgan.

White House. 2015. "Letter by President Obama Vetoing Keystone XL Bill." February 24, 2015. http://www.newsweek.com/here-text-president-obamas-veto-keystone-xl-pipeline-bill-309206.

White House. 2021. "Executive Order on Protecting Public Health and the Environment and Restoring Science to Tackle the Climate Crisis." January 20, 2021. https://www.whitehouse.gov/briefing-room/presidential-actions/2021/01/20/executive-order-protecting-public-health-and-environment-and-restoring-science-to-tackle-climate-crisis/

Whittington, Les. 2013. "Stephen Harper Endorses Energy East Pipeline Proposal." *Toronto Star*, August 2, 2013. https://www.thestar.com/news/canada/2013/08/02/stephen_harper_endorses_energy_east_pipeline_proposal.html.

Wiener-Bronner, D. 2014. "The World's Largest Solar Energy Plant Is Also a Massive Death Ray for Birds." *The Atlantic*, 2014. https://www.theatlantic.com/national/archive/2014/02/nevadas-massive-solar-plant-death-ray-birds/358244/.

Wilderness Committee. 2013. *Annual Report for the Year Ended 30 April 2013*. https://www.wildernesscommittee.org/sites/all/files/publications/2013_WC_annual-report-web.pdf.

Wilderness Committee. n.d. *Stop the Site C Dam*. https://www.wildernesscommittee.org/sitec.

Wilderness Committee and Tanker Free BC. 2012. "Oil Spills and Vancouver's Stanley Park: A Report on the Consequences of Oil Tanker Traffic in Burrard Inlet." July 2012. https://www.wildernesscommittee.org/sites/all/files/publications/sp-report-web.pdf.

Williams, Byron, and Jared Wheeler. 2017. *AMC Comments on Energy East: Draft List of Issues and Draft Factors and Scope of Factors for the Environmental Assessments*. File No. OF-Fac-Oil-E266-2014-01 02. https://apps.neb-one.gc.ca/REGDOCS/Item/View/3282518.

Willon, P. 2011. "Giant New Utility Poles Spark Controversy in Chino Hills." *Los Angeles Times*, November 27, 2011. http://articles.latimes.com/2011/nov/27/local/la-me-powerlines-20111128.

Wilsdon, James, and Rebecca Willis. 2004. "See-Through Science: Why Public Engagement Needs to Move Upstream." *Technology and Science, Policy-Making, Environment. Demos*. https://doi.org/10.13140/RG.2.1.3844.3681.

Wind Energy and Electric Vehicle Review. 2012. "The Ivanpah Solar Energy Project Named Concentrating Solar Power Project of the Year." https://www.evwind.es/2012/02/22/the-ivanpah-solar-energy-project-named-concentrating-solar-power-project-of-the-year/16757.

Winfield, Mark. 2013. "The Environment, 'Responsible Resource Development' and Evidence Based Policy-Making in Canada." In *Evidence Based Policy-Making in Canada*, edited by Shaun Young. Toronto: Oxford University Press.

Wingrove, Josh. 2014. "Memo Contradicts Harper's Stance on Emission Limits." *Globe and Mail*, May 22, 2014. http://www.theglobeandmail.com/news/politics/memo-contradicts-harpers-stance-on-emission-limits/article18818653/.

Wood, Tim. 2018. "Energy's Citizens: The Making of a Canadian Petro-Public." *Canadian Journal of Communication* 43:75–92. http://doi.org/10.22230/cjc.2017v43n1a3312.

Wood MacKenzie Inc. 2011. *A Netback Impact Analysis of West Coast Export Capacity*. Prepared for Alberta Department of Energy. http://www.energy.alberta.ca/Org/pdfs/WoodMackenzieWestCoastExport.pdf.

World Bank. 2013. *Turn Down the Heat: Climate Extremes, Regional Impacts, and the Case for Resilience*. Washington, DC: World Bank. June 1, 2013. http://documents.worldbank.org/curated/en/975911468163736818/pdf/784240WP0Full00D0CONF0to0June19090L.pdf.

Wotherspoon, Terry, and John Hansen. 2013. "The 'Idle No More' Movement: Paradoxes of First Nations Inclusion in the Canadian Context." *Social Inclusion* 1:21–36.

Wright, David. 2018. "Federal Linear Energy Infrastructure Projects and the Rights of Indigenous Peoples: Current Legal Landscape and Emerging Developments." *Review of Constitutional Studies* 23:175–223.

Wüstenhagen, Rolf, Maarten Wolsink, and Mary Jean Bürer. 2007. "Social Acceptance of Renewable Energy Innovation: An Introduction to the Concept." *Energy Policy* 35 (5): 2683–2691.

Yaffe, B. 2012. "U.S. Report Sounds Death Knell for Pipeline." *Vancouver Sun*, July 13, 2012.

YCharts. 2017. "Brent WTI Spread." August 7, 2017. https://ycharts.com/indicators/brent_wti_spread.

Zahariadis, N. 2014. "'Ambiguity and Multiple Streams." In *Theories of the Policy Process*, 3rd ed., edited by Paul Sabatier and Christopher Weible, 22–58. Boulder, CO: Westview Press.

Zeller, Tom Jr. 2013. "Cape Wind: Regulation, Litigation and the Struggle to Develop Offshore Wind Power in the U.S." *Renewable Energy World*, February 28, 2013. https://www.renewableenergyworld.com/2013/02/28/cape-wind-regulation-litigation-and-the-struggle-to-develop-offshore-wind-power-in-the-u-s/#gref.

Zickfeld, Kirsten, et al. 2009. "Setting Cumulative Emissions Targets to Reduce the Risk of Dangerous Climate Change." *Proceedings of the National Academy of Sciences* 106:16129–16134. https://doi.org/10.1073/pnas.0805800106.

Zolden, Evan. 2015. "Congressional Dysfunction, Public Opinion, and the Battle over the Keystone XL Pipeline." *Loyola University Chicago Law Review* 617–645.

Zussman, Richard. 2018. "British Columbians Could Be Facing Gas at $2 to $3 per Litre without Alberta Oil." *Global News*, March 8, 2018. https://globalnews.ca/news/4071934/british-columbians-oil-ban-trans-mountain/.

Index

American and Comparative Environmental Policy

Sheldon Kamieniecki and Michael E. Kraft, series editors

Russell J. Dalton, Paula Garb, Nicholas P. Lovrich, John C. Pierce, and John M. Whiteley, *Critical Masses: Citizens, Nuclear Weapons Production, and Environmental Destruction in the United States and Russia*

Daniel A. Mazmanian and Michael E. Kraft, editors, *Toward Sustainable Communities: Transition and Transformations in Environmental Policy*

Elizabeth R. DeSombre, *Domestic Sources of International Environmental Policy: Industry, Environmentalists, and U.S. Power*

Kate O'Neill, *Waste Trading among Rich Nations: Building a New Theory of Environmental Regulation*

Joachim Blatter and Helen Ingram, editors, *Reflections on Water: New Approaches to Transboundary Conflicts and Cooperation*

Paul F. Steinberg, *Environmental Leadership in Developing Countries: Transnational Relations and Biodiversity Policy in Costa Rica and Bolivia*

Uday Desai, editor, *Environmental Politics and Policy in Industrialized Countries*

Kent Portney, *Taking Sustainable Cities Seriously: Economic Development, the Environment, and Quality of Life in American Cities*

Edward P. Weber, *Bringing Society Back In: Grassroots Ecosystem Management, Accountability, and Sustainable Communities*

Norman J. Vig and Michael G. Faure, editors, *Green Giants? Environmental Policies of the United States and the European Union*

Robert F. Durant, Daniel J. Fiorino, and Rosemary O'Leary, editors, *Environmental Governance Reconsidered: Challenges, Choices, and Opportunities*

Paul A. Sabatier, Will Focht, Mark Lubell, Zev Trachtenberg, Arnold Vedlitz, and Marty Matlock, editors, *Swimming Upstream: Collaborative Approaches to Watershed Management*

Sally K. Fairfax, Lauren Gwin, Mary Ann King, Leigh S. Raymond, and Laura Watt, *Buying Nature: The Limits of Land Acquisition as a Conservation Strategy, 1780–2004*

Steven Cohen, Sheldon Kamieniecki, and Matthew A. Cahn, *Strategic Planning in Environmental Regulation: A Policy Approach That Works*

Michael E. Kraft and Sheldon Kamieniecki, editors, *Business and Environmental Policy: Corporate Interests in the American Political System*

Joseph F. C. DiMento and Pamela Doughman, editors, *Climate Change: What It Means for Us, OurUs, Our Children, and Our Grandchildren*

Christopher McGrory Klyza and David J. Sousa, *American Environmental Policy, 1990–2006: Beyond Gridlock*

John M. Whiteley, Helen Ingram, and Richard Perry, editors, *Water, Place, and Equity*

Judith A. Layzer, *Natural Experiments: Ecosystem-Based Management and the Environment*

Daniel A. Mazmanian and Michael E. Kraft, editors, *Toward Sustainable Communities: Transition and Transformations in Environmental Policy*, second edition

Henrik Selin and Stacy D. VanDeveer, editors, *Changing Climates in North American Politics: Institutions, Policymaking, and Multilevel Governance*

Megan Mullin, *Governing the Tap: Special District Governance and the New Local Politics of Water*

David M. Driesen, editor, *Economic Thought and U.S. Climate Change Policy*

Kathryn Harrison and Lisa McIntosh Sundstrom, editors, *Global Commons, Domestic Decisions: The Comparative Politics of Climate Change*

William Ascher, Toddi Steelman, and Robert Healy, *Knowledge in the Environmental Policy Process: Re-Imagining the Boundaries of Science and Politics*

Michael E. Kraft, Mark Stephan, and Troy D. Abel, *Coming Clean: Information Disclosure and Environmental Performance*

Paul F. Steinberg and Stacy D. VanDeveer, editors, *Comparative Environmental Politics: Theory, Practice, and Prospects*

Judith A. Layzer, *Open for Business: Conservatives' Opposition to Environmental Regulation*

Kent Portney, *Taking Sustainable Cities Seriously: Economic Development, the Environment, and Quality of Life in American Cities*, second edition

Raul Lejano, Mrill Ingram, and Helen Ingram, *The Power of Narrative in Environmental Networks*

Christopher McGrory Klyza and David J. Sousa, *American Environmental Policy: Beyond Gridlock*, updated and expanded edition

Andreas Duit, editor, *State and Environment: The Comparative Study of Environmental Governance*

Joseph F. C. DiMento and Pamela Doughman, editors, *Climate Change: What It Means for Us, Our Children, and Our Grandchildren*, second edition

David M. Konisky, editor, *Failed Promises: Evaluating the Federal Government's Response to Environmental Justice*

Leigh Raymond, *Reclaiming the Atmospheric Commons: Explaining Norm-Driven Policy Change*

Robert F. Durant, Daniel J. Fiorino, and Rosemary O'Leary, editors, *Environmental Governance Reconsidered: Challenges, Choices, and Opportunities*, second edition

James Meadowcroft and Daniel J. Fiorino, editors, *Conceptual Innovation in Environmental Policy*

Barry G. Rabe, *Can We Price Carbon?*

Kelly Sims Gallagher and Xiaowei Xuan, *Titans of the Climate: Explaining Policy Process in the United States and China*

Matto Mildenberger, *The Logic of Double Representation in Climate Politics*

Mary Alice Haddad, *Effective Advocacy: Lessons from East Asia's Environmentalists*

George Hoberg, *The Resistance Dilemma: Place-Based Movements and the Climate Crisis*